Lecture Notes in Artificial Intelligence     1809

Subseries of Lecture Notes in Computer Science
Edited by J. G. Carbonell and J. Siekmann

Lecture Notes in Computer Science
Edited by G.Goos, J. Hartmanis, and J. van Leeuwen

T0224447

*Berlin*
*Heidelberg*
*New York*
*Barcelona*
*Hong Kong*
*London*
*Milan*
*Paris*
*Singapore*
*Tokyo*

Susanne Biundo   Maria Fox   (Eds.)

# Recent Advances in AI Planning

5th European Conference on Planning, ECP'99
Durham, UK, September 8-10, 1999
Proceedings

 Springer

Series Editors

Jaime G. Carbonell, Carnegie Mellon University, Pittsburgh, PA, USA
Jörg Siekmann, University of Saarland, Saarbrücken, Germany

Volume Editors

Susanne Biundo
University of Ulm, Computer Science Department
89069 Ulm, Germany
E-mail: biundo@informatik.uni-ulm.de

Maria Fox
University of Durham, Science Laboratories
South Road, Durham DH1 3LE, UK
E-mail: Maria.Fox@durham.ac.uk

Cataloging-in-Publication Data applied for

Die Deutsche Bibliothek - CIP-Einheitsaufnahme

Recent advances in AI planning : proceedings / 5th European Conference
on Planning, ECP'99, Durham, UK, September 8 - 10, 1999. Susanne
Biundo ; Maria Fox (ed.). - Berlin ; Heidelberg ; New York ; Barcelona ;
Hong Kong ; London ; Milan ; Paris ; Singapore ; Tokyo : Springer, 2000
   (Lecture notes in computer science ; Vol. 1809 : Lecture notes in
   artificial intelligence)
   ISBN 3-540-67866-2

CR Subject Classification (1998): I.2, F.2, G

ISBN 3-540-67866-2 Springer-Verlag Berlin Heidelberg New York

Springer-Verlag Berlin Heidelberg New York
a member of BertelsmannSpringer Science+Business Media GmbH
© Springer-Verlag Berlin Heidelberg 2000
Printed in Germany

Typesetting: Camera-ready by author, data conversion by PTP-Berlin, Stefan Sossna
Printed on acid-free paper      SPIN: 10720246      06/3142      5 4 3 2 1 0

# Preface

The European Conferences on Planning (ECP) are a major forum for the presentation of new research in Artificial Intelligence Planning and Scheduling. They developed from a series of European workshops and became successfully established as international meetings. Previous conferences took place in St. Augustin (Germany) in 1991, Vadstena (Sweden) in 1993, Assisi (Italy) in 1995, and Toulouse (France) in 1997.

ECP-99 was held in Durham, United Kingdom. The conference received submissions from all over Europe, from the US, Canada, South America, and New Zealand.

This volume contains the 27 papers that were presented at the conference. They cover a variety of aspects in current AI Planning and Scheduling. Several prominent planning paradigms are represented, including planning as satisfiability and other model checking strategies, planning as heuristic state-space search, and Graphplan-Based approaches. Moreover, various new scheduling approaches and combinations of planning and scheduling methods are introduced.

In addition to the conference papers, three invited talks were presented by distinguished researchers of the field: Fausto Giunchiglia (IRST Trento, Italy) gave an introduction to *Planning as Model Checking*. The corresponding paper by Fausto Giunchiglia and Paolo Traverso is included in this volume. Claude Le Pape (Bouygues Telecom, France) presented *Constraint-Based Scheduling: Theory and Applications*, and Nicola Muscettola (NASA Ames, USA) talked about *Planning at 96 Million Kilometers from Earth*.

ECP-99 received support from PLANET, the *European Network of Excellence in AI Planning*, the University of Durham, United Kingdom, and the University of Ulm, Germany.

March 2000

Susanne Biundo
Maria Fox

# Program Committee

**Program Chair:** Susanne Biundo, Germany

**Local Arrangements:** Maria Fox, United Kingdom

Rachid Alami, *France*
Ruth Aylett, *United Kingdom*
Michael Beetz, *Germany*
Susanne Biundo, *Germany*
Jim Blythe, *USA*
Craig Boutilier, *Canada*
Amedeo Cesta, *Italy*
Berthe Choueiry, *USA*
Stephen Cranefield, *New Zealand*
Brian Drabble, *USA*
Mark Drummond, *USA*
George Ferguson, *USA*
Maria Fox, *United Kingdom*
Malik Ghallab, *France*
Robert Goldman, *USA*
Carla Gomes, *USA*
Joachim Hertzberg, *Germany*
Toru Ishida, *Japan*

Peter Jonsson, *Sweden*
Subbarao Kambhampati, *USA*
Jana Koehler, *Switzerland*
Richard Korf, *USA*
Claude LePape, *France*
Witold Lukaszewicz, *Poland*
Alfredo Milani, *Italy*
Dana Nau, *USA*
Anna Perini, *Italy*
Martha Pollack, *USA*
Louise Pryor, *United Kingdom*
Barry Richards, *United Kingdom*
Wolfgang Slany, *Austria*
Jon Spragg, *United Kingdom*
Sam Steel, *United Kingdom*
Patrick Taillibert, *France*
Sylvie Thiébaux, *Australia*
Paolo Traverso, *Italy*

# Additional Reviewers

Marco Baioletti, *Italy*
Mathias Bauer, *Germany*
Marcus Bjäreland, *Sweden*
Claudio Castellini, *Italy*
Marco Daniele, *Italy*
Olivier Despouys, *France*
Jürgen Dorn, *Austria*
Mario Girsch, *Austria*
Emmanuel Guere, *France*
Joakim Gustafsson, *Sweden*
Patrik Haslum, *Sweden*
Marcus Herzog, *Austria*

Felix Ingrand, *France*
Derek Long, *United Kingdom*
Amnon Lotem, *USA*
Atif Memon, *USA*
David Moriarty, *USA*
Nysret Muslija, *Austria*
Angelo Oddi, *Italy*
Nilufer Onder, *USA*
Roberto Sebastiani, *Italy*
Werner Stephan, *Germany*
Ioannis Tsamardinos, *USA*

# Table of Contents

# Planning as Model Checking

Fausto Giunchiglia[1,2] and Paolo Traverso[1]

[1] IRST, Istituto per la Ricerca Scientifica e Tecnologica, 38050 Povo, Trento, Italy
[2] Dipartimento di Informatica e Studi Aziendali, Universita' di Trento, Italy
`fausto@irst.itc.it, leaf@irst.itc.it`

**Abstract.** The goal of this paper is to provide an introduction, with various elements of novelty, to the Planning as Model Checking paradigm.

## 1 Introduction

The key idea underlying the Planning as Model Checking paradigm is that planning problems should be solved model-theoretically. Planning domains are formalized as semantic models. Properties of planning domains are formalized as temporal formulas. Planning is done by verifying whether temporal formulas are true in a semantic model.

The most important features of the proposed approach are:

- The approach is well-founded. Planning problems are given a clear and intuitive (semantic) formalization.
- The approach is general. The same framework can be used to tackle most research problems in planning, e.g., planning in deterministic and in non-deterministic domains, conditional and iterative planning, reactive planning.
- The approach is practical. It is possible to devise efficient algorithms that generate plans automatically and that can deal with large size problems.

The goal of this paper is to provide an introduction, with various elements of novelty, to the Planning as Model Checking paradigm. The core of the paper are Sections 2, 3, 4, and 5. Section 2 gives a brief introduction to the model checking problem. Section 3 shows how planning problems can be stated as model checking problems. Section 4 shows how our formalization can be used to tackle various planning problems. Section 5 shows how the approach can be extended to non-deterministic domains. Section 6 shows how Planning as Model Checking can be implemented. Our implementation relies heavily on existing work in the context of finite-state program verification (see [9] for an overview), and in particular on the work described in [8,2,20]. Section 7 discusses the related work.

S. Biundo and M. Fox (Eds.): ECP-99, LNAI 1809, pp. 1–20, 2000.

## 2  Model Checking

The Model Checking problem is the problem of determining whether a formula is true in a model. Model checking is based on the following fundamental ideas:

1. A domain of interest (e.g., a computer program, a reactive system) is described by a semantic model.
2. A desired property of the domain (e.g., a specification of a program, a safety requirement for a reactive system) is described by a logical formula.
3. The fact that a domain satisfies a desired property (e.g., the fact that a program meets its specifications, that a reactive system never ends up in a dangerous state) is determined by checking whether the formula is true in the model, i.e., by model checking.

We formalize domains as Kripke Structures. We restrict ourselves to the case of finite domains, i.e., domains which can be described by Kripke Structures with a finite number of states. A Kripke Structure $K$ is a 4-tuple $\langle W, W_0, T, L \rangle$, where

1. $W$ is a finite set of *states*.
2. $W_0 \subseteq W$ is a set of initial states.
3. $T \subseteq W \times W$ is a binary relation on $W$, the *transition relation*, which gives the possible transitions between states. We require $T$ to be total, i.e., for each state $w \in W$ there exists a state $w' \in W$ such that $(w, w') \in T$.
4. $L : W \mapsto 2^{\mathcal{P}}$ is a *labeling function*, where $\mathcal{P}$ is a set of atomic propositions. $L$ assigns to each state the set of atomic propositions true in that state.

A Kripke Structure encodes the possible evolutions of the domain (or behaviours) as *paths*, i.e., infinite sequences $w_0 w_1 w_2 \ldots$ of states in $W$ such that, for each $i$, $(w_i, w_{i+1}) \in T$. We require that paths start from an initial state $w_0 \in W_0$. By requiring that $T$ is total, we impose that all paths are infinite.

As a simple example of Kripke Structure, consider Figure 1. It depicts a simple domain, where an item can be loaded/unloaded to/from a container which can be locked/unlocked. The corresponding Kripke Structure is the following:

1. $W = \{1, 2, 3, 4\}$
2. $W_0 = \{2\}$
3. $T = \{(1, 2), (2, 1), (2, 2), (2, 3), (3, 2), (3, 4), (4, 3)\}$
4. $L(1) = \{Locked\}$, $L(2) = \emptyset$, $L(3) = \{Loaded\}$, $L(4) = \{Loaded, Locked\}$.

We formalize temporal properties of domains in Computation Tree Logic (CTL) [12]. Given a finite set $\mathcal{P}$ of atomic propositions, CTL formulas are inductively defined as follows:

1. Every atomic proposition $p \in \mathcal{P}$ is a CTL formula;
2. If $p$ and $q$ are CTL formulas, then so are
   a) $\neg p,\ p \vee q,$

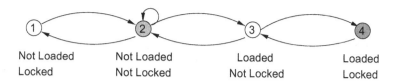

**Fig. 1.** An example of Kripke Structure

b) **AX**$p$, **EX**$p$,
c) **A**($p$**U**$q$), and **E**($p$**U**$q$))

**X** is the "next time" temporal operator; the formula **AX**$p$ (**EX**$p$) intuitively means that $p$ holds in every (in some) immediate successor of the current state. **U** is the "until" temporal operator; the formula **A**($p$**U**$q$) (**E**($p$**U**$q$)) intuitively means that for every path (for some path) there exists an initial prefix of the path such that $q$ holds at the last state of the prefix and $p$ holds at all the other states along the prefix. Formulas **AF**$p$ and **EF**$p$ (where the temporal operator **F** stands for "future" or "eventually") are abbreviations of **E**($\top$**U**$p$) and **A**($\top$**U**$p$) (where $\top$ stands for truth), respectively. **EG**$p$ and **AG**$p$ (where **G** stands for "globally" or "always") are abbreviations of $\neg$**AF**$\neg p$ and $\neg$**EF**$\neg p$, respectively. In the example in Figure 1, **EF***Loaded* holds in state 2, since *Loaded* holds eventually in a state of the path $2, 3, 4, 3, 4 \ldots$. Instead, **AF***Loaded* does not hold in state 2, since *Loaded* does not hold in any state of the path $2, 2, \ldots$.

CTL semantics is given in terms of Kripke Structures. We write $K, w \models p$ to mean that $p$ holds in the state $w$ of $K$. Let $p$ be a CTL formula. $K, w \models p$ is defined inductively as follows:

- $K, w \models p$ iff $p \in L(w)$, $p \in \mathcal{P}$
- $K, w \models \neg p$ iff $K, w \not\models p$
- $K, w \models p \vee q$ iff $K, w \models p$ or $K, w \models q$
- $K, w \models$ **AX**$p$ iff for all paths $\pi = w_0 w_1 w_2 \ldots$, with $w = w_0$, we have $K, w_1 \models p$
- $K, w \models$ **EX**$p$ iff there exists a path $\pi = w_0 w_1 w_2 \ldots$, with $w = w_0$, such that $K, w_1 \models p$
- $K, w \models$ **A**($p$**U**$q$) iff for all paths $\pi = w_0 w_1 w_2 \ldots$, with $w = w_0$, there exists $i \geq 0$ such that $K, w_i \models q$ and, for all $0 \leq j < i$, $K, w_j \models p$
- $K, w \models$ **E**($p$**U**$q$) iff there exists a path $\pi = w_0 w_1 w_2 \ldots$, with $w = w_0$, and $i \geq 0$ such that $K, w_i \models q$ and, for all $0 \leq j < i$, $K, w_j \models p$

We say that $p$ is true in $K$ ($K \models p$) if $K, w \models p$ for each $w \in W_0$. The Model Checking Problem for a CTL formula $p$ and a Kripke Structure $K$ is the problem of determining whether $p$ is true in $K$.

Algorithms for model checking exploit the structure of CTL formulas. For instance, an atomic formula $p$ is model checked by verifying that $p \in L(s)$ for all $s \in W_0$. As another example, model checking **AX**$p$ (**EX**$p$) is performed by model checking $p$ in all states (in some state) $s'$ such that $(s, s') \in T$, for each

```
1.  function MCHECKEF(p,K)
2.      CurrentStates:= ∅;
3.      NextStates := STATES(p,K);
4.      while NextStates ≠ CurrentStates do
5.          if (W₀ ⊆ NextStates)
6.              then return True;
7.          CurrentStates := NextStates;
8.          NextStates := NextStates ∪ ONESTEPMCHECK(NextStates,K);
9.      endwhile
10.     return False;
```

**Fig. 2.** Model Checking **EF**$p$.

$s \in W_0$. Finally, $\mathbf{A}(p\mathbf{U}q)$ or $\mathbf{E}(p\mathbf{U}q)$, can be model checked by exploiting the fact that

$$
\begin{aligned}
p\mathbf{U}q = q \ & \vee \\
(p \wedge \mathbf{X}q) \ & \vee \\
(p \wedge \mathbf{X}p \wedge \mathbf{XX}q) \ & \vee \\
& \cdots
\end{aligned}
$$

As a simple explanatory example, we show in Figure 2 a possible algorithm for model checking the CTL formula **EF**$p$, with $p \in \mathcal{P}$. Given a Kripke Structure $K = \langle W, W_0, T, L \rangle$, and a propositional formula $p$, the algorithm starts by computing the set of states where $p$ holds (line 3). We have in fact that:

$$
\text{STATES}(p, K) = \{s \in W : p \in L(s)\} \tag{1}
$$

Then, MCHECKEF explores the state space of $K$. It repeatedly accumulates in *NextStates* the states returned by ONESTEPMCHECK (line 8). Given a set of states *States* $\subseteq W$, ONESTEPMCHECK returns the set of states which have at least one immediate successor state in *States*:

$$
\text{ONESTEPMCHECK}(States, K) = \{s \in W : \exists s'. (s' \in States \wedge T(s,s'))\} \tag{2}
$$

Notice that **EF**$p$ always holds in each state in *NextStates*. The loop terminates successfully if *NextStates* contains all the initial states (termination condition at line 5). MCHECKEF returns *False* if *NextStates* does not contain all the initial states and there are no new states to explore, i.e., *NextStates* = *CurrentStates*. Termination is guaranteed by the monotonicity of the operator accumulating states and by the fact that the set of states is finite. When applied to the Kripke Structure in Figure 1 and to the formula **EF**(*Loaded* ∧ *Locked*), MCHECKEF starts with state 4 in *NextStates*, after the first step *NextStates* is {3,4}, after the second step *NextStates* is {2,3,4}, and then the algorithm stops returning *True*.

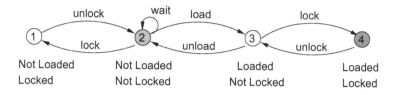

**Fig. 3.** An example of Planning Domain

## 3  Planning as Model Checking

The underlying idea of the Planning as Model Checking paradigm is to generate plans by determining whether formulas are true in a model. The fundamental ingredients are the following:

1. A planning domain is described by a semantic model, which defines the states of the domain, the available actions, and the state transitions caused by the execution of actions.
2. A planning problem is the problem of finding plans of actions given planning domain, initial and goal states.
3. Plan generation is done by exploring the state space of the semantic model. At each step, plans are generated by checking the truth of some suitable formulas in the model.

A *planning domain* $D$ is a 4-tuple $\langle F, S, A, R \rangle$ where

1. $F$ is a finite set of *fluents*,
2. $S \subseteq 2^F$ is a finite set of *states*,
3. $A$ is a finite set of *actions*,
4. $R : S \times A \mapsto S$ is a *transition function*.

The action $a \in A$ is said to be *executable* in $s \in S$ if $R(s, a)$ is defined.

In this section, we restrict ourselves to *deterministic actions*, which, given a state, lead to a single state. Notice that $R$ is a function. Notice also that planning domains are general enough to express actions with secondary effects [21], which can be described in ADL-like languages (e.g., PDDL) but not in pure STRIPS-like languages.

In Figure 3 we depict a simple planning domain. It is obtained from the example in Figure 1 by labeling transitions with actions. The corresponding planning domain is the following.

1. $F = \{Loaded, Locked\}$
2. $S = \{ \{\neg\ Loaded, Locked\}, \{\neg\ Loaded, \neg\ Locked\},$
   $\{ Loaded, \neg\ Locked\}, \{ Loaded, Locked\} \}$
3. $A = \{ lock, unlock, load, unload, wait\}$

```
1.   function PLAN(P)
2.       CurrentStates := ∅;
3.       NextStates := G;
4.       Plan := ∅;
5.       while (NextStates ≠ CurrentStates) do
6.           if I ⊆ NextStates
7.               then return Plan;
8.           OneStepPlan := ONESTEPPLAN(NextStates, D);
9.           Plan := Plan ∪ PRUNESTATES(OneStepPlan, NextStates);
10.          CurrentStates := NextStates;
11.          NextStates := NextStates ∪ PROJECTACTIONS(OneStepPlan);
12.  return Fail;
```

**Fig. 4.** A "Planning as Model Checking" Algorithm.

4. $R = \{$ $(\{\neg\ Loaded,\ Locked\}, unlock, \{\neg\ Loaded,\ \neg\ Locked\})$
   $(\{\neg\ Loaded,\ \neg\ Locked\},\ lock, \{\neg\ Loaded,\ Locked\})$
   $(\{\neg\ Loaded,\ \neg\ Locked\}, wait, \{\neg\ Loaded,\ \neg\ Locked\})$
   $(\{\neg\ Loaded,\ \neg\ Locked\}, load, \{Loaded,\ \neg\ Locked\})$
   $(\{Loaded,\ \neg\ Locked\}, unload, \{\neg\ Loaded,\ \neg\ Locked\})$
   $(\{Loaded,\ \neg\ Locked\}, lock, \{Loaded,\ Locked\})$
   $(\{Loaded,\ Locked\}, unlock, \{Loaded,\ \neg\ Locked\})$ $\}$

A *planning problem* $P$ for a Planning Domain $D = \langle F, S, A, R \rangle$ is a 3-tuple $\langle D, I, G \rangle$, where $I = \{s_0\} \subseteq S$ is the initial state, and $G \subseteq S$ is the set of goal states.

Notice that we are restricting ourselves to planning problems with a completely specified initial situation, namely problems with a single initial state. Notice also that planning problems allow for conjunctive and disjunctive goals, and, more in general, for goals that can be described by any propositional formula.

Intuitively, plans specify actions to be executed in certain states. More precisely, a *plan* $\pi$ for a planning problem $P = \langle D, I, G \rangle$ with planning domain $D = \langle F, S, A, R \rangle$ is defined as

$$\pi = \{\langle s, a \rangle :\ s \in S, a \in A\} \tag{3}$$

We call $\langle s, a \rangle$ a *state-action pair*.

We say that a plan is executable if all its actions are executable. In the following, we consider plans which have at least one state-action pair $\langle s, a \rangle$ with $s \in I$. A simple algorithm for plan generation is presented in Figure 4. PLAN searches backwards from $G$ to $I$. At each step (see line 8), given a set of states *States*, ONESTEPPLAN computes the set of state-action pairs $\langle s, a \rangle$ such that the action $a$ leads from $s$ to a state in *States*:

$$\textsc{OneStepPlan}(States, D) =$$
$$\{\langle s, a \rangle : s \in S, a \in A, \exists s'. \ (s' \in States \land s' = R(s, a))\} \tag{4}$$

$\textsc{PruneStates}(OneStepPlan, NextStates)$ eliminates from $OneStepPlan$ the state-action pairs the states of which are already in $NextStates$, and thus have already been visited.

$$\textsc{PruneStates}(\pi, States) = \{\langle s, a \rangle \in \pi : s \notin States\} \tag{5}$$

$\textsc{ProjectActions}$, given a set of state-action pairs, returns the corresponding set of states.

$$\textsc{ProjectActions}(\pi) = \{s : \langle s, a \rangle \in \pi\} \tag{6}$$

Consider the example in Figure 3, with 2 as initial state and 4 as goal state. $\textsc{Plan}$ starts from state 4, after the first step $Plan$ is $\{\langle 3, lock \rangle\}$, and after the second step $Plan$ is $\{\langle 2, load \rangle, \langle 3, lock \rangle\}$. Notice that the pair $\{\langle 4, unlock \rangle\}$ is eliminated by $\textsc{PruneStates}$. Therefore the algorithm stops returning $\{\langle 2, load \rangle, \langle 3, lock \rangle\}$.

Let us now see in which sense the planning problem is a model checking problem:

1. The planning domain is a semantic model.
2. The planning problem is specified through a set of goal states that corresponds to a formula representing a desired property of the domain.
3. Plan generation is done by checking whether suitable formulas are true in a semantic model.

More precisely, let $K = \langle W, W_0, T, L \rangle$ and $P = \langle D, I, G \rangle$ with $D = \langle F, S, A, R \rangle$ be a Kripke Structure and a planning problem, respectively. $W$, $W_0$ and $T$ correspond to $S$, $I$ and $R$, respectively. The set of atomic propositions $\mathcal{P}$ of the labeling function $L$ corresponds to the set of fluents $F$. We have the following differences:

1. The arcs defined by $R$ are labeled by actions.
2. $R$ is not required to be total. Indeed, in planning domains we may have states where no actions are executable.
3. $R$ is a function. Indeed, we are in deterministic domains. We extend to non-determinism in Section 5.
4. $I$ is a singleton. We extend to partially specified initial situations in Section 5.

Therefore, a planning problem corresponding to a Kripke Structure $K = \langle W, W_0, T, L \rangle$ is $P = \langle D, I, G \rangle$, where

1. $D = \langle F, S, A, R \rangle$ with
   a) $F = \mathcal{P}$,
   b) $S = W$,
   c) $A = \{u\}$,
   d) $R = \{(s, u, s') : (s, s') \in T\}$;
2. $I = W_0$.

It is now worthwhile to compare the algorithms MCHECKEF and PLAN. The basic routine of MCHECKEF, ONESTEPMCHECK (see (2)), can be defined in terms of the basic routine of PLAN, ONESTEPPLAN (see (4)), on the planning domain $D_k$ corresponding to the Kripke Structure $K$:

$$\text{ONESTEPMCHECK}(States, K) = \tag{7}$$
$$\text{PROJECTACTIONS}(\text{ONESTEPPLAN}(States, D_k))$$

Notice that MCHECKEF and PLAN are very similar. The main difference is that MCHECKEF returns either *True* or *False*, while PLAN returns either a plan or a failure. Let $p$ be a propositional CTL formula such that $K, w \models p$ for all $w \in G$ and $K, w \not\models p$ for all $w \notin G$. Then, $K \models \mathbf{EF}p$ iff there exists a plan satisfying the planning problem $P$ corresponding to $K$. Therefore, we have reduced planning to the model checking of the formula $\mathbf{EF}p$. This corresponds to an underlying assumption of classical planning: the requirement for a plan is merely on the final states resulting from its execution. However, previous papers on planning (see, e.g., [26,24,25]) have stated that the planning problem should be generalized to the problem of finding plans that satisfy some specifications on their execution paths, rather than on the final states of their execution. The Planning as Model Checking paradigm can be a good approach for extending the classical planning problem.

## 4   Situated Plans

The plans we construct are actually "situated plans" (see, e.g., [15,14]), namely plans that, at run time, are executed by a reactive loop that repeatedly senses the state, selects an appropriate action, executes it, and iterates, e.g., until the goal is reached. Indeed, a plan $\pi = \{\langle s, a \rangle : s \in S, a \in A\}$ can be viewed as the *iterative plan*

$$\textbf{while } s \in \{s : \langle s, a \rangle \in \pi\} \textbf{ do} \tag{8}$$
$$a \text{ such that } \langle s, a \rangle \in \pi$$

Plans as defined in Section 3 are rather general: they do not depend on the goal of the planning problem, and no condition is imposed on the fact that a plan should attempt to achieve a goal. A condition that a plan should satisfy is that, if a goal is achieved, then the plan stops execution. We call these plans, *goal preserving plans*.

A *goal preserving plan* is a set of state-action pairs
$\pi = \{\langle s, a \rangle : s \in S, a \in A, s \notin G\}$.

The goal preserving condition can be generalized. A plan, rather than "not acting in a goal state", can still act without abandoning the set of goal states. Consider for instance the task of a robot for surveillance systems: it may be required, after reaching a given location, e.g., an area of a building, to move inside that area without leaving it.

A *dynamic goal preserving plan* is a set of state-action pairs
$\pi = \{\langle s, a \rangle : s \in S, a \in A, s \in G \supset R(s,a) \in G\}$.

The goal preserving condition can be even weaker. Consider a surveillance robot which is required to repeatedly charge batteries in a given location in order to explore a building (the robot should eventually charge its batteries infinitely often). Let the goal states represent the charge battery location. This can be specified by requiring the plan to "pass through" the set of goal states infinitely often.

A *fair goal preserving plan* is a set of state-action pairs $\pi$ such that for each $\langle s_0, a_0 \rangle \in \pi$ with $s_0 \in G$, there exists $\{\langle s_1, a_1 \rangle, \ldots \langle s_n, a_n \rangle\} \subseteq \pi$ such that $s_{i+1} = R(s_i, a_i)$ for each $0 \leq i \leq n-1$, and $s_n \in G$

A condition that a plan should satisfy is that it should achieve the goal. We call such plans, *goal achieving plans*. Intuitively, for each state-action pair in a goal achieving plan, there should be a path leading from the state of the state-action pair to a goal state. More precisely, this requirement can be described as follows: "all the state-action pairs $\langle s, a \rangle$ contained in a plan $\pi$ should be such that, either $a$ leads from $s$ to the goal ($R(s,a) \in G$), or $a$ leads from $s$ to a state $s'$ such that $\pi$ contains $\langle s', a' \rangle$ and $a'$ leads from $s'$ to the goal ($R(s', a') \in G$), and so on. This informal requirement can be formalized as follows:

A *goal achieving plan* is defined inductively as follows.

1. $\pi = \{\langle s, a \rangle : s \in S, a \in A, R(s,a) \in G\}$ is a goal achieving plan
2. If $\pi'$ is a goal achieving plan, then $\pi = \pi' \cup \{\langle s, a \rangle\}$ such that $R(s,a) \in \{s' : \langle s', a' \rangle \in \pi'\}$ is a goal achieving plan

We can now define situated plans.

A *situated plan* is a goal preserving and goal achieving plan.

Situated plans can be "robust" to unexpected action outcomes, namely, when executed, they can achieve the goal in spite of the fact that some actions may have outcomes that are not modeled. Consider the plan $\{\langle 2, load \rangle, \langle 3, lock \rangle\}$ in the example in Figure 3. The plan is robust to the fact that the execution of *load* may lead from state 2 to state 2 (leave the item unloaded). The plan is not robust to the fact that the execution of *load* may lead from state 2 to state 1 (leave the item unloaded and accidentally lock the container).

Universal plans [23] are particular cases of situated plans. The intuitive idea is that a universal plan maps sets of states to actions to be executed for each possible situation arising at execution time.

A *universal plan* is a situated plan $\pi$ such that the states in $\pi$ are all the possible states of the domain, i.e., for each $s \in S$ there exists $\langle s, a \rangle \in \pi$.

The definition of universal plan can be generalized to include situations which are not modeled by states of the planning domain, as far as the situations can be described by the language of the planner. A plan of this kind is a situated plan $\pi$ such that for each $s \in 2^F$, there exists $\langle s, a \rangle \in \pi$.

As a very specific case, we can define plans which resemble classical plans, namely plans which consist of sequences of actions that lead from the initial state to a goal state.

A *quasi-classical plan* is a plan $\pi$ such that
1. $\pi = \{\langle s_1, a_1 \rangle, \ldots, \langle s_n, a_n \rangle\}$
2. $s_1 = s_0 \in I$,
3. $R(s_i, a_i) = s_{i+1}$ for each $1 \leq i \leq n$,
4. $s_{n+1} \in G$

At planning time, a quasi-classical plan is similar to a classical plan. However, a quasi-classical plan is executed like a classical plan just under the main hypothesis of classical planning that the model of the world is perfect. Consider again the quasi-classical plan $\{\langle 2, load \rangle, \langle 3, lock \rangle\}$ in the example in Figure 3. At planning time, the plan is similar to the classical plan *load, lock*. Suppose that, the execution of *load* leads twice from state 2 to state 2 (leaves the item unloaded). The execution of the classical plan *load, lock* does not achieve the goal (state 4). The execution of the quasi-classical plan $\{\langle 2, load \rangle, \langle 3, lock \rangle\}$ results in the execution of the action *load* three times followed by the execution of the action *lock*, and achieves the goal. In this case, the execution of the quasi-classical plan is equivalent to the execution of the classical plan *load, load, load, lock*.

The different behaviour of a classical plan and a quasi-classical plan at execution time is a consequence of the fact that a quasi-classical plan does not impose *a priori* before execution any partial/total order on the execution of actions. We see plans as *sets*, and as such, the order of the execution of actions is determined at execution time, depending on the actual situation of the world.

We have now enough terminology to state and discuss several interesting properties of the algorithm PLAN in Figure 4.

- Let us first consider *correctness*. PLAN returns situated plans. Actions are guaranteed to be executable since state-action pairs are constructed from the planning domain. The termination condition at line 6 guarantees that the initial state is in the returned plan. The returned plan is goal preserving since states-action pairs whose states are in $G$ are pruned away by PRUNESTATES. Finally, the returned plan is a goal achieving plan, since ONESTEPPLAN, at each step, when applied to the set of states *NextStates*, finds situated plans for the planning problems with initial states in *NextStates* and goal $G$.
- There may be different notions of *completeness*. A first notion is "if there exists a situated plan, PLAN finds it". The algorithm is complete in this sense, since it explores the state space exhaustively. A second notion is: "if there is no situated plan, PLAN terminates with failure". PLAN always terminates. It is guaranteed to terminate even in the case there is no solution. This follows from the fact that *NextStates* is monotonically increasing at each

step, and the number of states $S$ is finite. Termination in the case a solution does not exists is a property which is not guaranteed by several state of the art planners, e.g., SATplan [19], BlackBox [18] and UCPOP [21]. A third notion of completeness is: "the planning algorithm returns *all* the possible situated plans". PLAN is *not* complete in this sense, since it stops after that the initial state is reached. The algorithm can be easily modified to satisfy this notion of completeness by continuing state exploration after the initial state is reached. This can be done by eliminating the termination condition at lines 6 and 7, and by modifying the return statement at line 12 as follows:

12.     **if** $I \subseteq NextStates$
13.         **then return** *Plan*;
14.         **else return** *Fail*;

- Consider now *optimality*. We can define optimality from the initial state, or more in general, from all the states in the plan. Intutively, a plan is optimal if for each state in the plan, each sequence of states, obtained by traversing the planning domain according to the plan, is of minimal length. PLAN is optimal in this general sense. This follows from the fact that the search is breadth-first.
- Finally consider the combination of *optimality* and the third notion of *completeness*: "the planning algorithm returns *all* the possible optimal situated plans". PLAN satisfies this property. Indeed, at each iteration, ONESTEP PLAN accumulates all the state-action pairs which lead to the subgoal. As a consequence of this property, given a set of state-action pairs returned by PLAN, the planner can select one of the many possible plans for execution, and can switch freely from one plan to another during execution.

## 5   Non-determinism

Several realistic applications need non-deterministic models of the world. Actions are modeled as having possible multiple outcomes, and the initial situation is partially specified. Non-deterministic planning domains and non-deterministic planning problems are obtained from definitions given in Section 3 by extending them with transition relations and sets of initial states.

A *planning domain D* is a 4-tuple $\langle F, S, A, R \rangle$ where

1. $F$ is a finite set of *fluents*,
2. $S \subseteq 2^F$ is a finite set of *states*,
3. $A$ is a finite set of *actions*,
4. $R \subseteq S \times A \times S$ is a *transition relation*.

In Figure 5 we depict a simple non-deterministic planning domain. The action *load* can have two effects: either the item is correctly loaded, or it is misplaced in a wrong position. An action *adjust* can reposition the item.

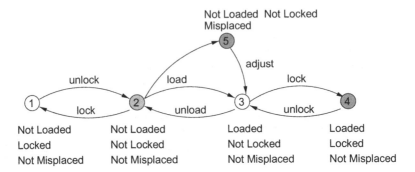

**Fig. 5.** An example of Non-deterministic Planning Domain

A *planning problem* $P$ for a planning domain $D = \langle F, S, A, R \rangle$ is a 3-tuple $\langle D, I, G \rangle$, where $I \subseteq S$ is the set of initial states, and $G \subseteq S$ is the set of goal states.

Along the lines described in Section 3, we can provide different specifications of plans as sets of state-action pairs for the non-deterministic case. However, in non-deterministic planning problems, we need to distinguish different kinds of solutions. Consider, for instance, the plan $\{\langle 2, load \rangle, \langle 3, lock \rangle\}$ in the planning domain in Figure 5. In general, we cannot say whether this plan does or does not achieve the goal state 4. It depends on the outcome of the non-deterministic action *load*. The plan *may* achieve the goal, and this is the case if *load* leads to state 3, or may not achieve the goal, in the case the outcome is state 5. Consider now the plan $\{\langle 2, load \rangle, \langle 3, lock \rangle, \langle 5, adjust \rangle\}$. It does achieve the goal, independently of the outcome of *load*. This plan specifies a conditional behaviour, of the kind "load; if load fails, then adjust; lock". We distinguish therefore between plans which may achieve the goal and those which are guaranteed to do so.

A *weak (strong) plan* is a set of state-action pairs which may (is guaranteed to) achieve the goal.

The distinction between weak and strong plans was first introduced in [7]. The algorithm PLAN (Figure 4), if applied to non-deterministic planning problems, finds weak plans. A Planning as Model Checking algorithm for weak plans (searching the state space forward rather than backward) was first presented in [4]. Planning as Model Checking for strong plans was first solved in [7], where the notion of plan as set of state-action pairs was first introduced. In [5], the authors exploit the Planning as Model Checking paradigm in order to generate strong plans as conformant plans, i.e., sequences of actions, rather than sets of state-action pairs.

Consider now the example in Figure 6. The action *load* has three possible outcomes: either it loads the item correctly, or it misplaces it, or it leaves the item unloaded (e.g., the robot simply fails to pick-up the item). In such a domain, there is no strong plan. The plan $\{\langle 2, load \rangle, \langle 3, lock \rangle, \langle 5, adjust \rangle\}$ may in principle

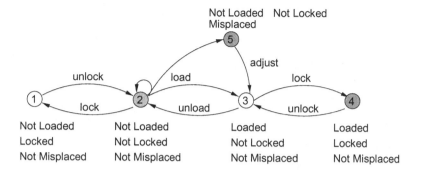

**Fig. 6.** An example of a domain for strong cyclic planning

loop forever, leaving the situation in state 2. This plan specifies an iterative behaviour, of the kind "load until the item is either loaded or misplaced; if it is misplaced, then adjust; lock". Plans encoding iterative trial-and-error strategies, like "pick up a block until succeed", are the only acceptable solutions in several realistic domains, where a certain effect (e.g., action success) might never be guaranteed *a priori* before execution. The planner should generate iterative plans that, in spite of the fact that they may loop forever, have good properties.

A *strong cyclic plan* is a plan whose executions always have a possibility of terminating and, when they do, they are guaranteed to achieve the goal.

The plan $\{\langle 2, load\rangle, \langle 3, lock\rangle, \langle 5, adjust\rangle\}$ is a strong cyclic plan for the goal state 4. Indeed, execution may loop forever in state 2, but it has always a possibility of terminating (*load* may non-deterministically lead to states 3 and 5), and, if it does, the plan achieves the goal both from state 3 and from state 5. The plan $\{\langle 2, load\rangle, \langle 3, lock\rangle\}$ is not a strong cyclic solution.

The problem of planning for strong cyclic solutions was first tackled in [6], where an efficient algorithm was proposed. The algorithm looks for strong plans, and, if it does not find one, iterates backward by applying ONESTEPPLAN, and then by computing the strongly connected components of the set of states produced by ONESTEPPLAN. A formal account for strong cyclic planning was first given in [10] where strong cyclic solutions are formalized as CTL specifications on the possible executions of plans. A strong cyclic plan is a solution such that "*for each* possible execution, *always* during the execution, there *exists* the possibility of *eventually* achieving the goal". More precisely, strong cyclic plans are plans whose executions satisfy the CTL formula **AGEF**$\mathcal{G}$, where $\mathcal{G}$ is a propositional formula representing the set of goal states. Strong and weak plans are plans whose executions have to satisfy the CTL formulas **AF**$\mathcal{G}$ and **EF**$\mathcal{G}$, respectively.

# 6   Planning via Symbolic Model Checking

We now consider the problem of implementing Planning as Model Checking. The problem is that realistic planning domains result most often in large state spaces. With this problem in mind we have defined planning domains and problems which are strictly related to Kripke Structures and CTL specifications. This has allowed us to exploit all the work done within the Computer Science community in the area of *symbolic model checking* [2,20], based on Ordered Binary Decision Diagrams (OBDD's) [1][1]. As a practical consequence, we have implemented the Model Based Planner (MBP), a planner built on top of NuSMV [3], a state of the art OBDD based symbolic model checker. MBP can deal efficiently with rather large size planning problems. For instance, it manages to find strong cyclic plans in non-deterministic domains with more than $10^7$ states in a few minutes [6].

In the rest of this section, in order to keep the paper self contained, we review the idea of OBDD-based symbolic model checking. We show how it can be applied to the Planning as Model Checking approach. As a matter of presentation, we keep distinct the description of symbolic model checking (Section 6.1) from the description of its OBDD-based implementation (Section 6.2).

## 6.1   Symbolic Representation

The fundamental ideas of *Planning via Symbolic Model Checking* are the following:

1. The Planning Problem is represented symbolically: the sets of states and the transitions of the semantic model are represented symbolically by logical formulas.
2. Plans are represented symbolically as formulas.
3. Planning is performed by searching through *sets* of states, rather than single states, by evaluating the assignments verifying (falsifying) the corresponding logical formulas.

Let $D = \langle F, S, A, R \rangle$ be a Planning Domain. Its symbolic representation is a boolean formula. We construct the boolean formula corresponding to $D$ as follows. We associate to each fluent in $F$ a boolean variable. Let $\underline{x} = x_1, \ldots, x_n$ be a vector of $n$ distinct boolean variables, where each $x_i$ corresponds to a distinct fluent in $F$ (let $n$ be the cardinality of $F$). $S$ and each subset $Q$ of $S$ can be represented by a boolean formula in the variables $\underline{x}$, that we write as $S(\underline{x})$ and $Q(\underline{x})$, respectively.

Consider the example in Figure 3. $\underline{x}$ is *Loaded, Locked*. Since $S = 2^F$, $S(\underline{x}) = \top$. State 1 ($\{\neg\ Loaded, Locked\}$) is represented by $\neg Loaded \land Locked$ and $Q = \{3, 4\} = (\{Loaded, \neg Locked\}, \{Loaded, Locked\})$ by the formula $Q(\underline{x}) = Loaded$.

---

[1]   Various alternatives of the notions used in this paper have been provided. For instance, models can be formalized as $\omega$-automata [27] and another common temporal logic is Linear Time Temporal Logic (LTL) [12]

We associate to each action in $A$ a boolean variable. Let $\underline{a} = a_1, \ldots, a_m$ be a vector of $m$ distinct boolean variables (also distinct from each $x_i$), where each $a_i$ corresponds to a distinct action in $A$ (let $m$ be the cardinality of $A$). For simplicity, in the following, we restrict ourselves to the case where formulas in the $m$ boolean variables $\underline{a}$ can be redefined as formulas in the variable $Act$, which can be assigned $m$ distinct values $a_1, \ldots a_m$. A boolean encoding of a formula in $Act$ into a boolean formula in $a_1, \ldots, a_m$ can be easily defined: e.g., $Act = a_1 \leftrightarrow a_1 \wedge \neg a_2 \wedge \ldots \neg a_m$. In order to keep the notation simple, from now on, when we write $a_1$ in a boolean formula we mean $a_1 \wedge \neg a_2 \wedge \ldots \neg a_m$, and similarly for the other $a_i$'s.

The transition function $R$ is represented symbolically by a formula $R(\underline{x}, \underline{a}, \underline{x}')$, where $\underline{x}' = x'_1, \ldots, x'_n$ is a vector of $n$ distinct boolean variables (also distinct from each $x_i$ and $a_i$). Intuitively, $\underline{x}'$ is used to represent the value of fluents after the execution of an action. For instance, let $R$ be the transition from state 1 to state 2 caused by *unlock* in Figure 3. Then

$$R = (\{\neg Loaded, Locked\}, unlock, \{\neg Loaded, \neg Locked\})$$

and

$$R(\underline{x}, \underline{a}, \underline{x}') = (\neg Loaded \wedge Locked \wedge unlock) \supset \neg Loaded' \wedge \neg Locked'$$

The symbolic representation of the Planning Domain in Figure 3 is the following.

1. $\underline{x} = \{Loaded, Locked\}$
2. $\underline{a} = \{lock, unlock, load, unload, wait\}$
3. $\underline{x}' = \{Loaded', Locked'\}$
4. $R(\underline{x}, \underline{a}, \underline{x}) = ((\neg Loaded \wedge Locked \wedge unlock) \supset \neg Loaded' \wedge \neg Locked') \wedge$
   $\qquad ((\neg Loaded \wedge \neg Locked \wedge lock) \supset \neg Loaded' \wedge Locked') \wedge$
   $\qquad ((\neg Loaded \wedge \neg Locked \wedge wait) \supset \neg Loaded' \wedge \neg Locked') \wedge$
   $\qquad ((\neg Loaded \wedge \neg Locked \wedge load) \supset Loaded' \wedge \neg Locked') \wedge$
   $\qquad ((Loaded \wedge \neg Locked \wedge unload) \supset \neg Loaded' \wedge \neg Locked') \wedge$
   $\qquad ((Loaded \wedge \neg Locked \wedge lock) \supset Loaded' \wedge Locked') \wedge$
   $\qquad ((Loaded \wedge Locked \wedge unlock) \supset Loaded' \wedge \neg Locked')$

Intuitively, symbolic representations of sets of states and transitions can be very compact: the number of variables in a formula does not depend in general on the number of states or transitions the formula represents. Given a Planning Domain with e.g., $10^6$ states in $S$, where the fluent *Loaded* is true in a subset $Q$ of e.g., $5 \times 10^5$ states, $S$ is represented by the formula $\top$, $Q$ by *Loaded*, and the empty set by $\bot$.

A symbolic representation of a Planning Problem $P = \langle D, I, G \rangle$ is obtained from the symbolic representation of the Planning Domain $D$, and from the boolean formulas $I(\underline{x})$ and $G(\underline{x})$. A symbolic plan for a symbolic planning domain $D$ is any formula $\phi(\underline{x}, \underline{a})$. For instance, the symbolic plan for the situated plan $\{(2, load), (3, lock)\}$ is $(\neg Loaded \wedge \neg Locked \supset load) \wedge (Loaded \wedge \neg Locked \supset lock)$.

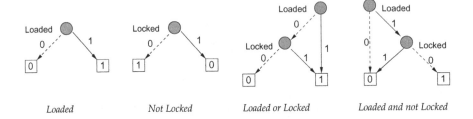

**Fig. 7.** OBDD's for *Loaded*, ¬ *Locked*, *Loaded* ∨ *Locked*, and *Loaded* ∧ ¬ *Locked*

Consider now the algorithm PLAN in Figure 4. Notice that it explores sets of states, rather than single states. The formula representing ONESTEPPLAN, can be written as a Quantified Boolean Formula (QBF).

$$\exists \underline{x}'(States(\underline{x}') \wedge R(\underline{x}, \underline{a}, \underline{x}'))) : \tag{9}$$

An equivalent propositional boolean formula can be easily obtained. For instance, $\exists y.\phi(x, y)$ can be rewritten to $\phi(x, \top) \vee \phi(x, \bot)$.

## 6.2 OBDD-Based Implementation

Planning via Symbolic Model Checking can still be implemented in different ways. A technique which has been successfully applied in the area of formal verification is that known as *Symbolic Model Checking* [2,20], which makes use of Ordered Binary Decision Diagrams (OBDD's) [1]. OBDD's are a compact representation of the assignments satisfying (and falsifying) a given boolean formula. Binary Decision Diagrams (BDD's) are rooted, directed, binary, acyclic graphs with one or two terminal nodes labeled 1 or 0 (for truth and falsity, respectively). A node in a BDD corresponds to a variable in the corresponding boolean formula. The two out-coming arcs from a BDD node represent the assignments of the corresponding variable to true (the arc is labeled with 1) and false (labeled with 0). OBDD's are BDD's with a fixed linear ordering on the propositional variables, which results in a corresponding order of the BDD nodes. Given a variable ordering, OBDD's are a canonical representation for boolean formulas. Figure 7 depicts the OBDD's for the formulas *Loaded*, ¬ *Locked*, *Loaded* ∨ *Locked*, and *Loaded* ∧ ¬ *Locked*, where the variable ordering is *Loaded* ≺ *Locked*. Figure 8 gives the OBDD's encoding the formulas

$$(\neg Loaded \wedge Locked \wedge lock) \supset \neg Loaded' \wedge \neg Locked' \tag{10}$$

$$(\neg Loaded \wedge \neg Locked \wedge lock) \supset \neg Locked' \tag{11}$$

where the variable ordering is *Loaded* ≺ *Locked* ≺ *lock* ≺ *Loaded'* ≺ *Locked'*.

Operations on two sets $S_1$ and $S_2$, e.g., the union $S_1 \cup S_2$ and the intersection $S_1 \cap S_2$, can be viewed as composing the corresponding formulas with corresponding connectives, e.g., $S_1(\underline{x}) \vee S_2(\underline{x})$ and $S_1(\underline{x}) \wedge S_2(\underline{x})$, and as a consequence, as

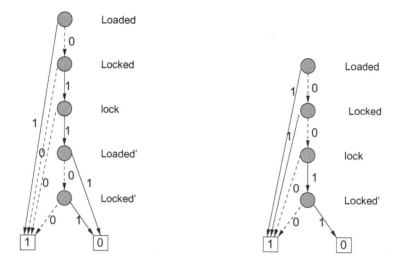

**Fig. 8.** OBDD's for the formulas (10) and (11)

operations on the corresponding OBDD's. For instance, in Figure 7, the OBDD for the formula *Loaded* ∧ ¬*Locked* (the rightmost one in the figure) is obtained from the the OBDD's for the formulas *Loaded* and ¬*Locked* (the two leftmost ones) by simply replacing the terminal node labeled "1" of the OBDD for *Loaded* with the OBDD for ¬*Locked*.

OBDD's retain the advantage of a symbolic representation: their size (the number of nodes) does not necessarily depend on the actual number of assignments (each representing, e.g., a state or a state transition). Furthermore, OBDD's provide an efficient implementation of the operations needed for manipulating sets of states and transitions, e.g., union, projection and intersection.

## 7  Related Work

The idea of a model-theoretic approach to planning is along the lines proposed in [17], which argues in favor of a model-theoretic approach to knowledge representation and reasoning in Artificial Intelligence.

Planning as Model Checking is a major conceptual shift w.r.t. most of the research done in planning so far, like STRIPS-like planning (see, e.g., [13,21]) and deductive planning (see, e.g., [24,25]). The framework is much more expressive than STRIPS-like planning, and is closer to the expressiveness of deductive planning frameworks.

The Planning as Model Checking approach has provided the possibility to tackle and solve planning problems which have never been solved so far, like strong planning [7] and strong cyclic planning [6,10] in non-deterministic domains. The experimental results reported in [7,6,5] show that Planning as Model Checking can be implemented efficiently.

Planning as propositional satisfiability [19,18] is conceptually similar to Planning as Model Checking (even if technically different), since it is based on the idea that planning should be done by checking semantically the truth of a formula. The framework of planning as propositional satisfiability has been limited so far to deterministic classical problems.

The work in [22] exploits the idea of Planning as Model Checking presented in [4,7,6] to build an OBDD-based planner. [11] proposes a framework similar to ours, which is based on an automata-theoretic approach to planning for LTL specifications. In [11] there is no notion of situated plans.

## 8 Conclusions and Acknowledgements

The goal of this paper has been to provide an introduction, with various elements of novelty, to the Planning as Model Checking paradigm. Various papers on this topic have been published – see the references. The three papers which introduce the two key intuitions are: [4] that first introduces the idea of seeing planning as a semantic problem, and [7,6] that first introduce the idea of seeing a plan as a set of state-action pairs. The first idea was envisaged by the first author of this paper as a follow on of the work, described in [16], where he first introduced and analyzed the graph of the states (the Kripke Structure) of a planning problem. The second idea was envisaged by the authors of [7,6] as an effective and practical way to represent conditional and iterative plans. The ideas reported in Sections 3 and 4 had never been written before. Many people in our group have given substantial contributions to the development of the approach, the most noticeable are: Alessandro Cimatti, Marco Roveri, Enrico Giunchiglia and Marco Daniele.

## References

1. R. E. Bryant. Graph-Based Algorithms for Boolean Function Manipulation. *IEEE Transactions on Computers*, C-35(8):677–691, August 1986.
2. J. R. Burch, E. M. Clarke, K. L. McMillan, D. L. Dill, and L. J. Hwang. Symbolic Model Checking: $10^{20}$ States and Beyond. *Information and Computation*, 98(2):142–170, June 1992.
3. A. Cimatti, E.M. Clarke, F. Giunchiglia, and M. Roveri. NuSMV: a new Symbolic Model Verifier. In N. Halbwachs and D. Peled, editors, *Proceedings Eleventh Conference on Computer-Aided Verification (CAV'99)*, number 1633 in Lecture Notes in Computer Science, pages 495–499, Trento, Italy, July 1999. Springer.
4. A. Cimatti, E. Giunchiglia, F. Giunchiglia, and P. Traverso. Planning via Model Checking: A Decision Procedure for $\mathcal{AR}$. In S. Steel and R. Alami, editors, *Proceeding of the Fourth European Conference on Planning*, number 1348 in Lecture Notes in Artificial Intelligence, pages 130–142, Toulouse, France, September 1997. Springer-Verlag. Also ITC-IRST Technical Report 9705-02, ITC-IRST Trento, Italy.

5. A. Cimatti and M. Roveri. Conformant Planning via Model Checking. In Susanne Biundo, editor, *Proceeding of the Fifth European Conference on Planning*, Lecture Notes in Artificial Intelligence, Durham, United Kingdom, September 1999. Springer-Verlag.

6. A. Cimatti, M. Roveri, and P. Traverso. Automatic OBDD-based Generation of Universal Plans in Non-Deterministic Domains. In *Proceeding of the Fifteenth National Conference on Artificial Intelligence (AAAI-98)*, Madison, Wisconsin, 1998. AAAI-Press. Also IRST-Technical Report 9801-10, Trento, Italy.

7. A. Cimatti, M. Roveri, and P. Traverso. Strong Planning in Non-Deterministic Domains via Model Checking. In *Proceeding of the Fourth International Conference on Artificial Intelligence Planning Systems (AIPS-98)*, Carnegie Mellon University, Pittsburgh, USA, June 1998. AAAI-Press.

8. E. Clarke, O. Grumberg, and D. Long. Model Checking. In *Proceedings of the International Summer School on Deductive Program Design*, Marktoberdorf, Germany, 1994.

9. E. M. Clarke and O. Grumberg. Research in automatic verification and finite-state concurrent systems. *Annual Review of Computer Science*, 2(1):269–289, 1987.

10. M. Daniele, P. Traverso, and M. Y. Vardi. Strong Cyclic Planning Revisited. In Susanne Biundo, editor, *Proceeding of the Fifth European Conference on Planning*, Lecture Notes in Artificial Intelligence, Durham, United Kingdom, September 1999. Springer-Verlag.

11. G. de Giacomo and M.Y. Vardi. Automata-theoretic approach to planning with temporally extended goals. In Susanne Biundo, editor, *Proceeding of the Fifth European Conference on Planning*, Lecture Notes in Artificial Intelligence, Durham, United Kingdom, September 1999. Springer-Verlag.

12. E. A. Emerson. Temporal and modal logic. In J. van Leeuwen, editor, *Handbook of Theoretical Computer Science, Volume B: Formal Models and Semantics*, chapter 16, pages 995–1072. Elsevier, 1990.

13. R. E. Fikes and N. J. Nilsson. STRIPS: A new approach to the application of Theorem Proving to Problem Solving. *Artificial Intelligence*, 2(3-4):189–208, 1971.

14. M. Georgeff. An embedded reasoning and planning system. In J. Tenenberg, J. Weber, and J. Allen, editors, *Proc. from the Rochester Planning Workshop: from Formal Systems to Practical Systems*, pages 105–128, Rochester, 1989.

15. M. Georgeff and A. L. Lansky. Reactive reasoning and planning. In *Proc. of the 6th National Conference on Artificial Intelligence*, pages 677–682, Seattle, WA, USA, 1987.

16. F. Giunchiglia. Abstrips abstraction – Where do we stand? Technical Report 9607-10, ITC-IRST, Trento, Italy, July 1996. To appear 1999 in the Artificial Intelligence Review.

17. J. Y. Halpern and M. Y. Vardi. Model Checking vs. Theorem Proving: A Manifesto. In J. Allen, R. Fikes, and E. Sandewall, editors, *Principles of Knowledge Representations and Reasoning: Proceedings fo the Second International Conference*, pages 325–334, 1991.

18. H. Kautz and B. Selman. BLACKBOX: A new approach to the application of theorem proving to problem solving. In *Working notes of the AIPS-98 Workshop on Planning as Combinatorial Search*, 1998.

19. Henry Kautz and Bart Selman. Pushing the Envelope: Planning, Propositional Logic, and Stochastic Search. In *Proc. AAAI-96*, pages 1194–1201, 1996.

20. K.L. McMillan. *Symbolic Model Checking*. Kluwer Academic Publ., 1993.

21. J. Penberthy and D. Weld. UCPOP: A sound, complete, partial order planner for ADL. In *Proc. of KR-92*, 1992.

22. R. Jensen and M. Veloso. Obdd-based universal planning: Specifying and solving planning problems for synchronized agents in non-deterministic domains. Technical report, CMU, Carnegie Mellon University, USA, 1999.

23. M. J. Schoppers. Universal plans for Reactive Robots in Unpredictable Environments. In *Proc. of the 10th International Joint Conference on Artificial Intelligence*, pages 1039–1046, 1987.

24. S. Steel. Action under Uncertainty. *J. of Logic and Computation, Special Issue on Action and Processes*, 4(5):777–795, 1994.

25. W. Stephan and S. Biundo. A New Logical Framework for Deductive Planning. In *Proc. of IJCAI93*, pages 32–38, 1993.

26. P. Traverso and L. Spalazzi. A Logic for Acting, Sensing and Planning. In *Proc. of the 14th International Joint Conference on Artificial Intelligence*, 1995. Also IRST-Technical Report 9501-03, IRST, Trento, Italy.

27. M. Y. Vardi and P. Wolper. An automata-theoretic approach to automatic program verification. In *Proc. of LICS86*, pages 332–344, 1986.

# Conformant Planning via Model Checking

Alessandro Cimatti[1] and Marco Roveri[1,2]

[1] ITC-IRST, Via Sommarive 18, 38055 Povo, Trento, Italy,
[2] DSI, University of Milano, Via Comelico 39, 20135 Milano, Italy
{cimatti,roveri}@irst.itc.it

**Abstract.** Conformant planning is the problem of finding a sequence of actions that is guaranteed to achieve the goal for any possible initial state and nondeterministic behavior of the planning domain. In this paper we present a new approach to conformant planning. We propose an algorithm that returns the set of all conformant plans of minimal length if the problem admits a solution, otherwise it returns with failure. Our work is based on the planning via model checking paradigm, and relies on symbolic techniques such as Binary Decision Diagrams to compactly represent and efficiently analyze the planning domain. The algorithm, called CMBP, has been implemented in the MBP planner. CMBP is strictly more expressive than the state of the art conformant planner CGP. Furthermore, an experimental evaluation suggests that CMBP is able to deal with uncertainties more efficiently than CGP.

## 1 Introduction

The planning via model checking [5,8,7,9] paradigm is based on the interpretation of a planning domain as a finite state automaton [5]. A high level action language, AR [10], is used to describe complex, nondeterministic domains with multiple initial states, and actions with conditional and uncertain effects. Symbolic representation and exploration techniques on the style of symbolic model checking [3,15], based on the use of Binary Decision Diagrams (BDDs) [2], allow for efficient planning in nondeterministic domains. The planning algorithm presented in [8] allows to find *strong* plans, i.e. conditional (contingent) plans which are guaranteed to achieve the goal for any initial state and any possible nondeterministic evolution of the domain. The algorithms defined in [7] and in [9] also allow for the generation of iterative trial-and-error strategies.

The work in [8,7,9] rely on the hypothesis of complete run-time observability. That is, the status of the world after the execution of a (possibly nondeterministic) action is assumed to be completely observable. The derived plans can be (heavily) conditioned to run-time observations. However, in many real world situations, sensorial information may be costly or unavailable, and techniques are needed to deal with incomplete run-time observability. In this work we extend the planning via model checking paradigm by proposing a new algorithm for conformant planning, i.e. the problem of finding a plan achieving the goal for any possible contingency in total absence of run-time information. Since no information is available at run time, the plan can not be conditioned to run-time observation, and thus it must be a sequence of actions, i.e. a classical plan. Differently from the classical planning problem, however, here a sequence of actions can result in (many) different executions, depending on the initial state and on

S. Biundo and M. Fox (Eds.): ECP-99, LNAI 1809, pp. 21–34, 2000.

the different uncertain outcomes of actions. This makes conformant planning much harder than classical planning.

The conformant planning algorithm is applicable to complex planning domains, with conditional actions, uncertainty in the initial state and in the outcomes of actions. The algorithm is complete, i.e. it returns with failure if and only if the problem admits no conformant solution. If a solution exists, it returns *all* conformant plans of minimal length. The algorithm has been implemented in MBP (Model Based Planner) [5,8,7], a planner developed on top of the NuSMV [4] model checker, and an experimental analysis has been carried out. The experimental results show that the algorithm can solve rather complex problems, and compares nicely with the state of the art conformant planner CGP [19]. In particular, it is able to express and solve problems with uncertain effects of actions, which can not be expressed in CGP. Furthermore, differently from CGP, our algorithm is not directly related to the *number* of initial states and uncertainties in action effects, and can plan rather efficiently in highly nondeterministic domains.

This paper is structured as follows. In section 2 we present some necessary background. In section 3 we describe the algorithm, and in section 4 we present the experimental results. In section 5 we draw the conclusions and discuss some future research.

## 2  Background

A planning domain is a 4-tuple $\mathcal{D} = (\mathcal{F}, \mathcal{S}, \mathcal{A}, \mathcal{R})$, where $\mathcal{F}$ is the (finite) set of fluents (atomic propositions), $\mathcal{S} \subseteq 2^{\mathcal{F}}$ is the set of states, $\mathcal{A}$ is the (finite) set of actions, and $\mathcal{R} \subseteq \mathcal{S} \times \mathcal{A} \times \mathcal{S}$ is the transition relation. Intuitively, a state is identified with the set of propositions holding in it. $\mathcal{R}(s, \alpha, s')$ holds iff when executing the action $\alpha$ in the state $s$ the state $s'$ is a possible outcome. An action $\alpha$ is not applicable in $s$ iff there is no state $s'$ such that $\mathcal{R}(s, \alpha, s')$ holds. An action $\alpha$ has an uncertain outcome in $s$ if there are two distinct states $s'$ and $s''$ such that $\mathcal{R}(s, \alpha, s')$ and $\mathcal{R}(s, \alpha, s'')$. In the following we assume a planning domain $\mathcal{D}$ is given. We say that an action $\alpha$ is applicable in the set of states $S$ if it is applicable to every state of $S$. The result of executing an action $\alpha$ in the set of states $S$ (also called the image of $S$ under $\alpha$), written $Exec[\alpha](S)$, is the set of all possible outcomes of the execution of $\alpha$ in any state of $S$, i.e.

$$Exec[\alpha](S) \doteq \{s' \mid \mathcal{R}(s, \alpha, s') \ with \ s \in S\}$$

If $s$ is a state, we write $Exec[\alpha](s)$ instead of $Exec[\alpha](\{s\})$. The *weak* preimage of a set of states $S$ under the action $\alpha$, written $WPreImage[\alpha](S)$, is the set of all states where the execution of $\alpha$ can lead to $S$. In symbols,

$$WPreImage[\alpha](S) \doteq \{s \mid \mathcal{R}(s, \alpha, s') \ with \ s' \in S\}$$

We call this set weak preimage to stress the fact that, for every state in it, reaching $S$ when executing $\alpha$ is possible but not necessary. The *strong* preimage of a set $S$ under the action $\alpha$, written $SPreImage[\alpha](S)$, is the set of all states where $\alpha$ is applicable and every possible execution is in $S$. I.e.,

$$SPreImage[\alpha](S) \doteq \{s \mid \emptyset \neq Exec[\alpha](s) \subseteq S\}$$

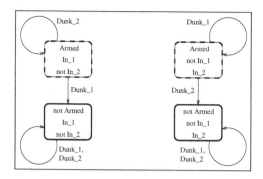

**Fig. 1.** The automaton for the BT domain

In this paper we consider plans to be sequences of actions. We use $\epsilon$ for the 0-length plan, $\alpha$ to denote an action, $\pi$ and $\rho$ to denote plans, and $\pi; \rho$ for plan concatenation. The applicability set of a plan is the set of states from which we can execute any prefix of the plan without ending up in a state where the rest of the plan is not applicable. The execution of a plan in a set of states is the set of "final" states of the possible execution traces from any of the initial states.

**Definition 1 (Applicability set of a Plan).** *Let $\pi$ be a plan. The applicability set of $\pi$, written $Appl[\pi]$, is a subset of $S$ defined as follows:*

1. $Appl[\epsilon] = S$;
2. $Appl[\alpha] = \{s \mid Exec[\alpha](s) \neq \emptyset\}$;
3. $Appl[\alpha; \rho] = \{s \mid s \in Appl[\alpha], \ and \ Exec[\alpha](s) \subseteq Appl[\rho]\}$;

**Definition 2 (Plan Execution).** *Let $S$ be a finite set of states. Let $\pi$ be a plan for $\mathcal{D}$. The execution of $\pi$ in $S$, written $Exec[\pi](S)$, is defined as:*

1. $Exec[\epsilon](S) = S$;
2. $Exec[\alpha](S) = \{s' \mid s \in S, \ and \ \mathcal{R}(s, \alpha, s')\}$;
3. $Exec[\alpha; \pi](S) = Exec[\pi](Exec[\alpha](S))$;

The classical example used to illustrate conformant planning is the bomb in the toilet (BT) problem. Figure 1 depicts the corresponding automaton. There are two packages, and one of them contains an armed bomb. It is possible to dunk either package in the toilet (actions $Dunk_1$ and $Dunk_2$). Dunking the package containing the bomb has the effect of disarming the bomb, while dunking the other package has no effect. Initially the bomb is armed, but there is uncertainty in the initial configuration since it is not known where the bomb is (dashed line states). We want to find a conformant solution to the problem of disarming the bomb, i.e. a sequence of actions that will disarm the bomb for all initial states. In this case, there are two possible conformant plans of length 2, namely dunking both packages in either order.

A planning probelm is a triple $(\mathcal{D}, Init, Goal)$, where $\mathcal{D}$ is the planning domain, and $Init$ and $Goal$ are nonempty sets of states of $\mathcal{D}$. In the following, when clear from the context, we omit the domain from a planning problem. A formal characterization of conformant planning can be given as follows.

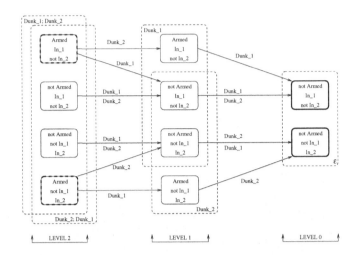

**Fig. 2.** Solving the BT problem

**Definition 3 (Conformant Plan).** *The plan $\pi$ is a conformant plan for (a conformant solution to) the planning problem $(\mathcal{D}, Init, Goal)$ iff $Init \subseteq Appl[\pi]$, and $Exec[\pi](Init) \subseteq Goal$.*

In words, a plan $\pi$ is a conformant solution to a planning problem $(Init, Goal)$ if two conditions are satisfied. First, $\pi$ must be applicable in $Init$, i.e. after executing any prefix of $\pi$ in any of the initial states, the remaining plan is always applicable. Second, all the states resulting from the execution of $\pi$ in $Init$ must be goal states.

## 3   The Conformant Planning Algorithm

The conformant planning algorithm uses as data structures states-plan (SP) tables, of the form $SPT = \{(S_1.\pi_1)\ldots(S_n.\pi_n)\}$ where, for $i = 1,\ldots,n$, $S_i$ is a set of states, $\pi_i$ is a sequence of actions, and $\pi_i \neq \pi_j$ for all $j\neq i$. We call $(S_i.\pi_i)$ a states-plan pair, and $S_i$ the set of states indexed by $\pi_i$. When no ambiguity arises, we write $SPT(\pi_i)$ for $S_i$. The intuition is that $\pi_i$ is a conformant solution for any planning problem $(S, Goal)$, with $S \subseteq S_i$. Thus we call $S_i$ *conformance set* of $\pi_i$ in $SPT$.

The algorithm proceeds backwards, from the goal to the initial states. It performs a breadth first search, building at each step conformant plans of increasing length. The status of the search (a level) is represented by a SP table, containing plans of the same length. The SP tables are stored in an array, $SPTarr$, $SPTarr[i]$ being the SP table corresponding to the $i$-th level of search.

Figure 2 describes how the algorithm solves the BT problem. The goal states are depicted with a thick solid line. A SP pair is depicted as states encircled by a dashed line, annotated by the indexing plan. The SP table at level 0, $SPTarr[0]$, is $\{(Goal.\epsilon)\}$, i.e. the set of goal states indexed by the 0-length plan $\epsilon$. (Notice that $\epsilon$ is a conformant solution to every problem with goal set $Goal$ and initial states contained in $Goal$.) The SP table at level 1, $SPTarr[1]$, contains two SP pairs with (overlapping) sets of states indexed by the length 1 plans $Dunk_1$ and

```
        function CONFORMANTPLAN(Init, Goal)
0       begin
1           i = 0;
2           SPTarr[0] := { (Goal. ε) };
3           Plans = GETPLANS(Init, SPTarr[0]);
4           while ((SPTarr[i] ≠ ∅) ∧ (Plans = ∅)) do
5               i := i + 1;
6               SPTarr[i] := CONFORMANTPREIMAGE(SPTarr[i-1]);
7               SPTarr[i] := CONFORMANTPRUNE(SPTarr, i);
8               Plans := GETPLANS(Init, SPTarr[i]);
9           done
10          if (SPTarr[i] = ∅) then
11              return Fail;
12              else return Plans;
13      end
```

**Fig. 3.** The conformant planning algorithm.

$Dunk_2$. The set indexed by $Dunk_1$, $SPTarr[1](Dunk_1)$, contains all states where $Dunk_1$ is applicable and all possible resulting states are in $Goal$. Notice that for *all* the states in this set, $Dunk_1$ leads to the goal. Thus, $Dunk_1$ is a conformant plan for every subset of $SPTarr[1](Dunk_1)$. However, neither of the SP pairs of length 1 corresponds to a conformant solution to our problem, because neither of the corresponding conformance sets contains all the initial states. Notice also that, under the hypothesis of complete observability, after one step the Strong Planning procedure presented in [8] would return a conditional plan specifying to execute only one action, i.e. dunk exactly the package containing the bomb. At level 2, $SPTarr[2]$ contains two SP pairs corresponding to the plans $Dunk_1; Dunk_2$ and $Dunk_2; Dunk_1$. Both the corresponding sets of states contain all the initial states (thick dashed line). This means that for all initial states either plan is applicable and will result in a goal state. Thus, we have found two conformant plans for the BT problem. These plans are conformant for any choice of the initial states. $SPTarr[2]$ does not contain all possible plans of length 2. $Dunk_1; Dunk_1$ and $Dunk_2; Dunk_2$ are not present. The reason is that, for each $i$, $SPTarr[2](Dunk_i; Dunk_i)$ would not differ from $SPTarr[1](Dunk_i)$. In other words, $Dunk_i; Dunk_i$ is subsumed by $Dunk_i$, and can be pruned. In general, if the expansion of a further level only results in no plans or plans which are subsumed by shorter plans, then the algorithm terminates concluding that the problem admits no conformant plan.

## 3.1   Set-Theoretic View

The conformant planning algorithm CONFORMANTPLAN($Init, Goal$), presented in Figure 3, takes in input a planning problem in form of the set of states $Init$ and $Goal$. It returns $Fail$ if and only if the problem admits no conformant solution. If a conformant plan exists, CONFORMANTPLAN($Init, Goal$) returns the set of all the conformant plans of minimal length.

The algorithm proceeds as follows, by filling the array of SP tables $SPTarr$. First it checks if there are plans of length 0, i.e. if $\epsilon$ is a solution. The function

GETPLANS, given a SP table and a set of (initial) states, computes the set of all possible conformant plans contained in the SP table.

$$\text{GETPLANS}(Init, SPT) \doteq \{\pi \mid there\ exists\ (S.\pi) \in SPT\ and\ Init \subseteq S\} \quad (1)$$

If no conformant plan of length $i$ exists (($Plans = \emptyset$) in line 4), then we enter the loop, and build conformant plans of increasing length (lines 5 to 8). The iteration terminates (line 4) when either a plan is found ($Plans \neq \emptyset$), or the space of conformant plans has been completely explored ($SPTarr[i] = \emptyset$).

At each iteration, the function CONFORMANTPREIMAGE is called to build a new SP table, containing conformant plans of length $i$, extending the conformant plans of length $i-1$ contained in $SPTarr[i-1]$.

$$\text{CONFORMANTPREIMAGE}(SPT) \doteq \quad (2)$$
$$\{(S\ .\ \alpha; \pi) \mid there\ exists\ (S'.\pi) \in SPT,\ and\ S = SPreImage[\alpha](S') \neq \emptyset\}$$

The resulting SP table is then stored in the $i$-th position of $SPTarr$. The function CONFORMANTPRUNE is responsible to remove from the newly generated SP table the plans which are either subsumed by other plans of the same length, or by plans present in the SP tables built at previous steps. It takes in input the array of SP tables $SPTarr$, and an index of the current step.

$$\text{CONFORMANTPRUNE}(SPTarr, i) \doteq$$
$$\{(S'.\pi') \in SPTarr[i] \mid$$
$$there\ is\ no\ (S.\pi) \in SPTarr[i]\ such\ that\ \pi \neq \pi'\ and\ S' \subsetneq S, \quad (3)$$
$$and\ for\ all\ j < i,\ there\ is\ no\ (S.\pi) \in SPTarr[j].(S' \subset S)\}$$

The termination of the algorithm follows from the calls to CONFORMANTPRUNE, which guarantee that the set of explored conformance sets is monotonically increasing, and thus a fix point is eventually reached when a plan does not exist (given the finiteness of the domain). The optimality of the algorithm follows from the breadth-first style of the search.

### 3.2   Symbolic Representation

From a conceptual point of view the algorithm of Figure 3 is rather simple. The problem is how to implement it efficiently. The basic idea, mutuated from symbolic model checking [15,3], is to represent the sets to be computed (e.g. sets of states, SP tables) symbolically, by means of propositional and quantified boolean formulae (QBF). These formulae, in turn, are represented and efficiently manipulated as BDDs. In the rest of this section we reinterpret the algorithm in terms of manipulation of propositional formulae. The issues related to BDDs are discussed in the next section.

We have a vector $\mathbf{x}$ of (distinct) boolean variables, called *state* variables, used to encode sets of states. For instance, for the BT problem, the variables in $\mathbf{x}$ could be $Armed$, $In_1$ and $In_2$. A state corresponds to a complete assignment to the variables in $\mathbf{x}$. The assignment $\{(Armed.\top)(In_1.\top)(In_2.\bot)\}$ (we write $\top$

and $\perp$ for the true and false truth values) corresponds to the state where the bomb is in package 1, and armed. We use formulae as representatives of the set of their models. Thus, a propositional formula in the variables in $\mathbf{x}$, written $\phi(\mathbf{x})$, represents the set of the states corresponding to the assignments which make $\phi$ true. For instance, the formula $\neg Armed$ represents the set of goal states, i.e. the states where the bomb is not armed. The formula $Armed \wedge (In_1 \leftrightarrow \neg In2)$ represents the set of initial states.

Another vector of *action* variables, $\boldsymbol{\alpha}$, is used to represent actions. For the BT problem, with a sequential encoding (i.e. assuming that only one action can be executed at each time), we can use one boolean variable $Act$, where the assignment $\{(Act.\top)\}$ represents the action $Dunk_1$, and the assignment $\{(Act.\perp)\}$ represents the action $Dunk_2$. A formula $\psi(\mathbf{x}, \boldsymbol{\alpha})$ represents a relation between states and actions (e.g., a universal plan, or an applicability condition). The formula $Armed \wedge Act$ specifies a relation holding between action $Dunk_1$, and every state where $Armed$ holds.

Transitions are 3-tuples containing a state (the initial state of the transition), an action (the action being executed), and a state (the resulting state of the transition). To represent the final state of transitions we use an additional vector of (next) state variables $\mathbf{x}'$. The transition relation of the automaton corresponding to the planning domain is thus represented by a formula $\mathcal{R}(\mathbf{x}, \boldsymbol{\alpha}, \mathbf{x}')$, each satisfying assignment of which represents a particular transition.

In order to represent SP tables, we need a way to represent plans. A plan of length $i$ is represented as an assignment to the vectors of plan variables, $\boldsymbol{\alpha}_1, \dots, \boldsymbol{\alpha}_i$, where each vector of variables $\boldsymbol{\alpha}_n$ ranges over actions, and represents the $n$-th action of a plan. For the BT problem, the assignment $\{(Act_1.\top) (Act_2.\perp)\}$ represents the plan $Dunk_1; Dunk_2$. The formula $\neg Act_1$ represents the set of the two plans of length 2 $Dunk_2; Dunk_1$ and $Dunk_2; Dunk_2$, since it imposes no constraint on the second action. In the following we assume that the variables in $\mathbf{x}, \mathbf{x}', \boldsymbol{\alpha}, \boldsymbol{\alpha}_1, \dots, \boldsymbol{\alpha}_i$ are all distinct. An SP table containing plans of length $i$ is represented by a formula in the state variables $\mathbf{x}$ and plan variables $\boldsymbol{\alpha}, \boldsymbol{\alpha}_1, \dots, \boldsymbol{\alpha}_i$.

Using a symbolic representation, we exploit the fact that if a variable $v$ does not occur in $\phi$, then it is irrelevant for the truth value of $\phi$: any satisfying assignment of $\phi$ where the truth value of $v$ is reversed is still a satisfying assignment. In general, the cardinality of the set represented by a given formula has a multiplying factor of two to the power of the number of variables which do not occur in the formula. This explains why a symbolic representation can have a dramatic improvement over an explicit-state (enumerative) representation.

In the following we describe in terms of propositional and QBF transformations some of the operations of the algorithm. The complete description can be found in [6]. We indicate with $\phi[\mathbf{v}'/\mathbf{v}]$ the parallel substitution (also called "shifting") in the formula $\phi$ of the variables in vector $\mathbf{v}$ with the (corresponding) variables in $\mathbf{v}'$. The computation of CONFORMANTPREIMAGE($SPT$), can be described as follows (where $SPT$ is the SP table in input, representing plans of length $i - 1$):

$$\text{CONFORMANTPREIMAGE}(SPT) \doteq \tag{4}$$
$$(\forall \mathbf{x}'.(\mathcal{R}(\mathbf{x}, \boldsymbol{\alpha}, \mathbf{x}') \to SPT(\mathbf{x}, \boldsymbol{\alpha}_{i-1}, \ldots, \boldsymbol{\alpha}_1)[\mathbf{x}'/\mathbf{x}]) \quad \wedge \quad \exists \mathbf{x}'.\mathcal{R}(\mathbf{x}, \boldsymbol{\alpha}, \mathbf{x}'))[\boldsymbol{\alpha}_i/\boldsymbol{\alpha}]$$

The free variables of the resulting formula are the current state variables $\mathbf{x}$ and the plan variables $\boldsymbol{\alpha}_i, \ldots, \boldsymbol{\alpha}_1$. The action variables $\boldsymbol{\alpha}$ in $\mathcal{R}$ are renamed to plan variables $\boldsymbol{\alpha}_i$. The next state variables in $\mathcal{R}$ and in $SPTarr$ (resulting from the shifting of $\mathbf{x}$ to $\mathbf{x}'$) are universally quantified away. Each set of assignments satisfying (4) and agreeing on the values assigned to plan variables represents a relation between a set of states and a plan of length $i$, i.e. a SP pair.

GETPLANS extracts the assignments to plan variables such that the corresponding set contains the initial states. In symbols,

$$\text{GETPLANS}(Init, SPT) \doteq \forall \mathbf{x}.(Init(\mathbf{x}) \to SPT(\mathbf{x}, \boldsymbol{\alpha}_i, \ldots, \boldsymbol{\alpha}_1)) \tag{5}$$

## 4   Experimental Results

In this section we discuss some implementational issues, and present some results of the experimental evaluation (all the details are given in [6]). The conformant planning algorithm was implemented in MBP. MBP is based on the NUSMV model checker, is written in C, and uses the CUDD [20] state-of-the-art BDD package. MBP takes in input planning domains described in AR [10], generates the corresponding symbolic representation, and can apply different planning algorithms to the specified planning problems. In the following we call CMBP the conformant planning algorithm implemented in MBP.

The conformant planners which are most significant for comparison with CMBP are CGP [19] and QBFPLAN [17]. CGP extends the ideas of GRAPHPLAN [1] to deal with uncertainty. Basically, a planning graph is built of every possible sequence of possible worlds, and constraints among planning graphs are propagated to ensure conformance. We consider CGP the state of the art in conformant planning. CGP was shown to outperform several other planners such as Buridan [16] and UDTPOP [14] (see [19] for a detailed comparison).

QBFPLAN is (our name for) the planning system by Rintanen. QBFPLAN generalizes the idea of SAT-based planning [12,13,11] to nondeterministic domains, by encoding problems in QBF. Given a bound on the length of the plan, first a QBF encoding of the problem is generated, and then a QBF solver [18] is called. If no solution is found, a new encoding for a longer plan must be generated and solved. QBFPLAN is interesting for comparison, since it relies on a symbolic representation based on QBF (although it differs from CMBP in many other ways).

Both CGP and QBFPLAN are incomplete, i.e. can not conclude that a planning problem has no conformant solutions. CMBP, on the other hand, thanks to the pruning step, is complete, i.e. it can discover whether no solution exists. In the experimental evaluation, for a fair comparison, CMBP was run by disabling the pruning primitives.

**Table 1.** Results for the BT and BTC problems.

| | | CMBP | | | CGP | |
|---|---|---|---|---|---|---|
| | \|P\| | #P. | \|BDD\| | Time | \|L\| | Time |
| BT(2) | 2 | 2 | 3 | 0.000 | 1 | 0.000 |
| BT(4) | 4 | 24 | 37 | 0.000 | 1 | 0.000 |
| BT(6) | 6 | 720 | 287 | 0.020 | 1 | 0.010 |
| BT(8) | 8 | 40320 | 1337 | 0.150 | 1 | 0.020 |
| BT(10) | 10 | 3628800 | 7919 | 1.330 | 1 | 0.020 |

| | | CMBP | | | CGP | |
|---|---|---|---|---|---|---|
| | \|P\| | #P. | \|BDD\| | Time | \|L\| | Time |
| BTC(2) | 3 | 2 | 11 | 0.010 | 3 | 0.000 |
| BTC(3) | 5 | 6 | 28 | 0.010 | 5 | 0.010 |
| BTC(4) | 7 | 24 | 102 | 0.010 | 7 | 0.030 |
| BTC(5) | 9 | 120 | 225 | 0.050 | 9 | 0.130 |
| BTC(6) | 11 | 720 | 483 | 0.160 | 11 | 0.860 |
| BTC(7) | 13 | 5040 | 1005 | 0.520 | 13 | 2.980 |
| BTC(8) | 15 | 40320 | 2773 | 1.850 | 15 | 13.690 |
| BTC(9) | 17 | 362880 | 5876 | 6.020 | 17 | 41.010 |
| BTC(10) | 19 | 3628800 | 12336 | 16.020 | 19 | 157.590 |

| QBFPLAN | | | |
|---|---|---|---|
| BTC(6) | | BTC(10) | |
| \|P\| | Time | \|P\| | Time |
| 1 | 0.00 | 1 | 0.02 |
| 2 | 0.01 | 2 | 0.03 |
| 3 | 0.26 | 3 | 0.78 |
| 4 | 0.63 | 4 | 2.30 |
| 5 | 1.53 | 5 | 4.87 |
| 6 | 2.82 | 6 | 8.90 |
| 7 | 6.80 | 7 | 22.61 |
| 8 | 14.06 | 8 | 52.72 |
| 9 | 35.59 | 9 | 156.12 |
| 10 | 93.34 | 10 | 410.86 |
| 11 | (+) 2.48 | 11 | 1280.88 |
| | | 13 | 3924.96 |
| | | 14 | — |
| | | ... | ... |
| | | 18 | — |
| | | 19 | (+) 16.84 |

CMBP is strictly more expressive than CGP, which can handle uncertainty only in the initial state (although [19] describes how the approach can be extended to actions with uncertain effects). The comparison with CGP was carried out only on the cases with uncertainty on the initial condition. QBFPLAN is able to handle actions with uncertain effects. This is done by introducing auxiliary (choice) variables, the assignments to which correspond to the different possible outcomes of actions. These variables need to be quantified universally to ensure conformance of the solution. However, the encoding generator of QBFPLAN has ML code as its input format. The comparison with QBFPLAN is limited to the (few) problems for which the encodings already existed.

For CMBP and CGP, all the examples were run by setting a limit to the depth of the search. Since MBP uses a serial encoding, the limit corresponds to the maximum length of the plan. In CGP, the limit is on the number of levels in the planning graph. The chosen limit was enough to find a solution for the tested problems in both systems. Differently from e.g. BLACKBOX [11], QBFPLAN does not have a heuristic to guess the "right" length of the plan. Given a limit in the length of the plan, it generates all the encodings up to the specified length, and repeatedly calls the QBF decider on encodings of increasing length until a plan is found. We specified as limit the length of the shortest solution. BDD based computations are known to be sensitive to a number of factors, such as the ordering of variables. For all the examples reported here, CMBP used a fixed ordering strategy: action variables were positioned at the top, then plan variables, and state variables. Variables of a given kind were interleaved with the corresponding auxiliary variables (e.g. $\mathbf{x}$ with $\mathbf{x}'$, $\alpha_i$ with $\beta_i$). Dynamic variable reordering was disabled. The tests were performed on an Intel 300MhZ Pentium-II, 512MB RAM, running Linux. CGP is implemented in LISP, and was compiled and run under Allegro CL 4.3 [Linux/X86;R1]. CPU time was limited to 7200 sec (two hours) for each test. In the following tables, unless otherwise specified, we write — for a test that was not completed within the above time limit.

The evaluation was performed by running the systems on a number of para-

**Table 2.** Results for the BMTC problems

| | Low Unc. | | | | | | | Mid Unc. | | | | High Unc. | | | |
|---|---|---|---|---|---|---|---|---|---|---|---|---|---|---|---|
| BMTC | | CMBP | | | | CGP | | CMBP | | CGP | | CMBP | | CGP | |
| (p,t) | IS | \|P\| | #P. | \|BDD\| | Time | \|L\| | Time | IS | Time | \|L\| | Time | IS | Time | \|L\| | Time |
| (2,2) | 2 | 2 | 4 | 15 | 0.000 | 1 | 0.000 | 4 | 0.000 | 2 | 0.010 | 8 | 0.000 | 2 | 0.030 |
| (3,2) | 3 | 4 | 48 | 70 | 0.010 | 3 | 0.020 | 6 | 0.010 | 3 | 0.040 | 12 | 0.020 | 4 | 13.560 |
| (4,2) | 4 | 6 | 768 | 268 | 0.040 | 3 | 0.030 | 8 | 0.060 | 4 | 0.460 | 16 | 0.090 | 4 | 145.830 |
| (5,2) | 5 | 8 | 15360 | 662 | 0.180 | 5 | 1.390 | 10 | 0.260 | 5 | 13,180 | 20 | 0.340 | 4 | — |
| (6,2) | 6 | 10 | 368640 | 1499 | 0.640 | 5 | 3.490 | 12 | 0.830 | 5 | — | 24 | 1.150 | | |
| (7,2) | 7 | 12 | 1.03e7 | 3250 | 2.100 | 7 | 508.510 | 14 | 2.780 | | | 28 | 3.390 | | |
| (8,2) | 8 | 14 | 3.30e8 | 8357 | 7.960 | 7 | 918.960 | 16 | 10.380 | | | 32 | 12.330 | | |
| (9,2) | 9 | 16 | 1.18e10 | 17944 | 22.820 | 7 | — | 18 | 30.370 | | | 36 | 35.510 | | |
| (10,2) | 10 | 18 | 4.75e11 | 37968 | 72.730 | | | 20 | 87.370 | | | 40 | 121.740 | | |
| (2,4) | 2 | 2 | 24 | 31 | 0.000 | 1 | 0.000 | 8 | 0.010 | 1 | 0.020 | 32 | 0.010 | 2 | 1.610 |
| (3,4) | 3 | 3 | 144 | 122 | 0.030 | 1 | 0.010 | 12 | 0.050 | 2 | 0.290 | 48 | 0.150 | 2 | 8.690 |
| (4,4) | 4 | 4 | 576 | 426 | 0.100 | 1 | 0.010 | 16 | 0.320 | 2 | 0.730 | 64 | 0.840 | 2 | 32.190 |
| (5,4) | 5 | 6 | 57600 | 1985 | 0.680 | 3 | 0.500 | 20 | 1.610 | 2 | — | 80 | 3.420 | 3 | — |
| (6,4) | 6 | 8 | 5806080 | 5905 | 3.350 | 3 | 1.160 | 24 | 6.900 | | | 96 | 12.650 | | |
| (7,4) | 7 | 10 | 6.58e08 | 14939 | 14.210 | 3 | 2.410 | 28 | 23.090 | | | 112 | 40.410 | | |
| (8,4) | 8 | 12 | 8.44e10 | 40237 | 77.420 | 3 | 8.540 | 32 | 232.150 | | | 128 | 932.820 | | |
| (9,4) | 9 | | — | — | — | 4 | — | 36 | — | | | 144 | — | | |
| (10,4) | 10 | | | | | | | | | | | 160 | | | |
| (2,6) | 2 | 2 | 60 | 56 | 0.010 | 1 | 0.010 | 16 | 0.010 | 1 | 0.200 | 128 | 0.090 | 2 | 337.604 |
| (3,6) | 3 | 3 | 720 | 423 | 0.090 | 1 | 0.010 | 24 | 0.080 | 1 | 0.830 | 192 | 1.040 | 2 | 1459.110 |
| (4,6) | 4 | 4 | 8640 | 1879 | 0.510 | 1 | 0.040 | 32 | 1.190 | 2 | 30.630 | 256 | 6.460 | 2 | 5643.450 |
| (5,6) | 5 | 5 | 86400 | 6137 | 3.080 | 1 | 0.060 | 40 | 12.260 | 2 | 30.140 | 320 | 40.770 | 2 | — |
| (6,6) | 6 | 6 | 518400 | 14265 | 17.490 | 1 | 0.100 | 48 | 118.600 | 2 | 57.300 | 384 | 1819.520 | | |
| (7,6) | 7 | 8 | 2.03e08 | 67489 | 5939.520 | 3 | 211.720 | 56 | — | 2 | — | 448 | — | | |
| (8,6) | 8 | | — | — | — | 3 | 1015.160 | 64 | | | | 512 | | | |
| (9,6) | 9 | | | | | 3 | 3051.990 | 72 | | | | 576 | | | |
| (10,6) | 10 | | | | | 2 | — | 80 | | | | 640 | | | |

meterized problem domains. The first class of problems we tackled is based on the classical bomb in the toilet problem, BT(p), where p is the parametric number of packages. The results for the BT problems are shown in Table 1 (upper left). The columns relative to CMBP are the length of the plan ($|P|$), the number of plans (#P.), the size of the BDD representing the set of conformant solutions ($|BDD|$), and the run time needed for searching the automaton (expressed in seconds). The columns relative to CGP are the number of levels in the planning graphs, and the computation time needed for the search. For the BT problem CGP is almost insensitive to the problem size, and outperforms CMBP. One reason for this is that CGP inherits from GRAPHPLAN the ability to deal with parallel actions efficiently, and the BT problem is intrinsically parallel (the depth of the planning graph is always one, i.e. all packages can be dunked in parallel).

We call BTC(p) the extension where dunking a package (always) clogs the toilet, and flushing can remove the clogging. The results for this problems are

**Table 3.** Results for the RING problems.

| | | CMBP | | | CGP | |
|---|---|---|---|---|---|---|
| | P | #P. | BDD | Time | L | Time |
| RING(2) | 5 | 2 | 10 | 0.010 | 3 | 0.070 |
| RING(3) | 8 | 2 | 23 | 0.030 | 4 | — |
| RING(4) | 11 | 2 | 35 | 0.060 | | |
| RING(5) | 14 | 2 | 47 | 0.320 | | |
| RING(6) | 17 | 2 | 59 | 1.460 | | |
| RING(7) | 20 | 2 | 71 | 7.190 | | |
| RING(8) | 23 | 2 | 83 | 35.380 | | |
| RING(9) | 26 | 2 | 95 | 167.690 | | |

| CGP on RING(5) | | | | |
|---|---|---|---|---|
| IS | L | Time | L | Time |
| 1 | 5 | 0.010 | 9 | 0.020 |
| 2 | 5 | 0.060 | 9 | 0.140 |
| 4 | 5 | 0.420 | 9 | 1.950 |
| 8 | 5 | 6.150 | 9 | 359.680 |
| 16 | 5 | — | 9 | — |

shown in Table 1. Since the BTC does not allow for parallel actions, the impact of the depth of the plan length becomes significant, and CMBP outperforms CGP. The performance of QBFPLAN is reported in the rightmost table, only for the 6 and 10 package problems. Notice that each line reports the time needed to decide whether there is a plan of length $i$. QBFPLAN is outperformed both by CGP and by CMBP. QBFPLAN does not exploit the computations performed to analyze previous levels, and thus needs to restart from scratch problems of increasing length. In the rest of the comparison we do not consider QBFPLAN.

The next class of problems, called BMTC(p,t), is the generalization of the BTC problem to the case of multiple toilets. The results are reported in Table 2. (IS is the number of initial states.) In the first class of tests ("Low Uncertainty" columns), the only uncertainty is the position of the bomb, while toilets are known to be not clogged. The basic feature of the problem is that it becomes more parallelizable when the number of toilets increases. CGP is able to fully exploit this feature, while CMBP suffers because of its serial encoding. With many toilets CGP outperforms CMBP. However, the behavior of CGP degrades as soon as more than 5 levels in the planning graph need to be explored. Consider the results for the BMTC(6,2) and BMCT(7,2) problems. Notice also that CMBP finds all the 10321920 conformant solutions to BMTC(7,2) in 2.100 seconds.

The "Mid" and "High" columns show the results in presence of more uncertainty in the initial state. In the second [third, respectively] class of tests, the status of every other [every, resp.] toilet can be either clogged or non clogged. This increases the number of possible initial states. The results show that CMBP is much less sensitive to the number of initial states, CGP is almost unable to solve what were trivial problems.

We considered another class of problems, where we have a ring of rooms, each of them with a window, which can be either open, closed or locked. The robot can move (either clockwise or counterclockwise), close the window of the room where it is, and lock it if closed. The goal is to have all windows locked. In the problem RING($r$), where $r$ is the number of rooms, the position of windows obeys the law of inertia, i.e. it remains unchanged unless changed by an action of the robot. The uncertainty in the initial states can be both in the position of the robot, and in the status of the windows. The maximum number of initial states is $r * 3^r$, corresponding to full uncertainty on the position of the robot and on the status of each window. The results, in the case of maximum uncertainty, are reported in on the left in Table 3. On the right, we plot (for the RING(5)

**Table 4.** Results for the BTUC and URING problems.

| | CMBP | | | | | | CMBP | | | |
|---|---|---|---|---|---|---|---|---|---|---|
| | \|P\| | #P. | \|BDD\| | Time | | | \|P\| | #P. | \|BDD\| | Time |
| BTUC(2) | 3 | 2 | 11 | 0.000 | | URING(2) | 5 | 2 | 10 | 0.000 |
| BTUC(3) | 5 | 6 | 28 | 0.000 | | URING(3) | 8 | 2 | 23 | 0.010 |
| BTUC(4) | 7 | 24 | 102 | 0.020 | | URING(4) | 11 | 2 | 35 | 0.030 |
| BTUC(5) | 9 | 120 | 225 | 0.050 | | URING(5) | 14 | 2 | 47 | 0.080 |
| BTUC(6) | 11 | 720 | 483 | 0.170 | | URING(6) | 17 | 2 | 59 | 0.200 |
| BTUC(7) | 13 | 5040 | 1005 | 0.530 | | URING(7) | 20 | 2 | 71 | 0.530 |
| BTUC(8) | 15 | 40320 | 2773 | 1.830 | | URING(8) | 23 | 2 | 83 | 1.370 |
| BTUC(9) | 17 | 362880 | 5876 | 6.020 | | URING(9) | 26 | 2 | 95 | 4.600 |
| BTUC(10) | 19 | 3628800 | 12336 | 17.730 | | URING(10) | 29 | 2 | 107 | 14.320 |

problem) the dependency of CGP on the number of initial states combined with the number of levels to be explored (different goals were provided which require the exploration of different levels).

Finally, we considered problems with full uncertainty in action effects, which can not be expressed in CGP. In the BTUC(p), clogging is an uncertain outcome of dunking a package. In the URING(r), at each time instant, each window can open or close nondeterministically if it is not locked. The results are reported in Table 4. The run times are lower than in the inertial cases, this is due to the fact that there is no need to represent the effects of the law of inertia.

## 5    Conclusions and Future Work

In this paper we presented a new algorithm for conformant planning. The algorithm is applicable to complex planning domains, with conditional actions, uncertainty in the initial state and in the outcomes of actions, and nondeterministic changes in the environment. The algorithm returns the set of all conformant plans of minimal length, if a solution to the planning problem exists. Otherwise, it terminates with failure. This work relies on and extends the planning via symbolic model checking paradigm presented in [5,8,7,9]. The algorithm has been designed to be implemented efficiently taking full advantage of the symbolic representation based on BDD. The experimental results show that the algorithm is able to solve rather complex problems, and compares nicely with the state of the art conformant planner CGP, and with QBFPLAN. First, CMBP is complete, i.e. it is able to decide whether a conformant plan exists. Second, CMBP is strictly more expressive than CGP, as it allows for uncertainty in the action effects. Furthermore, CGP suffers from the enumerative nature of its algorithm, and its qualitative behavior seem to depend heavily on the *number* of possible situations to be considered. The experimental evaluation suggests that CMBP is able to deal with uncertainties more efficiently than CGP.

A first direction of future activity is the investigation of parallel encodings. CMBP inherits from MBP a serial encoding, and is thus outperformed by CGP in problems with a high degree of parallelizability (e.g. when multiple toilets are available). Furthermore, optimization techniques typical of symbolic model checking, such as partitioning techniques [3], could be used to reduce the computational cost of relational products and pruning. We have also developed another algorithm for conformant planning, based on a forward (rather than backward) traversal of the state space. Another direction of future research includes its experimental evaluation, and its integration with the backward algorithm presented in this paper. Finally, conformant planning via model checking will be extended to deal with the general case of planning under partial observability.

# References

1. Avrim L. Blum and Merrick L. Furst. Fast planning through planning graph analysis. *Artificial Intelligence 1–2*, 90:279–298, 1997.
2. R. E. Bryant. Graph-Based Algorithms for Boolean Function Manipulation. *IEEE Transactions on Computers*, C-35(8):677–691, August 1986.
3. J. R. Burch, E. M. Clarke, K. L. McMillan, D. L. Dill, and L. J. Hwang. Symbolic Model Checking: $10^{20}$ States and Beyond. *Information and Computation*, 98(2):142–170, June 1992.
4. A. Cimatti, E.M. Clarke, F. Giunchiglia, and M. Roveri. NuSMV: a new Symbolic Model Verifier. In N. Halbwachs and D. Peled, editors, *Proceedings Eleventh Conference on Computer-Aided Verification (CAV'99)*, number 1633 in Lecture Notes in Computer Science, pages 495–499, Trento, Italy, July 1999. Springer.
5. A. Cimatti, E. Giunchiglia, F. Giunchiglia, and P. Traverso. Planning via Model Checking: A Decision Procedure for $\mathcal{AR}$. In S. Steel and R. Alami, editors, *Proceeding of the Fourth European Conference on Planning*, number 1348 in LNAI, pages 130–142, Toulouse, France, September 1997. Springer-Verlag.
6. A. Cimatti and M. Roveri. Conformant Planning via Model Checking. Technical Report 9908-02, ITC-IRST, Trento, Italy, August 1999.
7. A. Cimatti, M. Roveri, and P. Traverso. Automatic OBDD-based Generation of Universal Plans in Non-Deterministic Domains. In *Proceeding of the Fifteenth National Conference on Artificial Intelligence (AAAI-98)*, Madison, Wisconsin, 1998. AAAI-Press.
8. A. Cimatti, M. Roveri, and P. Traverso. Strong Planning in Non-Deterministic Domains via Model Checking. In *Proceeding of the Fourth International Conference on Artificial Intelligence Planning Systems (AIPS-98)*, Carnegie Mellon University, Pittsburgh, USA, June 1998. AAAI-Press.
9. M. Daniele, P. Traverso, and M. Y. Vardi. Strong Cyclic Planning Revisited. In Susanne Biundo, editor, *Proceeding of the Fifth European Conference on Planning*. Durham, UK, September 1999. Springer-Verlag.
10. E. Giunchiglia, G. N. Kartha, and V. Lifschitz. Representing action: Indeterminacy and ramifications. *Artificial Intelligence*, 95(2):409–438, 1997.
11. H. Kautz and B. Selman. BLACKBOX: A New Approach to the Application of Theorem Proving to Problem Solving. In *Working notes of the Workshop on Planning as Combinatorial Search*, Pittsburgh, PA, USA, June 1998.
12. Henry A. Kautz, David McAllester, and Bart Selman. Encoding Plans in Propositional Logic. In *Proc. KR-96*, 1996.
13. Henry A. Kautz and Bart Selman. Pushing the Envelope: Planning, Propositional Logic, and Stochastic Search. In *Proc. AAAI-96*, 1996.
14. Nicholas Kushmerick, Steve Hanks, and Daniel S. Weld. An algorithm for probabilistic planning. *Artificial Intelligence*, 76(1-2):239–286, September 1995.
15. K.L. McMillan. *Symbolic Model Checking*. Kluwer Academic Publ., 1993.
16. M. Peot. *Decision-Theoretic Planning*. PhD thesis, Dept. Engineering-Economic Systems — Stanford University, 1998.

17. J. Rintanen. Constructing conditional plans by a theorem-prover. *Journal of Artificial Intellegence Research*, 1999. Accepted for publication.
18. J. Rintanen. Improvements to the Evaluation of Quantified Boolean Formulae. In *16th Iinternational Joint Conference on Artificial Intelligence*. Morgan Kaufmann Publishers, August 1999. To appear.
19. David E. Smith and Daniel S. Weld. Conformant graphplan. In *Proceedings of the 15th National Conference on Artificial Intelligence (AAAI-98) and of the 10th Conference on Innovative Applications of Artificial Intelligence (IAAI-98)*, pages 889–896, Menlo Park, July 26–30 1998. AAAI Press.
20. F. Somenzi. CUDD: CU Decision Diagram package — release 2.1.2. Department of Electrical and Computer Engineering — University of Colorado at Boulder, April 1997.

# Strong Cyclic Planning Revisited

Marco Daniele[1,2], Paolo Traverso[1], and Moshe Y. Vardi[3]*

[1] IRST, Istituto per la Ricerca Scientifica e Tecnologica, 38050 Povo, Trento, Italy
[2] Dipartimento di Informatica e Sistemistica, Università "La Sapienza", 00198 Roma
[3] Department of Computer Science, Rice University, Houston TX 77251, USA
daniele@irst.itc.it, leaf@irst.itc.it, vardi@cs.rice.edu

**Abstract.** Several realistic non-deterministic planning domains require plans that encode iterative trial-and-error strategies, e.g., "pick up a block until succeed". In such domains, a certain effect (e.g., action success) might never be guaranteed *a priori* of execution and, in principle, iterative plans might loop forever. Here, the planner should generate iterative plans whose executions always have a possibility of terminating and, when they do, they are guaranteed to achieve the goal. In this paper, we define the notion of *strong cyclic plan*, which formalizes in temporal logic the above informal requirements for iterative plans, define a planning algorithm based on model-checking techniques, and prove that the algorithm is guaranteed to return strong cyclic plans when they exist or to terminate with failure when they do not. We show how this approach can be extended to formalize plans that are guaranteed to achieve the goal and do not involve iterations (*strong plans*) and plans that have a possibility (but are not guaranteed) to achieve the goal (*weak plans*). The results presented in this paper constitute a formal account for "planning via model checking" in non-deterministic domains, which has never been provided before.

## 1 Introduction

Classical planning [16,21] makes some fundamental assumptions: the planner has complete information about the initial state of the world, effects of the execution of actions are deterministic, and the solution to the planning problem can be expressed as a sequence of actions. These assumptions are unrealistic in several practical domains (e.g., robotics, scheduling, and control). The initial state of a planning problem may be partially specified and the execution of an action in the same state may have many possible effects. Moreover, plans as sequences of actions are bound to failure: non-determinism must be tackled by planning conditional behaviors, which depend on the information that can be gathered at execution time. For instance, in a realistic robotic application, the action "pick-up a block" cannot be simply described as a STRIPS-like operator [16] whose effect is that "the block is at hand" of the robot. "Pick-up a block" might result either in a success or failure, and the result cannot be known *a priori* of execution. A

---

* Supported in part by NSF grants CCR-9628400 and CCR-9700061.

S. Biundo and M. Fox (Eds.): ECP-99, LNAI 1809, pp. 35–48, 2000.

useful plan, depending on the action outcome, should execute different actions, e.g., try to pick-up the block again if the action execution has failed.

Most often, a conditional plan is not enough: plans encoding iterative trial-and-error strategies, like "pick up a block until succeed", are the only acceptable solutions. In several realistic domains, a certain effect (e.g., action success) might never be guaranteed *a priori* of execution and, in principle, iterative plans might loop forever, under an infinite sequence of failures. The planner, however, should generate iterative plans whose executions always have a possibility of terminating and, when they do, they are guaranteed to achieve the goal.

The starting point of the work presented in this paper is the framework of *planning via model checking*, together with the related system MBP, first presented in [7] and then extended to deal with non-deterministic domains in [10,9] (see also [18] for an introduction to Planning as Model Checking). [7] proposes the idea to use model checking techniques to do planning and proposes an algorithm for generating *weak plans*, i.e., plans that may achieve the goal but are not guaranteed to do so. [10] proposes an algorithm to generate *strong plans*, i.e., plans that are guaranteed to achieve a desired goal in spite of non-determinism. [9] extends [10] to generate *strong cyclic plans*, whose aim is to encode iterative trial-and-error strategies. However, no formal notion of strong cyclic plan is given in [9] and, as far as we know, in any other work.

In this paper we provide a framework for planning via model checking where weak, strong, and strong cyclic plans can be specified uniformly in temporal logic. In the paper, we focus on strong cyclic plans, since their formal specifications and the provision of a correct algorithm is still an open problem at the current state-of-the-art. Indeed, this paper builds on [9] making the following contributions.

– We provide a formal definition of strong cyclic plan based on the well-known Computation Tree Logic (CTL) [14]. The idea is that a strong cyclic plan is a solution such that "*for each* possible execution, *always* during the execution, there *exists* the possibility of *eventually* achieving the goal". The formalization is obtained by exploiting the universal and existential path quantifiers of CTL, as well as the "always" and "eventually" temporal connectives.
– We define a new algorithm for strong cyclic planning. It is guaranteed to generate plans that cannot get stuck in loops with no possibility to terminate. The algorithm in [9] did not satisfy this requirement. Moreover, the new algorithm improves the quality of the solutions by eliminating nonrelevant actions.
– We prove that the algorithm presented in this paper is correct and complete, i.e., it generates strong cyclic plans according to the formal definition while, if no strong cyclic solutions exist, it terminates with failure.

The results presented in this paper provide a formal account for planning via model checking that has never been given before. Indeed, after providing a clear framework for strong cyclic plans, we show how it can be easily extended to express weak and strong plans. Weak plans are such that *there exists* at least one execution that *eventually* achieves the goal, strong plans are such that *all* executions *eventually* achieve the goal.

The paper is structured as follows. We define the notion of planning problem in Section 2 and the notion of strong cyclic solutions in Section 3. The description of the planning algorithm is given in Section 4. Finally, in Section 5 we show how the framework can be extended to formalize weak and strong plans. We conclude the paper with a comparison with some related work.

## 2    The Planning Problem

A *(non-deterministic) planning domain* can be described in terms of *fluents*, which may assume different values in different *states*, *actions* and a *transition function* describing how (the execution of) an action leads from one state to possibly many different states.

**Definition 1 (Planning Domain).** *A planning domain D is a 4-tuple* $\langle F, S, A, R \rangle$ *where F is the finite set of fluents, $S \subseteq 2^F$ is the set of states, A is the finite set of actions, and $R : S \times A \mapsto 2^S$ is the transition function.*

Fluents belonging (not belonging) to some state $s$ are assigned to TRUE (FALSE) in $s$. Our definitions deal with Boolean fluents while examples are easier to describe through fluents ranging over generic finite domains[1]. $R(s, a)$ returns all the states the execution of $a$ from $s$ can lead to. The action $a$ is said to be *executable* in the state $s$ if $R(s, a) \neq \emptyset$.

A *(non-deterministic) planning problem* is a planning domain, a set of initial states and a set of goal states.

**Definition 2 (Planning Problem).** *A planning problem P is a 3-tuple* $\langle D, I, G \rangle$ *where D is the planning domain, $I \subseteq S$ is the set of initial states and $G \subseteq S$ is the set of goal states.*

Both $I$ and $G$ can be represented through two Boolean functions $\mathcal{I}$ and $\mathcal{G}$ over $F$, which define the sets of states in which they hold. From now on, we switch between the two representations, as sets or functions, as the context requires.

Non-determinism occurs twice in the above definitions. First, we have a set of initial states, and not a single initial state. Second, the execution of an action from a state is a set of states, and not a single state.

As an explanatory example, let us consider the situation depicted in Figure 1 (left). The situation is a very small excerpt from an application we are developing for the Italian Space Agency [4]. A tray (T) provides two positions in which two containers ($C_1$ and $C_2$) for solutions may be hosted. In addition, a kettle (K) may host one container for boiling its solution. The kettle is provided with a switch (S) that can operate only if the container is well positioned on the kettle. This situation can be formalized as shown in Figure 1 (right). The set $F$ of (non-Boolean) fluents is $\{C_1, C_2, S\}$. $C_1$ and $C_2$ represent the positions of the containers, and can be on-T (on tray), on-K-ok (on kettle, steady), or on-K-ko (on kettle, not

---

[1] For non-Boolean variables, we use a Boolean encoding similarly to [15].

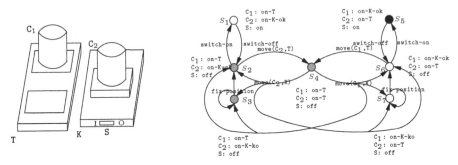

**Fig. 1.** An example (left) and its formalization (right).

steady). S represents the status of the kettle's switch (on or off). The set of states is represented by the nodes of the graph, which define fluents' values. The set of actions is represented by the edges' labels. Actions move($C_1$,T), move($C_2$,T), switch-on, and switch-off, are deterministic; move($C_1$,K), move($C_2$,K), and fix-position, are not. Indeed, when moving containers from the tray to the kettle, it can happen the containers are not correctly positioned. Moreover, it can be possible the wrong container is picked up and moved upon the kettle. Thus, $R(S_4, \text{move}(C_1,K)) = R(S_4, \text{move}(C_2,K)) = \{S_2, S_3, S_6, S_7\}$. Still, when trying to settle a container, it is possible getting no effect. Thus, $R(S_3, \text{fix-position}) = \{S_2, S_3\}$ and $R(S_7, \text{fix-position}) = \{S_6, S_7\}$. The planning problem is to boil the solution contained in $C_1$ starting from a situation where $C_1$ is on the tray and the kettle's switch is off, that is, $\mathcal{I}$ is $C_1 = \text{on-T} \wedge S = \text{off}$ (grey nodes, in Figure 1), and $\mathcal{G}$ is $C_1 = \text{on-K-ok} \wedge S = \text{on}$ (black node, in Figure 1).

A remark is in order. Non-deterministic planning problems can be expressed in different specification languages. For instance, in [7,10,9] the $\mathcal{AR}$ language [17] is used. Alternatively, we might use any language that allows us to express non-determinism, i.e., the fact that an action has multiple outcomes or, in other words, disjunctive postconditions. STRIPS-like [16] or ADL-like [21] languages (e.g., PDDL) are not expressive enough.

## 3   Strong Cyclic Plans

When dealing with non-determinism, plans have to be able to represent conditional and iterative behaviors. We define plans as *state-action tables* (resembling universal plans [22]) that associate actions to states. The execution of a state-action table can result in conditional and iterative behaviors. Intuitively, a state-action table execution can be explained in terms of a reactive loop that senses the state of the world and chooses one among the corresponding actions, if any, for the execution until the goal is reached.

**Definition 3 (State-Action Table).** *A state-action table $SA$ for a planning problem $P$ is a set of pairs $\{\langle s, a \rangle : s \in S \setminus G, a \in A, \text{ and } a \text{ is executable in } s\}$.*

The states of a state-action table may be any state, except for those in the set of goal states. Intuitively, this corresponds to the fact that when the plan

| State | Action |
|-------|--------|
| $S_1$ | `switch-off` |
| $S_3$ | `fix-position` |
| $S_2$ | `move(C_2,T)` |
| $S_6$ | `move(C_1,T)` |
| $S_4$ | `move(C_1,K)` |
| $S_4$ | `move(C_2,K)` |
| $S_7$ | `fix-position` |
| $S_6$ | `switch-on` |

**Fig. 2.** A state-action table.

achieves the goal no further action needs to be executed. Hereafter, we write STATES($SA$) for denoting the set of states in the state-action table $SA$, i.e., STATES($SA$)= $\{s : \exists a \in A.\langle s, a \rangle \in SA\}$.

**Definition 4 (Total State-Action Table).** *A state-action table $SA$ for a planning problem $P$ is* total *if, for all $\langle s, a \rangle \in SA$, $R(s,a) \subseteq$ STATES($SA$) $\cup G$.*

Intuitively, in a total state-action table, each state that can be reached by executing an action either is a goal state or has a corresponding action in the state-action table. The notion of total state-action table is important in order to capture strong (cyclic) plans, i.e., plans that must be specified for all possible outcomes of actions. In Figure 2, a total state-action table related to our example is shown.

Given a notion of plan as a state-action table, the goal is to formalize strong cyclic plans in terms of temporal logic specifications on the possible executions of state-action tables. A preliminary step is to formalize the notion of execution of a state-action table.

**Definition 5 (Execution).** *Let $SA$ be a state-action table for the planning problem $P$. An* execution *of $SA$ starting from the state $s_0 \in$ STATES($SA$) $\cup G$ is an infinite sequence $s_0 s_1 \ldots$ of states in $S$ such that, for all $i \geq 0$, either $s_i \in G$ and $s_i = s_{i+1}$, or $s_i \notin G$ and, for some $a \in A$, $\langle s_i, a \rangle \in SA$ and $s_{i+1} \in R(s_i, a)$.*

Executions are infinite sequences of states. Depending on non-determinism, we may have many possible executions corresponding to a state-action table. Each nongoal state $s_i$ has as successor a state $s_{i+1}$ reachable from $s_i$ by executing an action corresponding to $s_i$ in the state-action table; when the sequence reaches a goal state, the execution is extended with an infinite sequence of the same goal state. Of course, nontotal state-action tables may induce also executions *dangling* at nongoal states, i.e., executions reaching a nongoal state for which no action is provided.

The total state-action tables we are interested in, i.e., strong cyclic plans, are such that, informally, all their executions either lead to the goal or loop over a set of states from which the goal could be eventually reached. With respect to the state-action table of Figure 2, an example of the former case is executing `switch-on` when at $S_6$, which surely leads to the goal; while an example of the latter case is executing `fix-position` in $S_7$ that, even if looping at $S_7$, may lead to $S_6$ and, therefore, to the goal.

In order to capture the notion of strong cyclic plan, we need a formal framework that allows us to state temporal properties of executions. We have chosen the branching time logic CTL [14], which provides universal and existential path quantifiers and temporal operators like "eventually" and "always". CTL formulas are defined starting from a finite set $\mathcal{P}$ of propositions, the Boolean connectives, the temporal connectives X ("next-time") and U ("until"), and the path quantifiers E ("exists") and A ("for all"). Given a finite set $\mathcal{P}$ of propositions, CTL formulas are inductively defined as follows:

- Each element of $\mathcal{P}$ is a formula;
- $\neg\psi$, $\psi \vee \phi$, $\mathrm{EX}\psi$, $\mathrm{AX}\psi$, $\mathrm{E}(\phi\mathrm{U}\psi)$, and $\mathrm{A}(\phi\mathrm{U}\psi)$ are formulas if $\phi$ and $\psi$ are.

CTL semantics is given with respect to *Kripke structures*. A Kripke structure $K$ is a triple $\langle W, T, L \rangle$ where $W$ is a set of *worlds*, $T \subseteq W \times W$ is a total *transition relation*, and $L : W \mapsto 2^{\mathcal{P}}$ is a *labeling function*. A *path* $\pi$ in $K$ is a sequence $w_0 w_1 \ldots$ of worlds in $W$ such that, for $i \geq 0$, $T(w_i, w_{i+1})$. In what follows, $K, w \models \psi$ denotes that $\psi$ holds in the world $w$ of K. CTL semantics is then inductively defined as follows:

- $K, w_0 \models p$ iff $p \in L(w_0)$, for $p \in \mathcal{P}$
- $K, w_0 \models \neg\psi$ iff $K, w_0 \not\models \psi$
- $K, w_0 \models \psi \vee \phi$ iff $K, w_0 \models \psi$ or $K, w_0 \models \phi$
- $K, w_0 \models \mathrm{EX}\psi$ iff there exists a path $w_0 w_1 \ldots$ such that $K, w_1 \models \psi$
- $K, w_0 \models \mathrm{AX}\psi$ iff for all paths $w_0 w_1 \ldots$ we have $K, w_1 \models \psi$
- $K, w_0 \models \mathrm{E}(\phi\mathrm{U}\psi)$ iff there exist a path $w_0 w_1 \ldots$ and $i \geq 0$ such that $K, w_i \models \psi$ and, for all $0 \leq j < i$, $K, w_j \models \phi$
- $K, w_0 \models \mathrm{A}(\phi\mathrm{U}\psi)$ iff for all paths $w_0 w_1 \ldots$ there exists $i \geq 0$ such that $K, w_i \models \psi$ and, for all $0 \leq j < i$, $K, w_j \models \phi$

We introduce the usual abbreviations $\mathrm{AF}\psi \equiv \mathrm{A}(\mathrm{TRUE}\mathrm{U}\psi)$ (F stands for "future" or "eventually"), $\mathrm{EF}\psi \equiv \mathrm{E}(\mathrm{TRUE}\mathrm{U}\psi)$, $\mathrm{AG}\psi \equiv \neg\mathrm{EF}\neg\psi$ (G stands for "globally" or "always"), and $\mathrm{EG}\psi \equiv \neg\mathrm{AF}\neg\psi$.

The executions of a total state-action table $SA$ for the planning problem P can be encoded as paths of the Kripke structure $K^P_{SA}$ induced by $SA$.

**Definition 6 (Induced Kripke Structure).** *Let $SA$ be a total state-action table for the planning problem $P$. The Kripke structure $K^P_{SA}$ induced by $SA$ is defined as:*

- $W^P_{SA} = \text{STATES}(SA) \cup G$;
- $T^P_{SA}(s, s')$ iff $\langle s, a \rangle \in SA$ and $s' \in R(s, a)$, or $s = s'$ and $s \in G$;
- $L^P_{SA}(s) = s$.

The totality of $T^P_{SA}$ is guaranteed by the totality of $SA$. Strong cyclic plans can be specified through a temporal logic formula on their executions.

**Definition 7 (Strong Cyclic Plan).** *A strong cyclic plan for a planning problem* $P$ *is a total state-action table* $SA$ *for* $P$ *such that* $\mathcal{I} \subseteq W_{SA}^P$ *and, for all* $s \in \mathcal{I}$, *we have* $K_{SA}^P, s \models \text{AGEF}\mathcal{G}$.

That is, starting from the initial states, whatever actions we choose to execute and whatever their outcomes are, we always (AG) have a way of reaching the goal (EF$\mathcal{G}$). Notice that the state-action table in Figure 2 is a strong cyclic plan for the planning problem at hand.

# 4   The Strong Cyclic Planning Algorithm

The idea underlying our algorithm is that *sets* of states (instead of single states) are manipulated during the search. The implementation of the algorithm is based on OBDDs (Ordered Binary Decision Diagrams) [3], which allow for compact representation and efficient manipulation of sets. This opens up the possibility to deal with domains involving large state spaces, as shown by the experimental results in [9]. Our presentation is given in terms of the standard set operators (e.g., $\subseteq$, $\setminus$), hiding the fact that the actual implementation is performed through OBDD manipulation routines. In principle, however, the algorithm could be implemented through different techniques, provided that they make such set operations available. The algorithm is presented in two steps: first, algorithms computing basic strong cyclic plans are introduced (Figure 3 and 5), and then an algorithm for improving such basic solutions is given (Figure 7).

Given a planning problem $P$, STRONGCYCLICPLAN($P$) (Figure 3) generates strong cyclic plans. The algorithm starts with the largest state-action table in $SCP$ (line 2), and repeatedly removes pairs that either spoil $SCP$ totality or are related to states from which the goal cannot be reached (line 5). If the resulting $SCP$ contains all the initial states (line 7), the algorithm returns it (line 8), otherwise *Fail* is returned (line 9).

Pairs spoiling $SCP$ totality are pruned by function PRUNEOUTGOING (lines 14–20), which iteratively removes state-action pairs that can lead to nongoal states for which no action is considered. Its core is the function COMPUTEOUTGOING that, for a planning problem $P$ and a state-action table $SA$, is defined as $\{\langle s, a \rangle \in SA : \exists s' \notin \text{STATES}(SA) \cup G.s' \in R(s,a)\}$. With respect to the example shown in Figure 4 (left), during the first iteration, PRUNEOUTGOING removes $\langle S_4, e \rangle$ and, during the second one, it removes $\langle S_3, b \rangle$, giving rise to the situation shown in Figure 4 (middle).

Having removed the dangling executions results in disconnecting $S_2$ and $S_3$ from the goal, and give rise to a cycle in which executions may get stuck with no hope to terminate. This point, however, was not clear in the work presented in [9]. States from which the goal cannot be reached have to be pruned away. This task is accomplished by the function PRUNEUNCONNECTED (lines 21–27) that, when given with a planning problem $P$ and a state-action table $SA$, loops backwards inside the state-action table from the goal (line 25) to return the state-action pairs related to states from which the goal is reachable. Looping backward is realized through the function ONESTEPBACK that, when given with a planning

1.  **function** STRONGCYCLICPLAN($P$)
2.      $I := I \setminus G$; $SCP$:= $\{\langle s,a \rangle : s \in S \setminus G, a \in A,\ a$ is executable in $s\}$; $OldSCP$:=$\perp$
3.      **while** ($OldSCP{\neq}SCP$) **do**
4.          $OldSCP$:=$SCP$
5.          $SCP$:=PRUNEUNCONNECTED($P$, PRUNEOUTGOING($P$, $SCP$))
6.      **endwhile**
7.      **if** ($I \subseteq$ STATES($SCP$))
8.          **then return** $SCP$
9.          **else return** $Fail$

14. **function** PRUNEOUTGOING($P$, $SA$)
15.     $Outgoing :=$ COMPUTEOUTGOING($P$, $SA$)
16.     **while** ($Outgoing \neq \emptyset$) **do**
17.         $SA$:=$SA \setminus Outgoing$
18.         $Outgoing :=$ COMPUTEOUTGOING($P$, $SA$)
19.     **endwhile**
20.     **return** $SA$

21. **function** PRUNEUNCONNECTED($P$, $SA$)
22.     $ConnectedToG := \emptyset$; $OldConnectedToG := \perp$
23.     **while** $ConnectedToG \neq OldConnectedToG$ **do**
24.         $OldConnectedToG$:=$ConnectedToG$
25.         $ConnectedToG$:=$SA \cap$ ONESTEPBACK($P$, $ConnectedToG$)
26.     **endwhile**
27.     **return** $ConnectedToG$

**Fig. 3.** The algorithm.

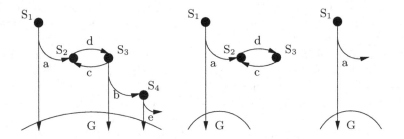

**Fig. 4.** Pruning the state-action table.

problem $P$ and a state-action table $SA$, returns all the state-action pairs possibly leading to states of $SA$ or $G$. Formally, ONESTEPBACK($P, SA$) = $\{\langle s, a \rangle : s \in S \setminus G, a \in A, \exists s' \in$ STATES($SA$) $\cup G.s' \in R(s,a)\}$. With respect to the example shown in Figure 4 (middle), PRUNEUNCONNECTED removes both $\langle S_2, d \rangle$ and $\langle S_3, c \rangle$, producing the situation shown in Figure 4 (right). Having removed the above pairs re-introduces dangling executions and, therefore, requires to apply the pruning phase once more, leading to the empty set. In general, the pruning phase has to be repeated until the putative strong plan $SCP$ is not changed either by PRUNEOUTGOING or by PRUNEUNCONNECTED (line 3).

1.   **function** STRONGCYCLICPLAN($P$)
2.      $I := I \setminus G$; $SCP := \emptyset$; $AccSA := \emptyset$; $OldAccSA := \perp$
3.      **while** ($I \not\subseteq$ STATES($SCP$) and $AccSA \neq OldAccSA$) **do**
4.         $OldAccSA := AccSA$; $AccSA :=$ ONESTEPBACK($P$, $AccSA$)
5.         $SCP := AccSA$; $OldSCP := \perp$
6.         **while** ($OldSCP \neq SCP$) **do**
7.            $OldSCP := SCP$
8.            $SCP :=$ PRUNEUNCONNECTED($P$, PRUNEOUTGOING($P$, $SCP$))
9.         **endwhile**
10.     **endwhile**
11.     **if** ($I \subseteq$ STATES($SCP$))
12.        **then return** $SCP$
13.        **else   return** *Fail*

**Fig. 5.** The incremental algorithm.

As an alternative (see Figure 5), rather than starting with the largest state-action table, one could start with an empty state-action table in *AccSA* (line 2) and incrementally extend it (line 4) until either a strong cyclic plan containing all the initial states is found, or *AccSA* is not extendible anymore (line 3).

The strong cyclic plans returned by STRONGCYCLICPLAN can be improved in two directions. Consider the example in Figure 6, where $S_3$ is the initial state. The strong cyclic plan returned by STRONGCYCLICPLAN for such example comprises all the possible state-action pairs of the planning problem. Note, however, that the pair $\langle S_1, a \rangle$ is absolutely useless, since it is unreachable from the initial state. Furthermore, the pair $\langle S_4, d \rangle$ is useless as well, because it moves the execution away from the goal. Indeed, when reaching $S_4$ from $S_3$, one does not want to go back to $S_3$ through $d$. The algorithm for getting rid of the above situations is shown in Figure 7.

Function PRUNEUNREACHABLE loops forward, inside the state-action table returned by the basic algorithm, collecting state-action pairs related to states that can be reached from the initial ones. Its core is the function ONESTEPFORTH (line 7) that, when given with a planning problem $P$ and a state-action table *ReachableFromI*, returns the set of pairs related to states reachable by executing actions in *ReachableFromI*. Formally, ONESTEPFORTH($P$, *ReachableFromI*) = $\{\langle s, a \rangle : s \in S, a \in A, a$ is executable in $s$ and $\exists \langle s', a' \rangle \in$ *ReachableFromI*$, s \in R(s', a')\}$. *ReachableFromI* is initialized with the pairs related to initial states

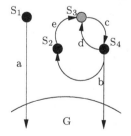

**Fig. 6.** Problems of the basic algorithm.

1.  **function** OPTIMIZE($P$, $SCP$)
2.     **return** SHORTESTEXECUTIONS($P$, PRUNEUNREACHABLE($P$, $SCP$))

3.  **function** PRUNEUNREACHABLE($P$, $SCP$)
4.     $ReachableFromI :=$ GETINIT($P, SCP$); $OldReachableFromI :=$ $\perp$
5.     **while** ($ReachableFromI \neq OldReachableFromI$) **do**
6.        $OldReachableFromI := ReachableFromI$
7.        $ReachableFromI := ReachableFromI \cup SCP \cap$ ONESTEPFORTH($P$, $ReachableFromI$)
8.     **endwhile**
9.     **return** $ReachableFromI$

10. **function** SHORTESTEXECUTIONS($P$, $SCP$)
11.    $Shortest := \emptyset$; $OldShortest := \perp$
12.    **while** ($Shortest \neq OldShortest$)
13.       $OldShortest := Shortest$
14.       $LastAdded := SCP \cap$ ONESTEPBACK($P$, $Shortest$)
15.       $Shortest := Shortest \cup$ PRUNEVISITED($LastAdded, Shortest$)
16.    **endwhile**
17.    **return** $Shortest$

**Fig. 7.** Optimization.

by GETINIT (line 4), defined as GETINIT($P, SCP$) $= \{\langle s, a \rangle \in SCP : s \in I\}$. With respect to Figure 6, this first optimization phase chops out the pair $\langle S_1, a \rangle$ while, with respect to the state-action table of Figure 2, $\langle S_1, \texttt{switch-off} \rangle$ is removed.

Function SHORTESTEXECUTIONS chops out all the pairs $\langle s, a \rangle$ that do not start one of the shortest executions leading from $s$ to the goal. Indeed, executions passing through $s$ can still reach the goal through one of the shortest ones. Shortest executions are gathered in $Shortest$ as a set of state-action pairs by looping backward (line 14) inside the (optimized through PRUNEUNREACHABLE) state-action table returned by the basic algorithm, and by introducing new pairs only when related to states that have not been visited yet (line 15). This latter task is performed by PRUNEVISITED, defined as PRUNEDVISITED($LastAdded$, $Shortest$) $= \{\langle s, a \rangle \in LastAdded : s \notin$ STATES($Shortest$)$\}$. With respect to Figure 6, this second optimization phase chops out the pair $\langle S_4, d \rangle$ while, with respect to the state-action table of Figure 2, $\langle S_6, \texttt{move(C}_1\texttt{,T)} \rangle$ is removed.

The algorithms for generating and optimizing strong cyclic plans are guaranteed to terminate, are correct and complete (the proofs can be found in [11]):

**Theorem 1.** *Let $P$ be a planning problem. Then*

1. OPTIMIZE($P$, STRONGCYCLICPLAN($P$)) *terminates.*
2. OPTIMIZE($P$, STRONGCYCLICPLAN($P$)) *returns a strong cyclic plan for $P$ if and only if one exists.*

# 5    Extensions: Weak and Strong Solutions

In this paper we focus on finding strong cyclic solutions, which has been an open problem at the current state-of-the-art for plan generation. However, strong cyclic plans are of course not the only interesting solutions. In some practical domains, it may be possible for the planner to generate strong plans, i.e., plans which are not iterative and guarantee goal achievement. In other applications, a plan may be allowed to lead to failures in very limited cases, i.e., some forms of weak solutions might be acceptable. A planner may be required to generate solutions of different "strength" according to the application domain.

Strong and weak plans have been introduced in [10]. We show here how they can be specified as temporal formulas on plan executions. This requires to generalize Definitions 5 and 6 for taking into account state-action tables that are not total. Given the state-action table $SA$ for the planning problem $P$, we first define $\text{CLOSURE}(SA) = \{s \notin \text{STATES}(SA) : \langle s', a' \rangle \in SA, s \in R(s', a')\} \cup G$.

**Definition 8 (Execution).** *Let $SA$ be a state-action table for the planning problem $P$. An execution of $SA$ starting from the state $s_0 \in \text{STATES}(SA) \cup \text{CLOSURE}(SA)$ is an infinite sequence $s_0 s_1 \ldots$ of states in $S$ such that, for all $i \geq 0$, either $s_i \in \text{CLOSURE}(SA)$ and $s_i = s_{i+1}$, or $s_i \notin \text{CLOSURE}(SA)$ and, for some $a \in A$, $\langle s_i, a \rangle \in SA$ and $s_{i+1} \in R(s_i, a)$.*

**Definition 9 (Induced Kripke Structure).** *Let $SA$ be a state-action table for the planning problem $P$. The Kripke structure $K_{SA}^P$ induced by $SA$ is defined as*

- $W_{SA}^P = \text{STATES}(SA) \cup \text{CLOSURE}(SA)$;
- $T_{SA}^P(s, s')$ *iff* $\langle s, a \rangle \in SA$ *and* $s' \in R(s, a)$, *or* $s = s'$ *and* $s \in \text{CLOSURE}(SA)$;
- $L_{SA}^P(s) = s$.

In the case of total state-action tables, since $\text{CLOSURE}(SA) = G$, these latter definitions collapse into the previous ones.

**Definition 10 (Weak Plan).** *A* weak plan *for a planning problem $P$ is a state-action table $SA$ for $P$ such that $\mathcal{I} \subseteq W_{SA}^P$ and, for all $s \in \mathcal{I}$, we have $K_{SA}^P, s \models \text{EF}\mathcal{G}$.*

**Definition 11 (Strong Plan).** *A* strong plan *for a planning problem $P$ is a total state-action table $SA$ for $P$ such that $\mathcal{I} \subseteq W_{SA}^P$ and, for all $s \in \mathcal{I}$, we have $K_{SA}^P, s \models \text{AF}\mathcal{G}$.*

# 6    Conclusions and Related Work

In this paper we have presented a formal account for strong cyclic planning in non-deterministic domains. We have formalized the notion of strong cyclic plans, i.e., plans encoding iterative trial-and-error strategies that always have a possibility of terminating and, when they do, are guaranteed to achieve the goal in spite of non-determinism. Strong cyclic plans are plans whose executions satisfy the CTL formula AGEF$\mathcal{G}$, where $\mathcal{G}$ is a propositional formula representing the set of goal states. We have shown how this approach can also embed "strong" and "weak" plans, whose executions have to satisfy the CTL formulas AF$\mathcal{G}$ and EF$\mathcal{G}$, respectively. We have defined an algorithm that is guaranteed to generate strong cyclic plans and to terminate, and have implemented it in MBP, a planner built on top of the symbolic model checker NuSMV [6]. MBP is currently used in an application for the "Italian Space Agency" (ASI) [4].

A future goal is to extend the planning task from the task of finding a plan which leads to a set of states (the goal) to the task of synthesizing a plan which satisfies some specifications in some temporal logic. This makes the planning task very close to controller synthesis (see, e.g., [1,20]), which considers both exogenous events and non-deterministic actions. From the controller synthesis perspective, in this paper we synthesize memoryless plans. Due to its generality, however, the work in [1,20] does not allow for concise solutions as state-action tables, and it is to be investigated how it can express and deal with strong cyclic plans. [19] proposes an approach to planning that has some similarities to the work on synthesis but abandons completeness for computational efficiency.

Most of the work in planning is focused on deterministic domains. Some works extend classical planners to "contingent" planners (see, e.g., [27]), which generate plans with conditionals, or to "conformant" planners [23,8], which try to find strong solutions as sequences of actions. Nevertheless, neither existing contingent nor existing conformant planners are able to generate iterative plans as strong cyclic solutions. Some deductive planning frameworks (see, e.g., [24, 25]) can be used to specify desired plans in non-deterministic domains. Nevertheless, the automatic generation of plans in these deductive frameworks is still an open problem. Some works propose an approach that is similar to planning via model checking. The TLplan system [2] (see also [12] for an automata-theoretic approach) allows for control strategies expressed as Linear Time Temporal Logic (LTL) [14] and implements a forward chaining algorithm that has strong similarities with LTL standard model checking [26]. However, the planner deals only with deterministic domains. Moreover, it is not clear how it could be extended to express strong cyclic solutions (where both a universal and existential path quantifiers are required) and to generate them. [5] proposes a framework based on process algebra and mu-calculus for reasoning about nondeterministic and concurrent actions. The framework is rather expressive, but it does not deal with the problem of plan generation. In planning based on Markov Decision Processes (MDP) (see, e.g., [13]), policies (much like state-action tables) are constructed from stochastic automata, where actions induce transitions with an associated probability, and states have an associated reward. The planning task is reduced to constructing optimal policies w.r.t. rewards and probability distributions. There is no explicit notion of weak, strong, and strong cyclic solution.

# References

1. E. Asarin, O. Maler, and A. Pnueli. Symbolic controller synthesis for discrete and timed systems. In *Hybrid System II*, volume 999 of *LNCS*. Springer Verlag, 1995.
2. F. Bacchus and F. Kabanza. Using temporal logic to express search control knowledge for planning. *Artificial Intelligence*, 1998. Submitted for pubblication.
3. R. E. Bryant. Graph-Based Algorithms for Boolean Function Manipulation. *IEEE Transactions on Computers*, C-35(8):677–691, August 1986.
4. A. Cesta, P. Riccucci, M. Daniele, P. Traverso, E. Giunchiglia, M. Piaggio, and M. Shaerf. Jerry: a system for the automatic generation and execution of plans for robotic devices - the case study of the Spider arm. In *Proc. of ISAIRAS-99*, 1999.
5. X.J. Chen and G. de Giacomo. Reasoning about nondeterministic and concurrent actions: A process algebra approach. *Artificial Intelligence*, 107(1):29–62, 1999.
6. A. Cimatti, E. Clarke, F. Giunchiglia, and M. Roveri. NUSMV: a reimplementation of SMV. Technical Report 9801-06, IRST, Trento, Italy, January 1998.
7. A. Cimatti, E. Giunchiglia, F. Giunchiglia, and P. Traverso. Planning via Model Checking: A Decision Procedure for $\mathcal{AR}$. In *ECP97*, pages 130–142, 1997.
8. A. Cimatti and M. Roveri. Conformant Planning via Model Checking. In *Proc. of ECP99*, 1999.
9. A. Cimatti, M. Roveri, and P. Traverso. Automatic OBDD-based Generation of Universal Plans in Non-Deterministic Domains. In *Proc. of AAAI98*, 1998.
10. A. Cimatti, M. Roveri, and P. Traverso. Strong Planning in Non-Deterministic Domains via Model Checking. In *Proc. of AIPS98*, 1998.
11. M. Daniele, P. Traverso, and M. Y. Vardi. Strong Cyclic Planning Revisited. Technical Report 9908-03, IRST, Trento, Italy, August 1999.
12. G. de Giacomo and M.Y. Vardi. Automata-theoretic approach to planning with temporally extended goals. In *Proc. of ECP99*, 1999.
13. T. Dean, L. Kaelbling, J. Kirman, and A. Nicholson. Planning Under Time Constraints in Stochastic Domains. *Artificial Intelligence*, 76(1-2):35–74, 1995.
14. E. A. Emerson. Temporal and modal logic. In J. van Leeuwen, editor, *Handbook of Theoretical Computer Science, Volume B: Formal Models and Semantics*, chapter 16, pages 995–1072. Elsevier, 1990.
15. M. Ernst, T. Millstein, and D. Weld. Automatic SAT-compilation of planning problems. In *Proc. of IJCAI-97*, 1997.
16. R. E. Fikes and N. J. Nilsson. STRIPS: A new approach to the application of Theorem Proving to Problem Solving. *Artificial Intelligence*, 2(3-4):189–208, 1971.
17. E. Giunchiglia, G. N. Kartha, and V. Lifschitz. Representing action: Indeterminacy and ramifications. *Artificial Intelligence*, 95(2):409–438, 1997.
18. F. Giunchiglia and P. Traverso. Planning as Model Checking. In *Proc. of ECP99*, 1999.
19. R. Goldman, D. Musliner, K. Krebsbach, and M. Boddy. Dynamic Abstraction Planning. In *Proc. of AAAI97*, 1998.
20. O. Kupferman and M.Y. Vardi. Synthesis with incomplete information. In *Proc. of 2nd International Conference on Temporal Logic*, pages 91–106, 1997.
21. J. Penberthy and D. Weld. UCPOP: A sound, complete, partial order planner for ADL. In *Proc. of KR-92*, 1992.
22. M. J. Schoppers. Universal plans for Reactive Robots in Unpredictable Environments. In *Proc. of IJCAI87*, pages 1039–1046, 1987.
23. D. Smith and D. Weld. Conformant Graphplan. In *AAAI98*, pages 889–896.

24. S. Steel. Action under Uncertainty. *J. of Logic and Computation, Special Issue on Action and Processes*, 4(5):777–795, 1994.
25. W. Stephan and S. Biundo. A New Logical Framework for Deductive Planning. In *Proc. of IJCAI93*, pages 32–38, 1993.
26. M Y. Vardi and P. Wolper. Reasoning about infinite computations. *Information and Computation*, 115(1):1–37, 15 November 1994.
27. D. Weld, C. Anderson, and D. Smith. Extending Graphplan to Handle Uncertainty and Sensing Actions. In *Proc. of AAAI98*, pages 897–904, 1998.

# Scaleability in Planning*

Vassilis Liatsos and Barry Richards

IC-Parc, William Penney Lab
Imperial College, London SW7 2AZ, UK
{vl,ebr}@icparc.ic.ac.uk

**Abstract.** This paper explores the performance of three planners, viz. *parc*PLAN, IPP and `Blackbox`, on a variant of the standard blocks-world problem. The variant problem has a restricted number of table positions, and the number of arms can vary (from 1 upwards). This type of problem is typical of many real world planning problems, where resources form a significant component. The empirical studies reveal that least commitment planning, as implemented in *parc*PLAN, is far more effective than the strategies in IPP and `Blackbox`. But the studies also reveal a serious limitation on the scaleability of *parc*PLAN's algorithm.

## 1   Introduction

In this paper we study the performance of *parc*PLAN [10] on a variant of the standard blocks-world problem, and for perspective compare it to IPP [9] and `Blackbox` [8]. In the standard problem the table is assumed to be indefinitely large; in the variant there is a fixed number of table positions. This restriction constitutes a resource limitation that significantly increases the difficulty of the problem. There is a further variation in regard to resources. In the standard problem there is just one arm available to move blocks; in the variant there may be several arms, the number varying from instance to instance. Problems of the variant type exhibit a critical dependency among three factors: the number of resources available, the degree of overlap among actions, and the length (duration) of individual actions. This dependency is typical of a large class of real-world planning problems, which is the reason for focusing on the variant blocks-world problems.

Any "viable" generic planner must be able to handle problems of this type successfully. The question is whether *parc*PLAN implements a scaleable generic strategy for these problems. This is the issue we investigate below. Here the comparative perspective provided by IPP and `Blackbox` is useful.

The *parc*PLAN system is a temporal planner. Actions, properties and goals are all indexed to intervals, which are represented in terms of their end-points. The search strategy is constraint-based and least commitment. This applies not only to temporal reasoning but to all forms of reasoning. The planning architecture is essentially "failure driven"; that is, it introduces an action into the

---

* This work was supported in part by a grant from the Engineering and Physical Sciences Research Council, Grant No. GR/L71919

plan only if it fails to find a feasible plan for the "current" set of actions. The solutions generated by *parc*PLAN use the minimal number of actions necessary to achieve all the goals.

There is a view that planning strategies like that in *parc*PLAN are "doomed" to be inefficient. Whilst this may well be true, we show that *parc*PLAN is better-suited than IPP and Blackbox in solving variant blocks-world problems. It is more efficient and scales much better.

This is not to imply, however, that *parc*PLAN is an effective generic planner. It too has significant limitations. The aim of the paper is twofold: first, to demonstrate that least commitment planning, as implemented in *parc*PLAN, has a great deal more "mileage" in it than has been recognised; second, to identify one of the critical limitations on the effectiveness of *parc*PLAN's algorithm. This limitation must be overcome if the system is to evolve into a "viable" generic planner. We think that this will require changing the search algorithm.

The paper is structured as follows. Section 2 presents the *parc*PLAN approach to representing planning problems and generating solutions. Section 3 reports the results of an empirical study of single-resource planning and Sect. 4 the results of a similar study of multiple-resource planning. Finally, Sect. 5 summarises the conclusions.

## 2   *parc*PLAN and the Planning Domain

The *parc*PLAN system falls broadly within the planning paradigm of Allen & Koomen [1]. That is, it is an interval-based temporal planner designed to support parallel action execution. But it differs from [1] in several respects. First, it treats intervals in terms of their end-points, and allows both indefinite and numerical reasoning. Second, it integrates temporal and resource reasoning, thereby supporting the requirements of reasoning about action overlap. This is realised in a module called *parc*TREC, which is described in [5]. Third, it implements a constraint-based strategy for introducing actions into a plan. This is essential to the process of generating optimal plans, which itself is another differentiating aspect of the *parc*PLAN architecture.

### 2.1   Representation

In *parc*PLAN properties and actions are uniformly indexed to intervals. This is realised in expressions of the form $Q@[start,end)$, which says that the property (action) $Q$ holds (occurs) over the time interval $[start,end)$. The interval $[start,end)$ is closed on the left and open on the right; that is, the left point belongs to the interval and the right is the first point not belonging to the interval. The reasons for this treatment need not be elaborated here. Below we illustrate how a planning problem is formulated in the variant blocks world.

The basic problem is depicted in part (i) of Fig. 1. Note that there are three table positions in the problem $(t1, t2, t3)$, and that the blocks are originally configured as shown in the "initial state". The task is to move the blocks into

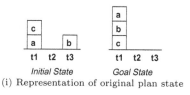

(i) Representation of original plan state

**Facts:**

> $on(a,t1)@[0,T1),\ on(b,t3)@[0,T2),$
> $on(c,a)@[0,T3),\ clear(b)@[0,T4),$
> $clear(c)@[0,T5),\ clear(t2)@[0,T6)$

**Goals:**

> $on(a,b)@[T7,20),\ on(b,c)@[T8,20),$
> $on(c,t1)@[T9,20)$

**Temporal Constraints:**

> $0 < T1, \ldots, 0 < T6, T7 < 20, \ldots, T9 < 20$

(ii) Specification of original plan state

**Fig. 1.** A blocks world example

the positions shown in the "goal state", using the arm(s) available and only the three table positions. The "initial state" and the "goal state" together reflect the original state of the planner.

One can see immediately that this problem is more difficult than the standard blocks-world formulation where the table is indefinitely large. Any standard problem can be solved in a maximum of $2n - 2$ moves using one arm, where $n$ is the number of blocks. Thus, a 3-block problem requires no more than 4 moves. The variant problem here requires 5 moves, if there is just one arm.

The informal characterisation of the original plan state in part (i) of Fig. 1 is specified more formally in part (ii). The specification has three components, viz. Facts, Goals and Temporal Constraints. The set of facts, together with some of the temporal constraints, reflect the situation in the "initial state", while the set of goals, together with other temporal constraints, reflect the situation in the "goal state". Note that there is no reference to the "initial state" in the facts, and no reference to the "goal state" in the goals. Here everything is indexed to intervals. Each of the facts begins to hold from time point 0, and is assumed to persist up to a certain unspecified time point, indicated by a variable. [1] Each of the goals must be established before time point 20, again indicated by a variable, and persists up to time 20. The constraints among the time points require that in the original plan state the facts are either earlier or co-temporal with the goals.

In general, the temporal constraints, which always relate two time points, are handled in the global constraint network Temporal Constraints. The variables here have the status of domain variables, with their domains consisting initially

---

[1] The Prolog convention is used to indicate variables: their names start with an uppercase letter.

| Action: | $move(Block,From,To,Arm)@[Start,End)$ |
|---|---|
| **Conditions:** | |
| 1. | $on(Block,From)@[T1,Start)$, |
| 2. | $clear(To)@[T2,End)$, |
| 3. | $clear(Block)@[T3,T4)$ |
| **Effects:** | |
| 4. | $on(Block,To)@[End,T5)$, |
| 5. | $clear(From)@[Start,T6)$ |
| **Constraints:** | |
| 6. | $designates(Block,block)$, |
| 7. | $designates(Arm,arm)$, |
| 8. | $designates(From,block\_table)$, $designates(To,block\_table)$, |
| 9. | $Block \neq From,\ Block \neq To$, |
| 10. | $T3 < Start,\ End < T4$ |

**Fig. 2.** Specification of the *move* action operator

of the integers 1 to 20. These domains will be pruned as the plan evolves from the original plan state and new temporal constraints are added.

The action operator *move* is a 4-place relation, specified in terms of Conditions, Effects, and Constraints. The relations among the conditions, effects and actions are less constrained than those in a state-based planner. Since actions occur over intervals, conditions need not always be true before the action begins, and effects may become true before it ends. This is the case for the *move* action, which is defined in Fig. 2.

While the Conditions (goals) and Effects (facts) are the familiar ones, the temporal constraints, given in line 10, allow for unfamiliar relationships. For example, the condition *clear(To)@[T2,End)* need not be true before the action starts. There is no constraint that $T2 < Start$ and hence the condition must only become true sometime before *End*. Similarly, the effect *clear(From)@[Start,T6)* becomes true before the action ends. In fact, it becomes true when the action starts and remains true up to *T6*, which may or may not be before the action ends; *T6* has no specific relation to *End*.

The non-temporal constraints are the usual ones. For example, the constraint in line 6, together with the Conditions and Effects, asserts that the object which is moved has to be a block and not a table position. The constraints in line 8 state that the source and destination of the block can be either a block or a table position. The two constraints in line 9, which are non-codesignation constraints, state that a block cannot be moved from itself, nor it can be moved on top of itself. Note, however, that there is no constraint *From ≠ To*, which would prevent moving a block from $X$ to $X$. When there is only one arm and hence actions must occur in a linear sequence, lifting up a block and putting it down in the same place is not productive. However, when there are multiple arms allowing actions to overlap, this type of action can be effective in obtaining an optimal solution.

There is a final component specifying the problem domain; this takes the form of rules, which state that certain relations between properties and actions cannot occur in a valid plan. In the case of properties, for example, two domain rules hold: (1) an object (whether a block or table position) cannot be clear

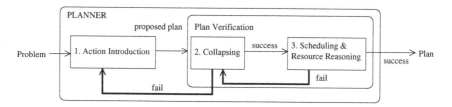

Fig. 3. *parc*PLAN's plan cycle

and have a block on top of it at the same time; (2) an object cannot have two different blocks on it at the same time. How these and other rules are captured need not detain us here.

## 2.2   Planning Algorithm

The planning algorithm in *parc*PLAN involves three basic steps: (1) action introduction, (2) collapsing and (3) scheduling and resource reasoning.[2] Action introduction and collapsing are two ways in which a goal can be achieved. Introducing an action to achieve a goal is realised by unifying the goal with one of the effects of the action operator and constraining the interval in the goal to be within the interval of that effect. Achieving a goal through collapsing is realised by unifying the goal with one of the facts in Facts and constraining the goal interval to be within the fact interval. If the resulting network of temporal constraints is consistent, collapsing is successful.[3] Figure 3 shows the planning cycle in *parc*PLAN, including the three basic steps and the failure points. These are explained in detail below.

**Step 1: Action Introduction** In the action introduction step *parc*PLAN looks for a goal which *does not* have a potential collapser. A fact $P@[S2,E2)$ is a *potential collapser* for a goal $Q@[S1,E1)$ if P and Q can codesignate, and $S1 \geq S2$ and $E2 \geq E1$ are consistent with the existing temporal constraints.[4] If there is a goal without a potential collapser, *parc*PLAN introduces an action into the plan by unifying an effect of the action operator with that goal, that is, by posting the appropriate codesignation and temporal constraints. It adds all the other effects of the action to the Facts, and all the conditions of the action to the Goals; and it removes the achieved goal from the Goals. Finally, it adds all the remaining constraints arising from the unification and the domain rules. The constraints are held in two separate networks, one for the codesignation constraints and the other for the temporal constraints.[5] *parc*PLAN continues to introduce actions in the plan until all the goals have at least one potential collapser; only then does the planner enter the collapsing step.

---

[2] The term "collapsing" originates from Allen and Koomen [1].

[3] A path consistency algorithm is used to detect inconsistency in the temporal network.

[4] Obviously, a goal can have more than one potential collapser.

[5] Note that all these constraints are resolved using least commitment.

**Step 2: Collapsing.** The task at this point is to collapse all the remaining goals while maintaining the consistency of the codesignation constraints and temporal constraints. Collapsing is handled as a constraint satisfaction problem where each goal is associated with a finite domain variable whose domain represents the potential collapsers for that goal. Choosing a value for such a variable collapses the associated goal with a particular potential collapser. To detect failures as early as possible the collapsing phase follows a strategy based on the *first-fail principle*; the most constrained goals are collapsed first, i.e. the ones with the least number of potential collapsers.[6]

**Step 3: Scheduling & Resource Reasoning.** If all the goals are successfully collapsed, *parc*PLAN enters the third step of the planning cycle. At this point the plan consists of the set of actions which have been introduced so far. Typically, these are partially specified (due to choices remaining in codesignation constraints) and partially ordered (due to choices remaining on the order of time points). The task now is to determine whether the actions can be made fully specific and scheduled subject to the available resources. This is essentially a disjunctive scheduling problem complicated by three interdependent factors, viz. the number of available resources, the degree of action overlap, and the duration of the actions. To solve this problem (the *resource feasibility problem*) efficiently requires special techniques for integrating temporal and resource reasoning. One such technique is presented in El-Kholy & Richards [4], where Boolean meta-variables are used to support the necessary integration. The technique here adopts the same search model as [4]; however the search strategy is based on the probe-backtracking technique [13,14]. Search terminates when *parc*PLAN succeeds in fully specifying the actions and scheduling them subject to the available resources. At this point, a plan has been found.

**Failure Points.** Clearly, there are two points where *parc*PLAN can fail. It can fail in the third step, because it cannot fully specify the actions consistently or cannot schedule them with the resources available. It can also fail in the second step, because it cannot successfully collapse the remaining goals. These failures represent points in the search where control needs to be returned to the previous step.

If failure is encountered in scheduling & resource reasoning, control is returned to the collapsing step, and *parc*PLAN searches for an alternative collapse. If it finds a successful collapse, control passes back to the third step of scheduling & resource reasoning; otherwise control is returned to the action-introduction step. To reduce the amount of "backtracking" between the last two steps, a global constraint is used to monitor the resource usage during collapsing. This constraint monitors the lower and upper bounds for resource requirements at each start point (see [4]). If at any point the lower bound exceeds the available resources, a failure is signalled. The idea here is that collapsing should not waste time looking at plans which exceed the resources available.

---

[6] Goals with only one potential collapser are collapsed immediately prior to any search.

Failure in the second step, i.e., failure to find a successful collapse, means that *at least one* of the goals needs to be achieved by introducing a new action. The problem is to find which goal that is; here search confronts a major challenge. This is not special to *parc*PLAN but is shared in different ways by all planners.[7]

Since it is not clear which goal needs to be select for action introduction, *parc*PLAN follows a least commitment strategy. The strategy adds an action to the plan but delays selecting a specific goal until the collapsing step has again been reached. This is implemented by introducing an action which can be used for *any* of the remaining goals. We call this a *generalised* action.[8] The limiting case is the wholly indefinite action $move(X, Y, Z, A)$, where $X$ is a block, $Y, Z$ are either a block or a table position, and $A$ is an arm. Once a generalised action has been introduced, control passes back to the collapsing step. Note that only one such action is required. If collapsing succeeds, one of the goals will be collapsed with an effect of the generalised action. At this point control passes to the third step. If, however, *parc*PLAN fails to find a successful collapse, it returns to the first step and introduces another generalised action into the plan. Clearly, this strategy introduces only those actions which are absolutely necessary to achieve the goals. This guarantees a plan with the minimal number of actions.

## 3   Experimental Study: Single Arm

In this section we focus on problems where there is just one arm available to move blocks. Here the third step in the planning cycle, viz. scheduling & resource reasoning, is simplified. Since there is only one arm, all the actions must be disjoint, i.e. the plan must be linear. As a result, when actions are introduced into the plan, a constraint is imposed to ensure that the actions are disjoint. This is sufficient to guarantee that if a successful collapse is found, the actions introduced so far can be scheduled to generate a plan.

For single-arm problems the burden of generating a plan is carried largely by the first two steps in the cycle. Our objective below is to examine this burden experimentally. We study a number of different problem sets, each containing 100 different instances. We name each set (benchmark) by indicating its size, i.e. the number of blocks, the number of table positions and the number of arms. For example, the set having 10 blocks, 4 table positions and 1 arm is denoted 10x4x1.

We explore each of the three planners, *parc*PLAN, IPP and Blackbox, on each problem set. We impose a uniform time-out of 600 seconds for each instance.[9] For this study we use Blackbox 3.4, and the PDDL-version of IPP used in the AIPS planning competition. These systems have a performance advantage over

---

[7] In IPP and Blackbox, for example, the same problem occurs when a plan is not found using $n$ time points; extending the plan to $n + 1$ time points generates a much larger search space.

[8] Joslin and Pollack use a similar approach in [7].

[9] The experiments were run on a Pentium II 300 MHz using Red Hat Linux 2.0.32. Benchmarks are available from [11]

**Table 1.** *parc*PLAN: single arm problems

| Benchmark | Solved | CPU time | Actions |
|-----------|--------|----------|---------|
| 4x4x1     | 100    | 0.20     | 0.08    |
| 6x6x1     | 100    | 0.70     | 0.02    |
| 8x8x1     | 100    | 1.53     | 0.08    |
| 10x10x1   | 100    | 4.65     | 0.25    |
| 12x12x1   | 100    | 8.49     | 0.45    |
| 14x14x1   | 100    | 14.10    | 0.64    |
| 16x16x1   | 100    | 26.27    | 0.93    |
| 18x18x1   | 100    | 36.52    | 0.96    |
| 20x20x1   | 100    | 64.82    | 1.32    |

*parc*PLAN since they are compiled for the particular architecture; *parc*PLAN runs on ECL$^i$PS$^e$ 4.1 [3].

As we mentioned above, *parc*PLAN will always generate a plan with the minimal number of actions. For the single-arm problems, the actions in the plan must be disjoint. On the assumption that their duration is some uniform constant length, *parc*PLAN yields plans with the shortest overall duration.

Note that finding such plans is not in general easy. Gupta & Nau [6] show that optimal planning in the standard blocks world is NP-complete.

### 3.1   Easier Problems

We first look at limiting case benchmarks, i.e. problem sets where the number of table positions is the same as the number of blocks. These are similar to, though not exactly like, the standard blocks-world problems. The aim here is to study the performance of the three planners as the size of the problem increases.

Table 1 shows the performance of *parc*PLAN measured on two dimensions, viz. the number of problems solved and the average CPU time (measured in seconds) to solve them. The problems range from 4 blocks 4 table positions, to 20 blocks 20 table positions. Recall that there is a cut-off of 600 seconds. *parc*PLAN solves all 100 instances of all the problems well within the cut-off. On the largest problem set it takes an average of about 65 seconds to solve each instance. It is interesting to note the effect of problem size on the average CPU time and the average number of generalised actions introduced to solve a particular instance. The rate at which CPU time rises is roughly similar to the rate at which the number of generalised actions is introduced.

To run IPP and Blackbox on the same benchmarks, we translate the problems into PDDL format [12]. We explore two alternative specifications. The Optimal version uses two actions pickup and putdown; this generates plans which are guaranteed to have the shortest plan length. The Feasible version uses one action move; here plans may contain superfluous actions and hence may not be the shortest.

Table 2 shows the performance of both Optimal and Feasible versions of IPP and Blackbox on the benchmarks above. Neither version of the two planners scales on these problems although the Feasible versions consistently solve more

**Table 2.** IPP and `Blackbox`: single arm problems

| Benchmark | IPP | | | | Blackbox | | | |
|---|---|---|---|---|---|---|---|---|
| | Optimal | | Feasible | | Optimal | | Feasible | |
| | Solved | CPU time | Solved | CPU time | Solved | CPU time | Solved | CPU time |
| 4x4x1 | 100 | 0.15 | 100 | 0.14 | 100 | 0.19 | 100 | 0.17 |
| 6x6x1 | 100 | 1.09 | 100 | 1.52 | 100 | 15.06 | 100 | 14.22 |
| 8x8x1 | 94 | 56.15 | 94 | 35.79 | 22 | 246.43 | 61 | 29.46 |
| 10x10x1 | 8 | 97.18 | 44 | 112.19 | 0 | – | 6 | 515.66 |
| 12x12x1 | 0 | – | 17 | 224.65 | n/a | n/a | n/a | n/a |

**Table 3.** *parc*PLAN: single arm, limited table problems

| Benchmarks | Solved | CPU time | Actions |
|---|---|---|---|
| 10x4x1 | 72 | 81.28 | 1.38 |
| 20x8x1 | 71 | 68.80 | 1.06 |
| 30x12x1 | 63 | 267.78 | 1.71 |

problems than the Optimal ones. The results clearly show that the performance of the Feasible version of IPP (which is the best of these) does not compare well to *parc*PLAN. The performance of IPP falls sharply, first between 6x6 and 8x8, and then between 8x8 and 10x10. Problem size here is a real challenge for both IPP and `Blackbox`.

### 3.2 Harder Problems

We now consider harder benchmarks. Here we fix the number of table positions to be 40% of the number of blocks. These problems are much harder than their counterparts, where the number of table positions matches the number of blocks. Table 3 shows that *parc*PLAN finds these far from trivial.

It is interesting that the success rate for *parc*PLAN seems relatively stable. It fails on 28 of the 10x4 instances, 29 of the 20x8 cases, and 37 of the 30x12. While the problems vary enormously in size, the success rate does not. As for the variation in CPU time, this correlates roughly to the number of generalised actions introduced.

We tried IPP and `Blackbox` on these benchmarks but the results were poor. Neither planner solved any of the 20x8 and 30x12 problems. Clearly IPP and `Blackbox` are not well-suited to these problems. Although *parc*PLAN is much better, we begin to see the limitations on its scaleability. There is sharp fall in performance on the 30x12x1 problem set.

## 4  Experimental Study: Multiple Arms

Here we consider problems where there are multiple arms to move blocks. In this case the third step in the planning cycle, viz. scheduling & resource reasoning, has real work to do. After finding a successful collapse, *parc*PLAN must determine whether the actions generated so far can be scheduled subject to the arms available. This requires search, which may not succeed. In the event, the system must "backtrack" to seek another successful collapse, and if it finds one, it re-enters the scheduling & resource reasoning step. Clearly, this can lead to a considerable amount of search.

## 4.1 Easier Problems

We first look at the performance of *parc*PLAN and IPP on benchmarks corresponding to the easier problem sets in the previous section. The difference here is in the number of arms available; we fix this uniformly at half the number of blocks. The smallest benchmark is 4x4x2 (4 blocks, 4 table positions, 2 arms) and the largest is 20x20x10 (20 blocks, 20 table positions, 10 arms).

For these problems we could test only the Optimal version of IPP since the Feasible version does not reason about resources. We present the results for both *parc*PLAN and IPP in Table 4.[10]

**Table 4.** Multiple arms problems

| Benchmark | *parc*PLAN | | | IPP | |
|-----------|--------|----------|---------|--------|----------|
|  | Solved | CPU time | Actions | Solved | CPU time |
| 4x4x2 | 100 | 0.14 | 0.14 | 100 | 0.17 |
| 6x6x3 | 100 | 0.45 | 0.28 | 100 | 1.44 |
| 8x8x4 | 100 | 0.71 | 0.04 | 92 | 15.99 |
| 10x10x5 | 100 | 5.63 | 0.12 | 44 | 87.52 |
| 12x12x6 | 100 | 12.24 | 0.04 | 12 | 96.49 |
| 14x14x7 | 100 | 5.88 | 0.01 | 0 | – |
| 16x16x8 | 100 | 5.68 | 0.00 | n/a | n/a |
| 18x18x9 | 100 | 8.04 | 0.00 | n/a | n/a |
| 20x20x10 | 100 | 10.93 | 0.00 | n/a | n/a |

The performance of IPP falls sharply on the 10x10x5 benchmarks; it can solve only 44 instances out of a 100 and takes an average of nearly 88 seconds to do so, which is an order of magnitude worse than *parc*PLAN. IPP's performance again falls sharply on the 12x12x6 problem set, where it solves only 12 instances. On the larger problems it cannot solve any instances within the 600 second cut-off.

In contrast, the performance of *parc*PLAN is impressive both on the number of instances solved and in the time required to solve them. It finds solutions for all instances of all benchmarks, with the worst average CPU time being just over 12 seconds on the 12x12x6 benchmark. Note that the relatively uniform performance profile reflects the relatively constant number of generalised actions introduced. Indeed such actions are very rarely used. For most of these problems there is no need to "backtrack" to the action-introduction step. This is why the strategy is so successful.

---

[10] We do not discuss `Blackbox` since its performance was poor relative to IPP.

## 4.2 Harder Problems

There are problems, however, where *parc*PLAN must resort to introducing a larger number of generalised actions; these it finds more difficult to solve, and some instances it cannot solve at all. We will show that the difficulty lies in the collapsing step, although the difficulty is not the same in all cases.

Table 5 profiles the performance of *parc*PLAN on problem sets consisting of 10 blocks and 10 table positions, with the number of arms varying from 5 down to 2. The first line of the table repeats the results on the 10x10x5 benchmarks from Table 4. While the results in table 5 are encouraging, the performance profile is curious. Note that *parc*PLAN is at its worst on the 10x10x4 and 10x10x3 problem sets. It solves fewer instances from these sets than it does from either the 10x10x2 set or the 10x10x5 set. The reason behind this behaviour is that the resource-monitoring constraint in the collapsing phase is able to prune more options and detect failures earlier when the resources are more limited. The more constrained the problem, the more effective the resource-monitoring constraint is in guiding the search. One can see that the constraint has the desired effect by looking at the reduction in the number of collapses as resources are reduced.

**Table 5.** *parc*PLAN: 10x10 problems

| | | Solved Cases | | | Failed Cases | |
|---|---|---|---|---|---|---|
| Benchmark | Solved | CPU time | Actions | Collapses | Actions | Collapses |
| 10x10x5 | 100 | 5.63 | 0.12 | 21.74 | – | – |
| 10x10x4 | 92 | 7.77 | 0.34 | 23.52 | 2.00 | > 1800 |
| 10x10x3 | 93 | 10.77 | 0.97 | 16.23 | 2.57 | > 1400 |
| 10x10x2 | 99 | 14.40 | 1.97 | 7.32 | 5.00 | 0 |

For the failed case of the 10x10x2 problem set, note that *parc*PLAN introduces 5 generalised actions but never finds a successful collapse. In effect, it never proceeds to the third step of the planning cycle; it times out looking for a successful collapse. For the failed cases of the 10x10x4 and 10x10x3 problem sets *parc*PLAN finds an enormous number of collapsers but in each case it discovers there exists no schedule for the actions subject to the arms available. Here there are too many successful collapses, all of which lead to a dead-end.

In general, the reason for failure here lies in collapsing. Either *parc*PLAN cannot find a successful collapse or it finds too many collapses, all of which are dead-ends. These collapsing problems constitute a major challenge for *parc*PLAN, one which is shared in different ways by all generic planners.

We now make the problem sets even harder. We reduce the number of table positions to 4, keeping the number of blocks at 10; and again we vary the number of arms from 5 down to 2. Table 6 shows the results. Here *parc*PLAN's performance degrades uniformly with the decrease in the arms available. Again note the situation in the failed cases. In the 10x10x4 and 10x10x3 problem sets, *parc*PLAN finds a large number of successful collapses (over 650 in one case and

**Table 6.** *parc*PLAN: 10x4 problems

| Benchmark | | Solved Cases | | | Failed Cases | |
|---|---|---|---|---|---|---|
| | Solved | CPU time | Actions | Collapses | Actions | Collapses |
| 10x4x5 | 100 | 4.46 | 0.11 | 8.39 | – | – |
| 10x4x4 | 98 | 9.72 | 0.40 | 9.46 | 2.00 | > 650 |
| 10x4x3 | 91 | 12.88 | 1.00 | 6.85 | 2.78 | > 250 |
| 10x4x2 | 84 | 52.73 | 2.30 | 13.10 | 4.50 | 29.50 |

**Table 7.** IPP: 10x10 and 10x4 problems

| 10x10 benchmarks | | | 10x4 benchmarks | | |
|---|---|---|---|---|---|
| Benchmark | Solved | CPU time | Benchmark | Solved | CPU time |
| 10x10x5 | 44 | 87.52 | 10x4x5 | 52 | 133.57 |
| 10x10x4 | 59 | 112.13 | 10x4x4 | 50 | 133.28 |
| 10x10x3 | 27 | 175.33 | 10x4x3 | 49 | 123.77 |
| 10x10x2 | 5 | 266.88 | 10x4x2 | 39 | 114.79 |

250 in the other); for each one it discovers that there is no viable schedule for the actions. In the 10x10x2 problem set it finds on average just under 30 successful collapses, and none lead to a viable schedule. Collapsing is the core issue.

Table 7 shows the performance of IPP on the same benchmarks.[11] The results indicate that the strategies of IPP and Blackbox are not well suited to solve problems where multiple resources are involved. This limitation has been noted by Srivastava and Kambhampati [15], who are exploring an extension to GraphPlan [2] with a view to handling such cases more efficiently.

## 5   Conclusions

The results of our experiments suggest that least commitment planning has more "mileage" in it than some have recognised [7]. The implementation of the least commitment strategy in *parc*PLAN is clearly effective for variant blocks-world problems, particularly in comparison with strategies in IPP and Blackbox. These strategies may be better than *parc*PLAN for other problems, although that remains to be investigated.

The main conclusion, however, relates to *parc*PLAN itself. It has serious limitations associated with collapsing. How these might be resolved is not immediately clear. What is clear is that this is one of the critical challenges for *parc*PLAN. Since the challenge appears in different forms in other generic planners, it seems to be critical to the future of generic planning.

---

[11]   Blackbox is ineffective for problems of this size and hence, we do not consider it further.

# References

[1] James F. Allen and J. A. Koomen. Planning Using a Temporal World Model. In *Proceedings of the Eighth International Joint Conference on Artificial Intelligence, IJCAI-83*, pages 741–747, Karlsruhe, Germany, 1983.

[2] Avrim L. Blum and Merrick L. Furst. Fast Planning Through Planning Graph Analysis. *Artificial Intelligence*, 90:281–300, 1997.

[3] ECRC. *ECL$^i$PS$^e$ User Manual*. European Computer Industry Research Centre, Munich, 4.1 edition, 1998.

[4] A. El-Kholy and B. Richards. Temporal and Resource Reasoning in Planning: the *parc*PLAN approach. In *Proceedings of the 12th European Conference on Artificial Intelligence (ECAI-96)*, pages 614–618, Budapest, Hungary, 1996. John Wiley and Sons, Ltd.

[5] Amin El-Kholy. *Resource Feasibility in Planning*. PhD thesis, Imperial College, University of London, 1996.

[6] Naresh Gupta and Dana S. Nau. On the Complexity of Blocks-World Planning. *Artificial Intelligence*, 56:223–254, 1992.

[7] D. Joslin and M. E. Pollack. Is "early commitment" in plan generation ever a good idea? In *Proceedings of the 13th National Conference on Artificial Intelligence (AAAI-96)*, pages 1188–1193, Portland, Oregon, 1996.

[8] H. Kautz and B. Selman. Blackbox: A New Approach to the Application of Theorem Proving to Problem Solving. *AIPS98 Workshop on Planning as Combinatorial Search*, pages 58–60, June 1998. Available from: `http://www.research.att.com/~kautz/blackbox/`.

[9] J. Koehler, B. Nebel, J Hoffmann, and Dimopoulos Y. Extending Planning Graphs to an adl Subset. In S. Steel and R. Alami, editors, *Fourth European Conference on Planning (ECP-97)*, number 1348 in Lectures Notes in Artificial Intelligence, pages 273–285, Toulouse, France, September 1997. Springer. Available from: `http://www.informatik.uni-freiburg.de/~koehler/ipp.html`.

[10] J. M. Lever and B. Richards. *parc*PLAN: a Planning Architecture with Parallel Actions, Resources and Constraints. In *Proceedings of the 9th International Symposium on Methodologies for Intelligent Systems*, pages 213–222, 1994.

[11] V. Liatsos and B. Richards. Planning benchmarks on a blocks world variant. Available from: `http://www.icparc.ic.ac.uk/parcPlan/ecp99/`.

[12] D. McDermott et al. PDDL – The Planning Domain Definition Language, version 1.2. Technical Report CVC TR-98-003, Yale Center for Computational Vision and Control, Yale University, October 1998. Available from: `http://www.cs.yale.edu/HTML/YALE/CS/HyPlans/mcdermott.html`.

[13] H. El Sakkout, T. Richards, and M. Wallace. Minimal Perturbation in Dynamic Scheduling. In H. Prade, editor, *13th European Conference on Artificial Intelligence. ECAI-98*. John Wiley & Sons, 1998.

[14] H. El Sakkout and M. Wallace. Probe Backtrack Search for Minimal Perturbation in Dynamic Scheduling. To appear in Constraints journal, 1999.

[15] B. Srivastava and S. Kambhampati. Scaling up Planning by teasing out Resource Scheduling. Technical Report ASU CSE TR 99-005, Department of Computer Science and Engineering, Arizona State University, Tempe, AZ 85287-5406, 1999. Available from: `http://rakaposhi.eas.asu.edu/yochan.html`.

# Exploiting Competitive Planner Performance

Adele E. Howe, Eric Dahlman, Christopher Hansen,
Michael Scheetz, and Anneliese von Mayrhauser

Computer Science Department, Colorado State University
Fort Collins, CO 80523 U.S.A.
e-mail: {howe, dahlman, hansenc, scheetz, avm}@cs.colostate.edu

**Abstract.** To date, no one planner has demonstrated clearly superior
performance. Although researchers have hypothesized that this should
be the case, no one has performed a large study to test its limits. In
this research, we tested performance of a set of planners to determine
which is best on what types of problems. The study included six planners and over 200 problems. We found that performance, as measured
by number of problems solved and computation time, varied with no one
planner solving all the problems or being consistently fastest. Analysis
of the data also showed that most planners either fail or succeed quickly
and that performance depends at least in part on some easily observable problem/domain features. Based on these results, we implemented
a meta-planner that interleaves execution of six planners on a problem
until one of them solves it. The control strategy for ordering the planners and allocating time is derived from the performance study data. We
found that our meta-planner is able to solve more problems than any
single planner, but at the expense of computation time.

## 1  Motivation

The implicit goal of planning research has been to create the best general purpose planner. Many approaches have been implemented (e.g., partial order, SAT
based, HTN) for achieving this goal. However, evidence from the AIPS98 competition [9] and from previous research (e.g., [14]) suggests that none of the
competing planners is clearly superior on even benchmark planning problems.

In this research, we empirically compared the performance of a set of planners
to start to determine empirically which works best when. No such study can
be comprehensive. To mitigate bias, we tested unmodified publically available
planners on a variety of benchmark and new problems. As a basis, we included
only problems in a representation that could be handled by all of the planners.

The AIPS98 planning competition facilitated comparison of planners through
a common representation language (PDDL [10]). Consequently, we had access to
a set of problems and planners that could accept them (Blackbox, IPP, SGP and
STAN). To expand the study, we added a domain that we had been developing
and two other planners with compatible representations (UCPOP and Prodigy).

Our study confirmed what we expected: no one planner excelled. Of the 176
problems solved by at least one of the planners, the most solved by a single

S. Biundo and M. Fox (Eds.): ECP-99, LNAI 1809, pp. 62–72, 2000.
© Springer-Verlag Berlin Heidelberg 2000

planner was 110. Computation times varied as well, with each planner posting the best time for some problem.

The data and PDDL provided a further opportunity. If no single planner is best, then an alternative is to build a meta-planner that incorporates several planners. The meta-planner tries the planners in sequence until a solution is found or a computational threshold is exceeded. Fortunately, we have found that if they can find any solution, most of the planners solve problems relatively quickly. The idea is a crude interpretation of the plan synthesis advocated in Kambhampati's IJCAI challenge [5], of the flexible commitment strategy advocated by Veloso and Blythe [14] and the interleaving of refinement strategies of UCP [6]; it was inspired by the Blackbox planner's inclusion of multiple SAT solving strategies[7].

We implemented a prototype meta-planner, called *BUS* for its role of communication and control. BUS allocates time to each of the planners until one solves it. As the basis of BUS's control strategy, we developed models of planner performance from the comparative study data. We tested BUS on the original problems and new problems. We expected that we would sacrifice some computation time to increase the overall problem solving rate. Although BUS provides a simple mechanism for averaging planner performance, its primary contribution is as a testbed for comparing performance of different planners and testing models of what works best when.

## 2   Planners

Because the AIPS98 competition required planners to accept PDDL, the majority (four) of planners used in this study were competition entrants, or are later versions thereof. The common language facilitated comparison between the planners without having to address the effects of a translation step. The two exceptions were UCPOP and Prodigy; however, their representations are similar to PDDL and were translated automatically. The planners represent four different approaches to planning: plan graph analysis, planning as satisfiability, partial order planning and state space planning with learning.

*STAN* [11] extends the Graphplan algorithm[2] by adding a preprocessor to infer type information about the problem and domain. This information is then used within the planning algorithm to reduce the size of the search space that the Graphplan algorithm would search. STAN can only handle problems using the STRIPS subset of PDDL.

*IPP* [8] extends Graphplan by accepting a richer plan description language, a subset of ADL. The representational capabilities are achieved via a preprocessor that expands more expressive constructs like quantification and negated preconditions into STRIPS constructs. The expanded domain is processed to remove parts of it that are irrelevant to the problem at hand. We used the AIPS98 version of IPP because the newer version no longer accepts PDDL.

*SGP* [16] also extends Graphplan to a richer domain description language. As with IPP, some of this transformation is performed using expansion techniques to remove quantification. SGP also *directly* supports negated preconditions and conditional effects. SGP tends to be slower (it is implemented in Common Lisp instead of C) but more robust than the other Graphplan based planners.

*Blackbox* [7] converts planning problems into boolean satisfiability problems, which are then solved using a variety of different techniques. In constructing the satisfiability problem, Blackbox uses the planning graph constructed as in Graphplan. We used version 2.5 of Blackbox because the newer versions were not able to parse some of the problems and in the parseable problems, we did not find a significant difference in performance.

*UCPOP* [1] is a Partial Order Causal Link planner. The decision to include UCPOP was based on several factors. First, it does not expand quantifiers and negated preconditions; for some domains, Graphplan-like expansion can be so great as to make the problem insolvable. Second, we had used UCPOP in developing an application which provides the third category of problems.

*Prodigy* [15] combines state-space planning with backward chaining from the goal state. A plan under construction consists of a head-plan of totally ordered actions starting from the initial state and a tail-plan of partially ordered actions related to the goal state. Informal results presented at the AIPS98 competition suggested that Prodigy performed well in comparison to the entrants.

## 3   Test Problems

Three criteria directed the compilation of the study test set; the problems should be: comprehensive, challenging and available in accepted representations. We included a wide set of problems, most of which were available to or had been used by the planner developers. We favored domains that were challenging, meaning that not all of the planners could solve the problems within a few minutes.

The common representation is the Planning Domain Definition Language (PDDL)[10]. PDDL is designed to support a superset of the features available in a variety of planners. At the minimum, the planners that are the subjects of our study all accept STRIPS representation. Although restricting problems to STRIPS reduced the test set significantly, we were concerned that our modifying the test problems could bias the results.

We included problems from three test suites: the AIPS98 competition set, the UCPOP benchmarks and a Software Testing domain. The first two sets are publically available. The third was developed over the past three years as an application of planning to software engineering; it has proven difficult for planners and had features not present in the other domains.

*Competition Domains.* The largest compendium of planning problems is probably the AIPS 98 competition collection [9]. Creators of competition planners knew their planners would be run on these problems and so had the opportunity to design to them. For UCPOP and Prodigy[1], the PDDL STRIPS representation was syntactically modified to match their requirements. The suite contained 155 problems in 6 domains.

*UCPOP Benchmarks.* The UCPOP distribution [13], includes 85 problems in 35 domains. The problems exploit a variety of representational capabilities. As a consequence, many of these problems could not be accepted by the chosen planners. From the distribution, we identified 18 problems from 7 domains that could be accepted by all of the planners with only minor syntactic modification.

*Our Addition: Software Testing.* Generating test cases for software user interfaces can be viewed as a planning problem. The commands to the interface can be represented as operators in the domain theory. The problems then describe changes that a user might wish to have happen to the underlying system as a consequence of executing a sequence of commands. The planner automates the process of generating test cases by constructing plans to solve these problems.

We developed a prototype system, based on the UCPOP planner, for a specific application: Storage Technology's Robot Tape Library interface [4]. The application involves moving tapes around, into and out of a large silo, reading them in a tape drive and monitoring the status of the system. We selected UCPOP initially (in 1995) because it was publically available and easy for software engineers to use. However, we were having trouble generating non-trivial test cases.

The basic domain theory contains 11 actions and 25 predicates[2]. The tape library can be configured as connected silos with a tape drive in each and positions for tapes designated by panels, rows and columns. The configuration is described in the initial conditions of problems along with identifiers and initial positions for the tapes.

We created three variants on the basic domain theory to recode some problematic aspects of the original (i.e., conditional effects and disjunctive preconditions) and six core problems whose goals required use of different actions on two tapes. We then extended the set of problems by varying the size of the library configuration; the configurations always included an equal number of panels, rows and columns of values 4, 8, 12, 16 and 20. These positions tested the vulnerability of planners to extraneous objects. These combinations produced 90 different problems.

## 4   Empirical Comparison of Planners

We hypothesized that the best planner for a given domain would vary across the domains. To support our design of the meta-planner, we hypothesized that

---

[1] We thank Eugene Fink for translation code from PDDL to Prodigy.

[2] We will be making the basic domain theory available on a web site.

**Table 1.** Summary of Planners' Performance: by counts of superior performance on problems and by comparing computation times for success and failure

| Planner | # Fastest Suite | Software | # Solved Suite | Software | Computation Time Comparison $\mu$ Success | $\mu$ Fail | T | $P <$ |
|---|---|---|---|---|---|---|---|---|
| Blackbox | 15 | 2 | 89 | 11 | 23.06 | 210.81 | -4.80 | 0.0001 |
| IPP | 4 | 1 | 71 | 2 | 156.28 | 821.08 | -1.36 | 0.178 |
| SGP | 13 | 0 | 91 | 0 | 1724.58 | – | – | – |
| STAN | 59 | 0 | 79 | 0 | 67.89 | 12.29 | 0.33 | 0.741 |
| UCPOP | 14 | 69 | 41 | 69 | 20.68 | 387.45 | -15.81 | 0.001 |
| Prodigy | 3 | 0 | 48 | 12 | 52.27 | 2828.43 | -15.12 | 0.0001 |

planners would exhibit significantly different times between success and failure (would either fail or succeed quickly) and that easily measurable features of problems and domains would be somewhat predictive of a planner's performance. The purpose of the empirical comparison is to test these three hypotheses.

We ran the six planners on 263 problems from the three sets. For each problem/domain combination, we recorded five easily observed features. For the domain, we counted the number of actions (Act) and the number of predicates (Pred). For the problem, we counted the number of objects (Obj), the number of predicates in the initial conditions (Init) and the number of goals (Goal).

We measured performance by whether the planner successfully found a solution and how much computation time was required to fail or succeed. We counted time-outs, core dumps and planner flagged failures as unsuccessful runs. We allocated up to 15 hours for most of the planners. UCPOP was given a search limit of 100000. All runs were conducted on the same platform: Sun Ultra 1 workstations with 128M of memory. For analysis, we filtered the results to include only those problems that were solved by some planner: 176 problems remained of the 263 in the tested set.

The first hypothesis is that the best planner varies across problems. For each planner, Table 1 lists the number of problems on which that planner posted the fastest time and the total number of problems solved. These results are subdivided according to the problem collections: "suite" includes the competition and UCPOP problems, and "software" is our software testing domain. No planner solved all of the problems or even more quickly solved all of its problems. STAN was fastest in general. IPP generally lagged the others, but did solve a few problems that the others did not. The competition planners solved more of the problems from the benchmark suite. We had hoped that some of the other planners would excel on the software testing domain, but UCPOP dominated for that domain.

The second hypothesis was that the computation time would depend on success: the time to succeed would differ significantly from the time to fail. To test this, we partitioned the data based on planner and success and performed two tailed, two sample T-tests for each planner (see Table 1). All of SGP's failures were time-outs with identical times. Blackbox, UCPOP and Prodigy

**Table 2.** ANOVA and Chi-square test results for dependence of performance on planner and domain/problem features

| Planner | Feature | Time F | P | Success Chi | P | Planner | Feature | Time F | P | Success Chi | P |
|---|---|---|---|---|---|---|---|---|---|---|---|
| Blackbox | Init | 2.94 | .023 | 7.62 | .054 | IPP | Init | 8.11 | .001 | 15.98 | .003 |
| | Obj | 0.64 | .632 | 7.92 | .048 | | Obj | 3.54 | .017 | 8.48 | .075 |
| | Goal | 52.34 | .001 | 32.17 | .001 | | Goal | 0.74 | .567 | 27.25 | .001 |
| | Act | 3.55 | .010 | 38.60 | .001 | | Act | 1.20 | .317 | 9.26 | .055 |
| | Pred | 17.75 | .001 | 36.76 | .001 | | Pred | 0.17 | .917 | 31.85 | .001 |
| SGP | Init | 4.54 | .005 | 2.45 | .656 | STAN | Init | 1.34 | .268 | 13.59 | .009 |
| | Obj | 1.06 | .380 | 0.71 | .950 | | Obj | 0.33 | .802 | 4.26 | .372 |
| | Goal | 3.26 | .016 | 3.03 | .552 | | Goal | 24.46 | .001 | 4.27 | .370 |
| | Act | 2.25 | .070 | 1.99 | .738 | | Act | 1.06 | .372 | 19.57 | .001 |
| | Pred | 2.51 | .048 | 14.49 | .006 | | Pred | 7.48 | .001 | 8.02 | .046 |
| UCPOP | Init | 5.79 | .001 | 16.39 | .002 | Prodigy | Init | - | - | 9.04 | .06 |
| | Obj | 53.63 | .001 | 44.12 | .001 | | Obj | - | - | 22.79 | .0001 |
| | Goal | 49.66 | .001 | 41.24 | .001 | | Goal | 6.747 | .0001 | 37.86 | .0001 |
| | Act | 4.01 | .004 | 30.65 | .001 | | Act | 37.014 | .0001 | 55.67 | .0001 |
| | Pred | 20.44 | .001 | 61.43 | .001 | | Pred | 61.51 | .0001 | 66.48 | .0001 |

show significant differences: succeeding quickly and failing slowly. Success and failure times for STAN and IPP were not significantly different.

Finally, we hypothesize that the performance of planners depend on observable problem and domain features. To test this, we partitioned the data according to each of the five features and six planners. We then tested the relationship between the features and time by running a set of one way ANOVAs with computation time as the dependent variable and the feature as the independent variable. We tested the dependence between feature and success using Chi-Squared tests with counts for successful and total runs. Because cells were sparse, we coded each feature into five bins because the lowest number of values for a feature was 10. Some Prodigy features were missing too many cells to be analyzed with five bins.

Table 2 shows that performance of each planner depends on some of the features (statistically significant at $P < .05$). The number of predicates is significant in almost every case. Each planner showed significant results on from one to all five features on both performance metrics. This suggests that a subset of the features can be used to predict success and required computation time; which of them will vary for each planner.

# 5   Meta-planner

BUS is a meta-planner that schedules problem solving by six other planners. The key idea is that if one planner can solve a subset of planning problems then multiple planners should be able to solve more. The current version is a prototype for exploring combinations of planners and control strategies. More importantly, BUS is a platform for empirically comparing planner performance.

To solve a problem, BUS first determines the order in which the planners should be tried. It calculates an expected run time for each planner to solve the problem and an expected probability of success. The control strategy is to order algorithms by $\frac{P(A_i)}{T(A_i)}$ where $P(A_i)$ is the expected probability of success of algorithm $A_i$ and $T(A_i)$ is the expected run time of algorithm $A_i$. In the general case, this strategy minimizes the expected cost of trying a sequence of $n$ algorithms until one works [12].

In our prototype, the models of expected time and success are linear regression models of the performance data. As described in the last section, we analyzed the data to determine the problem/domain features upon which each planner's performance most depends. The best features were incorporated in a regression model, which provides an intercept and slopes for each of our features. Fortunately, all the features were interval metrics. We created one model for each of the planners on each of the two performance metrics.

The time models are composed of four features. Models of all the features tended to produce either negative times or high times for some problems. Four appeared to provide the best balance between enough information to predict and not too much to lead it astray.

For the success models, we computed five separate regression models for each planner, one for each feature. The dependent variable was *success rate per feature*; because it is relative to a specific feature, we could not compute a multiple linear regression model. Instead, we combined the statistically significant linear regression models using a uniform weighting. Another complication is that probabilities vary only from 0 to 1.0. To compensate, we added a ceiling and floor at these values for the probability models.

These models are linear, which, based on visualizing the data, is not the most accurate model. The $R^2$ values for the time models were: 0.46 for Blackbox, 0.19 for IPP, 0.26 for SGP, 0.35 for Stan, 0.76 for UCPOP and 0.51 for Prodigy. The $R^2$ values for the individual success models varied from 0.04 to 0.51. Still for the prototype, we relied on the linear regression models as they were easily obtained and implemented and could be justified by the data.

The core of BUS is a process manager. Each planner is run as a separate process. Planners are run in a round robin like scheme ordered by the control strategy. A planner is pulled off the front of the queue and allocated a time slice. The duration of the time slice is the expected run time needed for the particular planner to solve the proposed problem. If the planner solves the problem, the planning halts. If the planner fails, then it is removed from the round robin. If the time slice finishes without solution or failure, the process manager checks whether the current planner has taken as much time as the next planner in the queue requires. If not, the current planner as well as the proceeding planners are each allocated additional time slices until either one of them solves the problem or exceeds the time expected for the next planner. When control passes to the next planner, the current one is suspended, and its computation time so far is

**Table 3.** Number of problems solved and solution times for individual planners and BUS on new problems

| Planner | # Solved | $\mu$ time solved | Planner | # Solved | $\mu$ time solved |
|---------|----------|-------------------|---------|----------|-------------------|
| Blackbox | 26 | 37.87 | IPP | 22 | 10.62 |
| SGP | 24 | 364.85 | Stan | 8 | 2.26 |
| UCPOP | 21 | 3.98 | Prodigy | 12 | 5.94 |
| BUS | 32 | 70.83 | | | |

recorded. Computation time is accumulated until the overall amount exceeds a threshold (30 minutes for our trials). BUS translates the PDDL representations for UCPOP and Prodigy.

## 6    Performance of Meta-planner

To test the efficacy of BUS, we ran it on a subset of problems solved in the comparison study plus some new problems. The comparison problems were all of the benchmark suite problems and the software testing problems on just one domain; these were included to set a baseline for performance. The new problems were included to determine how well the control strategy generalizes and to show that BUS can solve problems with representations that were not in the intersection of all of the planners. The new problems came from the AIPS98 competition test generators and the UCPOP distribution. We generated 10 each new logistics, mystery, and mprime problems. The UCPOP problems were from the travel, ferry, tire, and get-paid domains. We recorded the total computation time used by all the planners that BUS ran.

In expectation, BUS should be able to solve all of the problems albeit somewhat slower than the fastest single run times. The additional time accounts for overhead and for trying several planners before one solves it. BUS solved 133 of 151 problems from the benchmark suite and 11 of 25 problems from the software testing set. On the software problems, the strategy appeared to allocate too much time to planners other than UCPOP.

On the study test suite, we compared BUS's times to the best, average and worst times for each problem for the single planners. BUS performed better than average on 54 problems. BUS took longer than the best posted times on average: the mean for BUS was 72.83, the mean for the best solutions was 11.36. However, BUS required less time than the average and worst times across the planners ($T = -1.86, P < .066$ for comparison to average times).

For the new problems, 19 of the 30 generated problems were unsolvable, meaning that no individual planner could solve them in under 4 hours. The individual planners were allotted up to 4 hours, but SGP is the only one which required more than 30 minutes to solve two of its problems. As Table 3 shows, BUS did well in comparison to the other planners. It solved more problems than any individual planner albeit at the cost of extra computation time.

# 7  Observations

Although the current control strategy is simplistic, BUS demonstrated the efficacy of combining planner approaches in a meta-planner. As expected, it solved more problems from the comparison set than any single planner had done. Somewhat surprisingly, the computational cost for BUS was significantly lower than the average computation required by individual planners to solve the same problems. On new problems, BUS solved more problems than any individual planner.

While the evaluation of BUS so far is favorable, BUS needs both more work and more evaluation. The current control strategy was a first attempt, but does not adequately model the relationship between problems and planner performance. We have identified four key deficiencies: coverage of feature values, limited predictability of features, mismatch in the underlying modeling technique and effect of representation. First, even the current control strategy would be much improved by simply having more data; the current set of problems was uneven in the distribution of the feature values. Second, while statistical tests showed a dependency between the features and performance, the relationship is not strong enough to be adequately predictive. The features are shallow and do not relate directly to planner functionality. Finally, clearly, the linear regression models are not the most appropriate for the data; the features did not appear to be linear in predictability. Alternative control strategies could be based on machine learning, smoothing and non-linear regression techniques. The most promising available option is Eugene Fink's method for selecting problem solving methods [3], which might be adaptable to this task. His method has performed well on a domain similar to that used in some of our problems.

Researchers have long acknowledged the effect of representation on performance. In this study, planners were forced to use the STRIPS versions of problems. Additionally, the current version includes a translator for converting PDDL to UCPOP and Prodigy representations. Although the translation process is syntactic, we observed that the planners seemed to do better on the problems that were originally coded in their representation. The next version of BUS will keep a database of problems in their original representations, translating where possible and filtering when not. Also, we will be investigating whether the translator can perform some semantic manipulations as well to address this issue.

To further its utility, BUS needs a wider variety of planners. Initially, the different representations will be handled by the database. Later, we will work on automatic translations of the representations. The next version will also include a checker for filtering out problem types that do not work for some planners.

BUS is a prototype. We tested the assumptions underlying its design in a study on 176 planning problems. From that study, we determined that most planners either succeed or recognize failure quickly and that problem features can be used to predict likelihood of success and expected computation cost, two characteristics necessary for supporting the current design. We also derived a control strategy based on the data from the study. Although it tends to incur additional computational overhead, its performance was shown to be competitive with current state of the art planners. BUS serves both as a vehicle for exploring planner performance (we can determine which planner ultimately solved each problem) and for exploiting different planning approaches.

**Acknowledgments.** This research was supported by grants from the National Science Foundation: grant number CCR-9619787 and Career award number IRI-9624058. The U.S. Government is authorized to reproduce and distribute reprints for Governmental purposes notwithstanding any copyright notation thereon. We also thank the reviewers for their suggestions of clarifications and extensions.

# References

1. A. Barrett, D. Christianson, M. Friedman, K. Golden, S. Penberthy, Y. Sun, and D. Weld. UCPOP user's manual. Technical Report TR 93-09-06d, Dept of Computer Science and Engineering, University of Washington, Seattle, WA, November 1996. Version 4.0.
2. A. Blum and M. Furst. Fast planning through planning graph analysis. *Artificial Intelligence*, 90:281–300, 1997.
3. E. Fink. How to solve it automatically: Selection among problem-solving methods. In *Proceedings of the Fourth International Conference on Artificial Intelligence Planning Systems*, June 1998.
4. A.E. Howe, A. von Mayrhauser, and R.T. Mraz. Test case generation as an AI planning problem. *Automated Software Engineering*, 4(1), 1997.
5. S. Kambhampati. Challenges in bridging plan synthesis paradigms. In *Proceedings of the Fifteenth International Joint Conference on Artificial Intelligence*, 1997.
6. S. Kambhampati and B. Srivastava. Universal Classical Planning: An algorithm for unifying state-space and plan-space planning. In *Current Trends in AI Planning: EWSP '95*. IOS Press, 1995.
7. H. Kautz and B. Selman. Blackbox: A new approach to the application of theorem proving to problem solving. In *Working notes of the AIPS98 Workshop on Planning as Combinatorial Search*, Pittsburgh, PA, 1998.
8. J. Koehler, B. Nebel, J. Hoffmann, and Y. Dimopoulos. Extending planning graphs to an ADL subset. In *Fourth European Conference in Planning*, 1997.
9. D. McDermott. Aips98 planning competition results. http://ftp.cs.yale.edu/pub/mcdermott /aipscomp-results.html, June 1998.
10. D. McDermott, M. Ghallab, A. Howe, C. Knoblock, A. Ram, M. Veloso, D. Weld, and D. Wilkins. *The Planning Domain Definition Language*, May 1998.
11. M.Fox and D.Long. The automatic inference of state invariants in TIM. *JAIR*, 9:367–421, 1998.
12. H. A. Simon and J. B. Kadane. Optimal problems-solving search: All-or-none solutions. *Artificial Intelligence*, 6:235–247, 1975.
13. UCPOP Group. The UCPOP planner. http://www.cs.washington.edu/research/ projects/ai/ www/ucpop.html, 1997.
14. M. Veloso and J. Blythe. Linkability: Examining causal link commitments in partial-order planning. In *Proceedings of the Second International Conference on AI Planning Systems*, June 1994.

15. M. M. Veloso, J. Carbonell, M. A. Pérez, D. Borrajo, E. Fink, and J. Blythe. Integrating planning and learning: The prodigy architecture. *Journal of Experimental and Theoretical Artificial Intelligence*, 7(1):81–120, 1995.
16. D. Weld, C. Anderson, and D. Smith. Extending graphplan to handle uncertainty and sensing actions. In *Proc. of 16th National Conference on AI*, 1998.

# A Parallel Algorithm for POMDP Solution

Larry D. Pyeatt[1] and Adele E. Howe[2]

[1] Computer Science Department, Texas Tech University, Lubbock, TX 79409
pyeatt@cs.ttu.edu
[2] Computer Science Department, Colorado State University, Fort Collins, CO 80523
howe@cs.colostate.edu

**Abstract.** Most exact algorithms for solving partially observable Markov decision processes (POMDPs) are based on a form of dynamic programming in which a piecewise-linear and convex representation of the value function is updated at every iteration to more accurately approximate the true value function. However, the process is computationally expensive, thus limiting the practical application of POMDPs in planning. To address this current limitation, we present a parallel distributed algorithm based on the Restricted Region method proposed by Cassandra, Littman and Zhang [1]. We compare performance of the parallel algorithm against a serial implementation Restricted Region.

## 1 Introduction

The process of planning can be viewed as finding a *policy* for a sequential decision task. The policy is a mapping from states to actions that maximizes some reward function. For plan execution in this framework, the agent determines its current state and chooses an action from its policy at each time step. The effects of actions need not be deterministic, since the current state is all that is necessary to select an action from the policy. If the current state can be determined at each time step, then the planning problem can be modeled as a Markov decision process. However, the agent may not be equipped to determine the current state with certainty and can only make sensor observations that hint at the true state. In this case, the planning problem is more closely modeled as a partially observable Markov decision process (POMDP) [12,2].

A POMDP formulation of the planning problem is particularly well suited to mobile robotics, where the environment can be modeled as a state space, but the sensors may be inadequate to determine the current state. There are several methods for exactly solving a POMDP problem to find a policy, but current solution methods do not scale well and are practical only for small problems. This paper presents a parallel version of Restricted Region (RR), which is currently the best known exact solution algorithm for POMDP problems [12,1].

A POMDP problem is specified by a set of states $\mathcal{S}$, a set of actions $\mathcal{A}$, a set of observations $\mathcal{Z}$, a reward function $r(s, a)$, a set of transition probabilities $\Pr(s'|s, a)$, and a set of observation probabilities $\Pr(z|s', a)$. The transition probabilities specify the probability that the state is $s'$ given that the previous

S. Biundo and M. Fox (Eds.): ECP-99, LNAI 1809, pp. 73–83, 2000.

state was $s$ and action $a$ was taken. The observation probabilities specify the probability of seeing observation $z$ given that the current state is $s'$ and action $a$ was taken in the previous state.

The POMDP model assumes that true state cannot always be determined. Instead, a probability distribution $b$ over $\mathcal{S}$ is maintained. This probability distribution is known as the *belief state*. At each time step an action $a$ is taken, and an observation $z$ is received. The belief in state $s'$ is then updated using

$$b_z^a(s') = \frac{\Pr(z|s',a)\sum_{s\in\mathcal{S}}\Pr(s'|s,a)b(s)}{\sum_{s'\in\mathcal{S}}\left[\Pr(z|s',a)\sum_{s\in\mathcal{S}}\Pr(s'|s,a)b(s)\right]} \tag{1}$$

for all $s' \in \mathcal{S}$, and the resulting new belief state is used to select the next action from the policy. An observation is received and the process repeats.

In the next section, we will review the best known algorithm for finding POMDP policies. Section 3 describes our parallel version of the algorithm. In Section 4, we present our experiments and results, and Section 5 discusses future work.

## 2   Solving POMDPs Using Incremental Pruning

The goal for a POMDP solution method is to construct a policy $\pi$ that specifies the action to take for every possible belief state $b \in \mathcal{B}$ so as to maximize discounted future rewards. Most POMDP solution algorithms generate the policy by finding the value function, which specifies the value of taking each possible action from every belief state. The value function can always be represented as a piecewise linear and convex surface over the belief space [10,7]. A POMDP policy is a set of labeled vectors that are the coefficients of the linear segments that make up the value function.

In value-iteration algorithms, a new value function $\mathcal{V}'$ is calculated in terms of the current value function $\mathcal{V}$ at each iteration [1]. $\mathcal{V}'$ is the result of adding one more step of rewards to $\mathcal{V}$. The function $\mathcal{V}'$ is defined by:

$$\mathcal{V}'(b) = \max_{a\in\mathcal{A}}\left(\sum_{s\in\mathcal{S}}r(s,a)b(s) + \gamma\sum_{z\in\mathcal{Z}}\Pr(z|b,a)\mathcal{V}(b_z^a)\right). \tag{2}$$

Equation 2 defines the value for a belief state $b$ as the value of the best action that can be taken from $b$ where the best value is given by the expected immediate reward for that action plus the expected discounted value of the resulting belief state.

Equation 2 can be decomposed into combinations of simpler value functions:

$$\mathcal{V}'(b) = \max_{a\in\mathcal{A}}\mathcal{V}^a(b) \tag{3}$$

$$\mathcal{V}^a(b) = \sum_{z\in\mathcal{Z}}\mathcal{V}_z^a(b) \tag{4}$$

$$\mathcal{V}_z^a(b) = \frac{\sum_{s\in\mathcal{S}}r(s,a)b(s)}{|\mathcal{Z}|} + \gamma\Pr(z|b,a)\mathcal{V}(b_z^a). \tag{5}$$

Because these functions involve linear combinations, the transformations preserve piecewise linearity and convexity [10,4].

Since the value functions $V_z^a$, $V^a$, and $V'$ are PWL, we can represent them as sets of vectors composed of the coefficients for each PWL segment. The sets have a unique minimum size representation [4], denoted by $S_z^a$, $S^a$, and $S'$ respectively. We use the notation purge($\mathcal{X}$) to represent the operation of reducing a set $\mathcal{X}$ to its minimum size form.

A vector of all ones is denoted as $\mathbf{1}$, a vector of all zeros is denoted as $\mathbf{0}$, and we define a vector $e_s$ for every $s \in \mathcal{S}$ where $e_s$ is a vector of all zeros except for one element $e_s(s) = 1$. The equivalence of vectors is defined as $\alpha = \beta \iff \alpha(s) = \beta(s) \, \forall s \in \mathcal{S}$. The lexicographical ordering of vectors is defined as $\alpha > \beta \iff \alpha \neq \beta$ and $\alpha(s) > \beta(s)$ where $s = \operatorname{argmin}_{i \in \mathcal{S}} \alpha(i) \neq \beta(i)$. Set subtraction is defined by $\mathcal{X} \backslash \mathcal{Y} = \{\chi \in \mathcal{X} | \chi \notin \mathcal{Y}\}$. The *cross sum* of two vector sets is defined as $\mathcal{X} \oplus \mathcal{Y} = \{\chi + \psi | \chi \in \mathcal{X}, \psi \in \mathcal{Y}\}$.

Using the vector set notation, the value function is described as

$$S' = \mathsf{purge}\left(\bigcup_{a \in \mathcal{A}} S^a\right) \tag{6}$$

$$S^a = \mathsf{purge}\left(\bigoplus_{z \in \mathcal{Z}} S_z^a\right) \tag{7}$$

$$S_z^a = \mathsf{purge}\left(\{\tau(\alpha, a, z) | \alpha \in S\}\right) \tag{8}$$

where $\tau(\alpha, a, z)$ is the $|\mathcal{S}|$-vector given by

$$\tau(\alpha, a, z) = \frac{1}{|\mathcal{Z}|} r^a(s) + \gamma \sum_{s'} \alpha(s') \Pr(z|s', a) \Pr(s'|s, a). \tag{9}$$

Equations 6 and 7 come directly from Equations 3 and 4 and basic properties of piecewise linear convex functions. Equation 8 is derived by substituting Equation 1 into Equation 5. Calculating $\tau(\alpha, a, z)$, performing the cross sum, and performing the union are straightforward; so the problem is now reduced to purging sets of vectors. In order to get a solution quickly, the purge operation must be efficient.

## 2.1  Purging Vector Sets

Given a set of $|\mathcal{S}|$-vectors $A$ and a single $|\mathcal{S}|$-vector $\alpha$, we define the set of belief states in which vector $\alpha$ is dominant compared to all the other vectors in $A$ as

$$R(\alpha, A) = \{b | b \cdot \alpha > b \cdot \alpha', \forall \alpha' \in A \backslash \{\alpha\}\}. \tag{10}$$

The set $R(\alpha, A)$ is called the *witness region* of vector $\alpha$ because $b$ can testify that $\alpha$ is needed to represent the piecewise linear convex function given by $A \cup \{\alpha\}$. The witness region is empty if and only if there is no part of the belief space where $\alpha$ has a better value than any other vector in $A$.

```
purge (F,𝒮)
    W ← ∅
    – Select the obvious winners at each corner of the belief space and put them in W
    for s ∈ 𝒮
        max ← 0
        for φ ∈ F
            if ((e_s · φ > max) or ((e_s · φ = max) and (ω_s < φ))) then
                ω_s ← φ
                max ← e_s · φ
    for s ∈ 𝒮
        if (ω_s ∈ F) then
            W ← W ∪ {ω_s}
            F ← F\{ω_s}
    – Test the remaining vectors for domination
    while F ≠ ∅
        φ ← first(F)
        x ← dominate(φ, W)
        if (x = ⊥) then F ← F\{φ}
        else
            max ← 0
            for (φ ∈ F)
                if ((x · φ > max) or ((x · φ = max) and (ω < φ))) then
                    ω ← φ
                    max ← x · φ
            W ← W ∪ {ω}
            F ← F\{ω}
    return W
```

**Fig. 1.** The algorithm for purging a set of vectors.

Given this definition for the witness region $R$, we can define the purge function as

$$\mathsf{purge}(A) = \{\alpha | \alpha \in A, R(\alpha, A) \neq \emptyset\}, \qquad (11)$$

which simply returns the set of vectors in $A$ that have non-empty witness regions. This is the minimum size set that represents the piecewise linear convex value function.

Figure 1 shows the algorithm of $\mathsf{purge}(F)$. This operation, developed by White [11], returns the vectors in $F$ that do not have empty witness regions. It should be noted that dot product ties *must* be broken lexicographically [4]. The purge algorithm starts with an empty set, $W$, and fills it with vectors from $F$ that have non-empty witness regions. The algorithm first checks all of the vectors at the corners of belief space. The dominant vector at each corner is obviously dominant over some region, so these vectors are added to $W$ and removed from $F$ immediately. The algorithm then checks each vector $φ$ in $F$ to determine if it is dominant somewhere in $W$. If a point $x$ is found on vector $φ$

that dominates all the vectors in $W$, then $x$ is a *witness* to the fact that there is another vector that belongs in $W$. The correct vector $\phi \in F$ is the one that maximizes the dot product with $x$. The vector $\phi$ is added to $W$ and removed from $F$, and the loop repeats. When $F$ is empty, $W$ contains the parsimonious set of dominant vectors.

The dominate function called by purge finds the belief state $x$ that gives the largest value on $\phi$ that is not dominated. The dominate function creates a linear program and adds a constraint for each hyperplane in $W$. If no solution exists or if the value of the objective function is non-positive, then a null vector is returned. Otherwise, the solution to the linear program is the desired point $x$.

## 2.2 Incremental Pruning

Incremental pruning [1] sequences the purge operations to reduce the number of linear programs that have to be solved and to reduce the number of constraints in the linear programs themselves. Incremental pruning can exhibit superior performance and greatly reduced complexity compared to other methods [12,1].

From the associative property of addition, $(A \oplus B) \oplus C = A \oplus (B \oplus C)$, and it also holds that $\mathsf{purge}((A \oplus B) \oplus C) = \mathsf{purge}(\mathsf{purge}(A \oplus B) \oplus C)$. Therefore, Equation 7 can be rewritten as

$$S^a = \mathsf{purge}(\ldots \mathsf{purge}(\mathsf{purge}(S^a_{z_1} \oplus S^a_{z_2}) \oplus S^a_{z_3}) \ldots \oplus S^a_{z_{|\mathcal{Z}|}}), \tag{12}$$

which is the approach taken by the incremental pruning algorithm. All of the calls to purge in Incremental Pruning are of the form purge $(A \oplus B)$. This formulation allows purge to exploit the considerable regularity in the vector set passed to it.

We replace the call to dominate in Figure 1 with

$$x \leftarrow \mathsf{dominate}(\phi, D \backslash \{\phi\}) \tag{13}$$

where $D$ is a set of vectors that satisfy the following set of properties:

1. $D \subseteq (A \oplus B)$
2. Let $\alpha + \beta = \phi$ for $\alpha \in A$ and $\beta \in B$. For every $\alpha' \in A$ and $\beta' \in B$, if $\alpha' + \beta' \in W$, then either $\alpha' + \beta' \in D$, or $\alpha + \beta' \in D$, or $\alpha' + \beta \in D$.

Many choices for $D$ will satisfy the given properties, such as:

$$D = A \oplus B \tag{14}$$
$$D = (\{\alpha\} \oplus B) \cup (\{\beta\} \oplus A) \tag{15}$$
$$D = W \tag{16}$$
$$D = (\{\alpha\} \oplus B) \cup \{\alpha' + \beta | (\alpha' + \beta) \in W\} \tag{17}$$
$$D = (\{\beta\} \oplus A) \cup \{\beta' + \alpha | (\beta' + \alpha) \in W\} \tag{18}$$

Any such choice of $D$ allows us to use the domination check of Equation 13 to either remove $\phi$ from consideration or to find a vector that has not yet been added to $W$.

```
purge (F,S)
      W ← ∅
      – Select the obvious winners at each edge of the belief space and put them in W
      for s ∈ S
            max ← 0
            for φ ∈ F
                  if ((e_s · φ > max) or ((e_s · φ = max) and (ω_s < φ))) then
                        ω_s ← φ
                        max ← e_s · φ
      for s ∈ S
            if (ω_s ∈ F) then
                  W ← W ∪ {ω_s}
                  F ← F\{ω_s}
      – Test the remaining vectors for domination
      current ← 0
      for i ∈ NUMSERVERS
            child_i ← CreateThread(ThreadProc, F, W, current, i)
      for i ∈ NUMSERVERS
            join(child_i)
      for (ω ∈ F)
            if (marked(ω)) then
                  W ← W ∪ ω
      return W
```

**Fig. 2.** Parallel algorithm for purging a set of vectors. The algorithm splits into NUMSERVERS threads, which mark the vectors in $F$ that belong in $W$.

Incremental pruning is actually a *family* of algorithms. The particular choice for $D$ determines how efficient the algorithm is. In general, small sets make a better choice, since this reduces the number of constraints in the linear program created by the **dominate** function. Incremental Pruning using a combination of Equations 17 and 18 is referred to as *restricted region* (RR) and is currently deemed to be the best POMDP solution algorithm [1].

## 3   Parallel Restricted Region

Incremental Pruning decomposes the problem into independent parts. The easiest way to parallelize RR would be to perform the cross sum and filter operations on the $S_z^a$ sets in parallel. The number of threads using this approach would be limited to $|\mathcal{A}|$.

Our approach is to perform the individual domination tests in parallel. Each domination test is performed by a domination server which may reside on another computer. The sequential algorithm shown in Figure 1 assumes that $W$ and $F$ will be updated at the end of each domination check and that $W$ and $F$ do not change during the domination check. However, this cannot be guaranteed when

```
ThreadProc(F, W, current, i)
    SendToServer(i, "NewProblem")
    SendToServer(i, F, W)
    while current < |F|
        lock current
        φ ← F_current
        current ← current + 1
        unlock current
        SendToServer(i, "TestDomination")
        SendToServer(i, φ)
        x ← RecieveFromServer(i)
        if (x ≠ ⊥) then
            max ← 0
            for (φ ∈ F)
                if ((x · φ > max) or ((x · φ = max) and (ω < φ))) then
                    ω ← φ
                    max ← x · φ
            lock F
            F_ω ← mark(F_ω)
            unlock F
```

**Fig. 3.** Each thread communicates with one server. The thread sends $F$ and $W$ to the server, then begins processing vectors in $F$.

several domination tests are occurring simultaneously, and synchronizing the data between servers would be a significant problem. We change the algorithm slightly, so that the vectors are not moved from $F$ to $W$, but are simply marked as dominant. After all of the vectors in $F$ have been checked, the ones marked as dominant are moved to $W$. This approach allows the initial sets to be sent to all of the servers at the beginning of each parallel merge which minimizes subsequent communication.

Our goal is to distribute the linear programs over as many processors as possible and do as many in parallel as we can. In order to do so, we must modify the purge function. We modified the cross sum operator $\alpha \oplus \beta$ so that the resulting vector is "tagged" to indicate $\alpha$ and $\beta$. We then added an operation taggedα$(\alpha, X)$ to return the set of vectors in $X$ that are tagged with the given $\alpha$ and an operation taggedβ$(\beta, X)$ to return the set of vectors in $X$ that are tagged with the given $\beta$. The complete $F$ and $W$ sets are sent to the servers, which use these operations to extract the subsets required for each application of Equations 17 and 18. We then modified the purge algorithm to perform domination checks in parallel.

The parallel purge algorithm is shown in Figure 2. As with RR, the algorithm first finds the dominant vectors at each corner of the belief space and moves those vectors into $W$. Then it starts a number of threads to perform the domination checks on the remaining vectors in $F$. The threads all execute the same `ThreadProc` code (shown in Figure 3) locally and each thread is responsible for communicating with one server. Each thread sends $F$ to the remote server and then enters a loop. On each iteration, the thread takes the next available vector in $F$ and instructs the remote server to test that vector for domination. If the remote server returns a witness point, then the thread finds the largest vector in $F$ at that point and marks it for addition to $W$. When all of the vectors in $F$ have been tested, the threads terminate, and the main program continues.

```
Server()
    loop
        ReceiveFromThread(command)
        if (command = "NewProblem") then
            ReceiveFromThread(F, W)
        elseif (command = "TestDomination") then
            ReceiveFromThread(φ)
            α = taga(φ)
            A = taggedα(α, F)
            A^W = taggedα(α, W)
            β = tagb(φ)
            B = taggedβ(β, F)
            B^W = taggedβ(β, W)
            if (|A| + |B^W| < |B| + |A^W|) then
                D = A ∪ B^W
            else
                D = B ∪ A^W
            x = dominate(φ, D\{φ})
            SendToThread(x)
```

**Fig. 4.** Server algorithm for parallel purge.

Figure 4 shows the server algorithm for the parallel purge algorithm. The server understands two commands. The `NewProblem` command instructs it to receive the following $F$ and $W$ sets. The `TestDomination` command instructs the server to test a given vector in $F$ for domination. When a `TestDomination` command arrives, it extracts the $\alpha$ and $\beta$ tags from the vector and extracts the sets $A = \{\alpha\} \oplus B$, $B = \{\beta\} \oplus A$, $B^W = \{\alpha' + \beta | (\alpha' + \beta) \in W\}$, and $A^W = \{\beta' + \alpha | (\beta' + \alpha) \in W\}$. $D$ is calculated as $A \cup B^W$ or $B \cup A^W$, whichever is smaller. Finally, $\phi$ is tested for domination over $D$, and the result is returned.

## 4   Experiments and Results

We implemented the parallel RR algorithm in C++ and ran it on a network of Sun Ultra I workstations. For comparison, we also implemented a serial version of the RR algorithm derived from the same code base. Thus, the only difference in the two versions is in the incprune algorithm as outlined in this paper.

**Table 1.** Test problem parameters.

| Problem | $|\mathcal{S}|$ | $|\mathcal{A}|$ | $|\mathcal{Z}|$ | Iterations | Reference |
|---|---|---|---|---|---|
| Tiger | 2 | 3 | 2 | 273 | Cassandra, Kaelbling and Littman [3] |
| Shuttle | 8 | 3 | 5 | 283 | Chrisman [5] |
| Part Painting | 4 | 4 | 2 | 100 | Kushmerick, Hanks, and Weld [6] |
| 4×3 | 11 | 4 | 6 | 100 | Parr and Russell [8] |
| Robot | 64 | 3 | 16 | 25 | Pyeatt and Howe [9] |

**Table 2.** Results of experiments with run times given in seconds. The column headings for the Parallel RR algorithm indicate the number of servers.

| Problem | RR | Parallel RR | | | | |
|---|---|---|---|---|---|---|
| | | 1 | 2 | 3 | 4 | 5 |
| Tiger | 209 | 220 | 177 | 139 | 103 | 70 |
| Shuttle | 415 | 566 | 473 | 453 | 448 | 444 |
| Part Painting | 137 | 156 | 103 | 89 | 81 | 76 |
| 4×3 | 1359 | 1474 | 1156 | 1100 | 1079 | 1083 |
| Robot | >86400 | >86400 | >86400 | 83465 | 62034 | 41648 |

We selected four problems from the literature and one of our own. The problems from the literature are representative of the problems that can be solved with current POMDP solution methods. Our Robot problem is significantly more difficult, and the advantage of our parallel algorithm is more pronounced on this problem. The problems are listed in Table 1. Descriptions of the Tiger, Part Painting, and 4×3 problems are available from the references. Robot is a small robot navigation problem where the goal is to reach a specified location.

Table 2 shows the performance of the two algorithms on the selected problems. The parallel IP algorithm was run with a varying number of servers; the algorithm was terminated when run time exceeded 24 hours. This occurred in three cases on the Robot problem. On the Tiger and Part Painting problems, each additional processor results in successively smaller performance increase. On the 4×3 problem, the performance actually degraded when a fifth processor was added. This diminishing return results from the communications overhead increasing in relationship to the amount of computation that is performed. On the Shuttle problem, the serial algorithm outperformed the parallel algorithm due to the time required to send the $S_z^a$ sets over the network. On the Robot problem, which requires considerably more computation, the relationship of run-time to number of processors is more linear. Note that this relationship does not appear to correlate with the number of states, actions, or observations in the POMDP problem, but appears to be related to the size of the $S_z^a$ sets that are created. Thus, $|\mathcal{S}|$, $|\mathcal{A}|$, and $|\mathcal{Z}|$ are not good indicators of the difficulty of POMDP problems.

## 5    Future Work

Our current implementation is really a distributed algorithm that spreads the work across several computers connected with 10Mbit/s Ethernet. The major drawback to the current implementation is the network bottleneck created when all of the threads send the $F$ set to the servers at once. This bottleneck could be alleviated in several ways. The main program could broadcast the $F$ set to all servers with a single transmission. This approach was not taken because TCP streams do not support broadcast. UDP supports broadcast, but implements unreliable communication using datagrams instead of streams. Broadcasting large vector sets using UDP would require breaking the data into packets and providing some mechanism to ensure reliability. This approach, though feasible, would require an added layer of complexity. Another way to reduce the bottleneck is to use multiple high speed network interfaces instead of a single Ethernet. Linux Beowulf and IBM SP systems both provide multiple high speed networks and would be ideal for this application. Third, and most obviously, the parallel IP algorithm could be implemented on a shared memory multiprocessor system.

This algorithm only parallelizes Equation 4, but could be extended to perform the incprune in Equation 3 and the cross sum operation in Equation 5 in parallel as well. Currently, these operations account for the majority of the time used by the parallel algorithm on the Tiger and Robot problems.

The parallel algorithm is most useful on problems with large $S_z^a$ sets, where the overhead involved in sending the data to the servers is much less than the time required for computation in the incprune operation. Thus, for small problems, parallel solution does not present an advantage, while the most difficult POMDP problems can benefit greatly from parallelization.

**Acknowledgments.** This research was supported in part by National Science Foundation Career Award IRI-9624058. The United States Government is authorized to reproduce and distribute reprints for governmental purposes notwithstanding any copyright notation herein. We thank the reviewers for catching some minor errors in the notation and description.

## References

1. Anthony Cassandra, Michael L. Littman, and Nevin L. Zhang. Incremental pruning: A simple, fast, exact algorithm for partially observable markov decision processes. In *Proceedings of the Thirteenth Annual Conference on Uncertainty in Artificial Intelligence*, 1997.

2. Anthony R. Cassandra. A survey of POMDP applications. In Michael Littmann, editor, *Working Notes: AAAI Fall Symposium on Planning with Partially Observable Markov Decision Processes*, pages 17–24. AAAI, October 1998.
3. Anthony R. Cassandra, Leslie Pack Kaelbling, and Michael L. Littman. Acting optimally in partially observable stochastic domains. In *Proceedings of the Twelfth National Conference on Artificial Intelligence*, Seattle, WA, 1994.
4. Anthony R. Cassandra, Michael L. Littman, and Leslie Pack Kaelbling. Efficient dynamic-programming updates in partially observable Markov decision processes. Technical Report CS-95-19, Brown University, Providence, RI, 1996.
5. Lonnie Chrisman. Reinforcement learning with perceptual aliasing: The perceptual distinctions approach. In *Proceedings of the Tenth National Conference on Artificial Intelligence (AAAI-92)*, pages 183–188. AAAI, AAAI Press/MIT Press, July 1991.
6. N. Kushmerick, S. Hanks, and D. S. Weld. An algorithm for probabilistic planning. *Artificial Intelligence*, 76(1–2):239–286, 1995.
7. G. E. Monahan. A survey of partially observable Markov decision processes. *Management Science*, 28(1):1–16, 1982.
8. R. Parr and Stuart Russell. Approximating optimal policies for partially observable stochastic domains. In *Proceedings of the 1995 International Joint Conference on Artificial Intelligence*, Montreal, Quebec, August 1995.
9. Larry D. Pyeatt and Adele E. Howe. Integrating POMDP and reinforcement learning for a two layer simulated robot architecture. In *Third International Conference on Autonomous Agents*, Seattle, Washington, May 1999.
10. Richard D. Smallwood and Edward J. Sondik. The optimal control of partially observable Markov processes over a finite horizon. *Operations Research*, 21:1071–1088, 1973.
11. Chelsea C. White III. A survey of solution techniques for the partially observed Markov decision process. *Annals of Operations Research*, 32:215–230, 1991.
12. Nevin L. Zhang and Wenju Liu. Planning in stochastic domains: Problem characteristics and approximation. Technical Report HKUST-CS96-31, Department of Computer Science, Hong Kong University of Science and Technology, Hong Kong, 1996.

# Plan Merging & Plan Reuse as Satisfiability *

Amol Dattatraya Mali

Dept. of Elect. Engg. & Computer Science, P.O.Box 784
University of Wisconsin, Milwaukee, WI 53201, USA
mali@miller.cs.uwm.edu

**Abstract.** Planning as satisfiability has hitherto focused only on purely generative planning. There is an evidence in traditional refinement planning that planning incrementally by reusing or merging plans can be more efficient than planning from scratch (sometimes reuse is not more efficient, but becomes necessary if the cost of abandoning the reusable plan is too high, when users are charged for the planning solution provided). We adapt the satisfiability paradigm to these scenarios by providing a framework where reusable or mergeable plans can be either contiguous or partially ordered and their actions can be removed and new actions can be added. We report the asymptotic sizes of the propositional encodings for several cases of plan reuse and plan merging. Our empirical evaluation shows that the satisfiability paradigm can scale up to handle plan reuse and plan merging.

## 1 Introduction

Impressive results have been obtained by casting planning problems as propositional satisfiability in [Kautz & Selman 96]. In planning as satisfiability, a propositional encoding is generated by fixing the number of plan steps (say $K$). If the number of steps in the solution is more than $K$, the value of $K$ is increased and an encoding is regenerated. An encoding contains all action sequences of length $K$ and solving it can be viewed as extracting a plan from it. To ensure that each model of an encoding is a valid plan, several constraints about the satisfaction of the pre-conditions of actions and resolution of conflicts between the actions are included in the encoding.

This paradigm has become highly popular, as can be seen from the work that followed [Kautz & Selman 96] such as [Ernst et al 97] [Giunchiglia et al 98] [Kautz et al 96]. However this work is limited to purely generative style of planning where each problem is solved from scratch without reusing or merging plans. On the other hand, there is a considerable evidence and argument in the previous literature that reusing plans can provide improvements in the plan synthesis time [Hanks & Weld 95] [Kambhampati & Hendler 92] [Veloso 94]. Many times, plans have to reused even if there is no efficiency gain, because

---

* I thank Subbarao Kambhampati and the anonymous referees of the ECP-99 for useful comments on the previous draft of this paper. This work was performed when the author was a graduate student at Arizona State University.

S. Biundo and M. Fox (Eds.): ECP-99, LNAI 1809, pp. 84–96, 2000.

the cost of abandoning them is too high (because the users are charged for the planning solution provided). [Britanik & Marefat 95] and [Foulser et al 92] show how plans can be merged by resolving interactions between them and how this kind of plan synthesis can be sometimes faster than planning from scratch.

To examine how these arguments apply to planning as satisfiability, we set up several propositional encodings that contain the constraints from the reusable or mergeable plans. Since there are different forms in which plans can be stored (for further use, to solve a new problem) and there are different ways of encoding them in propositional logic (discussed in [Kautz et al 96]), a variety of ways of casting plan reuse and plan merging as satisfiability exist. These ways are of interest to us because of the different sizes of the encodings that they yield. Our encodings for plan merging and plan reuse are synthesized by adapting the encodings of [Kautz et al 96] to handle these scenarios and we assume some familiarity of the readers with their encodings. We identify the constraints that dominate the asymptotic sizes of the encodings for reuse and merging. Treating the number of actions and fluents as domain constants, we use the power of the variable (number of steps in an encoding) to compare the encoding sizes.

Our work makes the following contributions.

- An automated synthesis of propositional encodings for several cases of plan merging and plan reuse and a report of their asymptotic sizes.
- We show that the size of the state-based encodings which are the shown to be the smallest (have the fewest number of clauses, sum of the clause lengths and number of variables) in the generative planning scenario [Mali & Kambhampati 99], approaches the size of the causal encodings, in the plan merging scenario, when the order preservation restriction (defined in section 3) is enforced.
- We show that the causal encodings for solving planning problems by merging or reusing causal plans are smaller and also faster to solve, than the causal encodings for solving the same problems in a generative style.
- We show that the state-based encodings for merging or reusing contiguous plans are generally neither smaller nor faster to solve, than the state-based encodings that do not reuse or merge plans.
- We show that the causal encodings for merging causal plans can be smaller (and also faster to solve) than the state-based encodings for merging contiguous plans.

In section 2, we explain the notation used for representing the constraints in the plans. In section 3, we describe the semantics of plan merging and show how different cases of merging can be encoded as satisfiability. In section 4, we revisit these cases, but for reusing plans. In section 5, we report empirical results on problems in several benchmark domains[1] and discuss the insights obtained from them. We present conclusions in section 6.

---

[1] Available at http://www.cs.yale.edu/HTML/YALE/CS/HyPlans/ mcdermott.html

## 2  Notation

$p_i$ denotes a plan step. $o_j$ denotes a ground action. $U$ is the set of ground pre-condition and effect propositions $u_j$ in the domain. $u_j(t)$ denotes that $u_j$ is true at time $t$. $\phi$ denotes the null action (no-op) which has no pre-conditions or effects. $O$ is the set of ground actions in the domain, it also includes $\phi$. $\phi$ occurs at a time $j$ only when no other non null action occurs at $j$. That is, $\phi$ is mutually exclusive with every other non null action. $(p_i = o_j)$ denotes the step→action mapping and $o_j$ is called the binding of $p_i$. $p_i$ then inherits the pre-conditions and effects of $o_j$. $p_i \xrightarrow{f} p_j$ denotes a causal link where $p_i$ (contributor) adds the condition $f$ and $p_j$ (consumer) needs it and $p_i$ precedes $p_j$. $o_i(j)$ denotes that the action $o_i$ occurs at time $j$ (when this happens, the pre-conditions of $o_i$ are true at time $j$ and its effects are true at time $(j+1)$). $p_i \prec p_j$ denotes that $p_i$ precedes $p_j$. $k_i$ denotes the number of steps in $i$ th plan. $m$ denotes the number of plans being merged. $o_{ij}$ denotes $j$ th action ($j \in [1, k_i]$) from $i$ th plan, $i \in [1, m]$. $K$ denotes the sum of the number of steps from the $m$ plans being merged or the number of steps in a reusable plan. If there are $K$ time steps in a state-based encoding, they range from 0 to $(K - 1)$. $K'$ denotes the number of new steps added during plan reuse. A causal link $p_i \xrightarrow{f} p_j$ is said to be threatened if there is a step $p_q$ that deletes $f$, such that $p_i \prec p_q, p_q \prec p_j$.

In sections 3 and 4, we do not list all the constraints that the encodings contain (since many are similar to the constraints in the encodings of [Kautz et al 96]), but we rather focus on the constraints that either capture the representation of the mergeable and reusable plans or dominate the asymptotic size of an encoding. When plans are to be reused or merged, it is necessary to convert the constraints of these plans into clauses that can be included in the encodings. We call this process "the translation/ adaptation of constraints in the plans". We assume that the plans that are reused or merged may be stored as either a contiguous sequence of actions or as a causal structure (a set of causal links and precedence constraints). It should be noted that contiguous plans are not necessarily found by state space planners and causal plans are not necessarily found by partial order planners. We refer to an encoding for the non-incremental planning scenario (where plans are generated from scratch, without any merging or reuse) as a "naive" encoding.

## 3  Plan Merging

Plan merging consists of uniting separately generated plans into one global plan, obeying the constraints due to interactions within and between individual plans [Britanik & Marefat 95]. Our notion of merging is close to the one in [Kambhampati et al 96]. As per their definition of *mergeability*, a plan $P'$ found by merging the plans $P_1$ and $P_2$ contains all the constraints from $P_1$ and $P_2$. We allow the actions from the mergeable plans to be removed while the merging process takes places, however, if the actions are not removed, then the orderings between them (if the plans are contiguous) and both the orderings and causal links between them (if the plans are causal) are preserved in the final plan (which is an outcome of the merging process) (we refer to this criterion as the *order preservation restriction*).

We consider four cases of plan merging obtained by varying the following two factors - **1.** The plans that are merged may be either contiguous or partially ordered (partial order plans are also referred to as *causal* plans ). **2.** The encoding itself may be based on state space planning or partial order planning.

We assume that the specification of the merging problem contains initial state and goal state of the problem to be solved and the plans to be merged. For all four cases of merging, we make the following assumptions -
**1.** No new actions are added, that is, an action $o_j$ will not belong to the final plan if it does not appear in any of the $m$ plans that are merged. **2.** A contiguous plan to be merged is a contiguous action sequence $[o_{i_1} o_{i_2} .....]$. **3.** $i$ th causal plan to be merged has $k_i$ step$\longrightarrow$action bindings, $c_i$ causal links and $r_i$ precedence relations on the steps. **4.** Actions from the mergeable plans can be removed just by mapping the corresponding steps to the null action.

## 3.1 Merging Contiguous Plans Using the State-Based Encoding[2]

Since the steps in this encoding are contiguous and actions from the plans may need to be interleaved, we make $K$ copies of each plan. We concatenate the individual plans in a particular order to create a sequence of $K$ steps and then we concatenate this sequence with itself $(K - 1)$ times to create a sequence of $K^2$ steps. Then we convert this sequence into the structure in Figure 1 where each action $o_i$ in the sequence is replaced by the disjunction $(o_i \vee \phi)$ to allow the removal of $o_i$ while synthesizing the final plan. For example, if we want to merge the two contiguous plans $[o_1 o_2]$ and $[o_3 o_4]$, then the step$\longrightarrow$action mapping structure shown in Figure 1 is used. It can be verified that this structure represents all possible orderings of the actions in these two plans. As we show later, making $K$ copies of each plan allows us to preserve the orders of the actions in the original plans, in the final plan.

Consider the state-based encoding with explanatory frame axioms from [Kautz et al 96] to handle the merging process. Since the truth of each of the $| U |$ propositions is stated at each time step (to compute the world state), there are $O(K^2 * | U |)$ variables in the encoding. Since explanatory frame axioms explaining the change in the truth of each of the propositions from $U$ are stated for each of the $K^2$ time steps, there are $O(K^2 * | U |)$ clauses in the encoding. (An explanatory frame axiom states that if a proposition $f$ is true at time $t$ and false at time $(t + 1)$, some action having the effect of deleting $f$ must have occurred at $t$). Note that if a $K$ step encoding is generated to solve the planning problem from scratch, without merging the plans, it will have $O(K * (| O | + | U |))$

---

[2] Kautz et al. 96 use the term state-based encoding to refer to an encoding based on state space planning, with the action variables eliminated, since the occurrence of an action is equivalent to the truth of its pre-conditions and effects. We do not insist on the elimination of action variables.

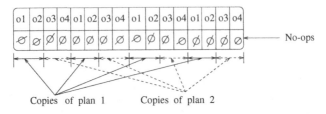

**Fig. 1.** Merging two contiguous plans by making 4 copies of each. K = 4. Thus 16 steps are used.

variables and $O(K * (| U | + | O |))$ clauses (because one has to consider the possibility of any action from the domain occurring at any time step (which requires $K* | O |$ variables) and because an action occurring at time $t$ implies the truth of its pre-conditions at $t$ and the truth of its effects at $(t + 1)$ which requires $O(K* | O |)$ clauses and $O(K* | U |)$ clauses are required to state the explanatory frame axioms).

Though the encoding has $K^2$ steps, we are looking for a plan of only $K$ steps. To reduce the potential non-minimality in the plans, we specify extra constraints. If a particular occurrence of an action from a plan is used (that is, it is not removed by choosing $\phi$), then all other $(K-1)$ copies of this occurrence of this action must be removed and those steps must be mapped to $\phi$, to reduce non-minimality in the plans (e.g. repeatedly loading and unloading the same package at the same location). This is stated in the following clauses.

$$\wedge_{q=1}^{K} \wedge_{i=1}^{m} \wedge_{j=1}^{k_i} (o_{ij}((q-1) * K + \textstyle\sum_{p=1}^{i-1} k_i + j - 1) \Rightarrow (\wedge_{q_1=1, q_1 \neq q}^{K} \neg o_{ij}((q_1 - 1) * K + \textstyle\sum_{p=1}^{i-1} k_i + j - 1)))$$

This specification requires $O(K^3)$ clauses. We also need clauses to respect the order preservation restriction. Since the $i$ th copy of any plan precedes its $j$ th copy (when $i < j$), if actions $o_a$ and $o_b$ are present in the plan such that $o_a$ precedes $o_b$, then if $o_b$ from $i$ th copy of the plan appears in the final plan, the copies of $o_a$ in all the remaining $j$ th copies of the plan, $j \in [i+1, K]$ must be discarded and $\phi$ should occur at those time steps. If this does not happen, the ordering relations between the actions from the original plan will be violated. The following clauses enforce the order preservation restriction. The number of such clauses required will be $O(K^3 * \alpha)$, $\alpha = max(\{k_i \mid i \in [1, m]\})$.

$$\wedge_{q=1}^{K} \wedge_{i=1}^{m} \wedge_{j=1}^{k_i} (o_{ij}((q-1)*K + \textstyle\sum_{p=1}^{i-1} k_i + j - 1) \Rightarrow (\wedge_{q_1=q+1}^{K} \wedge_{j_1=1}^{j-1} \neg o_{ij_1}((q_1 - 1) * K + \textstyle\sum_{p=1}^{i-1} k_i + j_1 - 1)))$$

Thus the number of variables and clauses in the encoding are $O(K^2* | U |)$ and $O(K^2* | U | + K^3 * \alpha)$ respectively.

## 3.2   Causal Plans Merged Using the Causal Encoding

We explain below how each constraint in a causal plan is represented in the encoding, to handle the merging scenario. The step⟶action binding $(p_i = o_j)$

is modified to $((p_i = o_j) \vee (p_i = \phi))$. We need the mutual exclusion constraint $\neg((p_i = o_j) \wedge (p_i = \phi))$ as well. Let us consider the causal links. A causal link $p_i \xrightarrow{f} p_j$ is modified to the constraints $((p_i \xrightarrow{f} p_j) \Rightarrow ((p_i = o_s) \wedge (p_j = o_q) \wedge (p_i \prec p_j)))$ and $(((p_i = o_s) \wedge (p_j = o_q)) \Rightarrow (p_i \xrightarrow{f} p_j))$ (for order preservation restriction) where $o_s, o_q$ are the actions to which $p_i, p_j$ are mapped in the plan being merged. This means that if the causal link is used, the step$\longrightarrow$action binding of the contributing and consuming steps must be used as well. Also, if the contributor and consumer steps are not mapped to $\phi$, then the causal links must be used as well. Some of the $r_i$ precedence relations in $i$ th plan might have been introduced to resolve the threats. In such a case we assume that the plan contains a record indicating that the ordering was introduced to resolve the threat. Such an ordering is also represented. For example, let the ordering $p_1 \prec p_2$, $p_1 = o_1$ be introduced to resolve the threat to the link $p_2 \xrightarrow{f} p_3$ posed by $p_1$. Then $p_1 \prec p_2$ is translated into the constraint $(((p_2 \xrightarrow{f} p_3) \wedge (p_1 = o_1)) \Rightarrow (p_1 \prec p_2))$. Note that since new causal links may be established between steps from different plans and new threats can arise, $O(K^2 * \mid U \mid)$ (all possible causal links) variables and $O(K^3 * \mid U \mid)$ threat resolution clauses are needed (to resolve the potential threats to each of the $O(K^2 * \mid U \mid)$ potential causal links by each of the $O(K)$ steps). These also determine the size of the encoding. Note that the actual encoding will be much smaller since the knowledge of the step$\longrightarrow$action binding can be propagated to generate fewer causal links and fewer threat resolution axioms. In this merging-based encoding, the size of the domain of the binding of each step is 2 (because it is either retained or removed and not mapped to a new action) rather than $\mid O \mid$, which is the case for the naive causal encoding. Thus the causal encoding for merging causal plans will be significantly smaller than the naive causal encoding. This is also confirmed by the sizes in our experiments reported in Fig. 6.

Though we do not discuss the cases of merging causal plans using the state-based encoding and merging contiguous plans using the causal encoding, we report their asymptotic sizes in Figure 3.

## 4   Plan Reuse

We assume that the specification of the reuse problem contains the plan to be reused and the initial and goal state of the problem to be solved, along with the ground actions in the domain. We focus on generating $(K+K')$ step encodings in which the reusable plan is represented, along with the $K'$ new steps. (A related issue is that of ensuring that some portion (even the maximum applicable) of the reusable plan is reused. In section 4.3, we discuss some strategies to ensure that the reusable plan is maximally recycled.) Similar to plan merging, we consider 4 cases of plan reuse. Our assumptions for reuse are stated below -
1. Actions in the reusable plan are either used in the final plan or removed by replacing them with $\phi$. 2. If the reusable plan is causal, it has $c$ causal links

and $r$ precedence constraints. **3.** A new step may be mapped to any of the $\mid O \mid$ actions in the domain.

If a contiguous plan is to be reused, we represent only the actions from the plan in an encoding. We do not represent the orderings of the actions from the plan because the actions in a contiguous plan are not ordered with the least commitment, that is, they may be ordered even if there is no reason to do so. On the other hand, if a causal plan is to be reused, we represent the actions, orderings and causal links from it into an encoding.

## 4.1   Reusing a Causal Plan Using the Causal Encoding

We discuss here how the constraints in a reusable causal plan are represented in an encoding. Each step⟶action binding $(p_i = o_j)$ in the plan is replaced by the constraint

$$((p_i = o_j) \vee (p_i = \phi)) \wedge \neg((p_i = o_j) \wedge (p_i = \phi))$$

The size of the domain of binding of each step $p_i$ from the reusable plan is 2 ($\{o_j, \phi\}$ here) rather than $\mid O \mid$ and this reduction can be propagated to reduce the size of the encoding, as confirmed by the results in Fig. 5. A causal link $p_i \xrightarrow{f} p_j$ is represented by the constraint $((p_i \xrightarrow{f} p_j) \Rightarrow ((p_i = o_s) \wedge (p_j = o_q) \wedge (p_i \prec p_j)))$ where $o_s, o_q$ are the actions to which $p_i, p_j$ are mapped (in the reusable plan). If the ordering $p_1 \prec p_2$, $p_1 = o_1$ was introduced to resolve the threat to the link $p_2 \xrightarrow{f} p_3$ posed by $p_1$, then $p_1 \prec p_2$ is translated into the constraint $(((p_2 \xrightarrow{f} p_3) \wedge (p_1 = o_1)) \Rightarrow (p_1 \prec p_2))$. Assuming that most of the causal links between the $K$ steps from the reusable plan will be present in it, the number of new variables in the encoding will be dominated by the new causal links that will be established between new $K'$ steps and the new $K'$ steps and the old $K$ steps. Hence the number of variables in the encoding will be $O((K'^2 + K' * K) * \mid U \mid + c)$. Assuming that the reusable plan is free of threats (since such threat resolving information is present in the reusable plan itself and is translated into appropriate constraints that are added to the encoding), since the steps from the reusable plan may threaten new causal links and the new steps may threaten both the new causal links as well as the causal links from the reusable plan, $O((K'^3 + K'^2 * K) * \mid U \mid + K' * c)$ clauses will be required to resolve the threats.

## 4.2   Reusing a Contiguous Plan Using the State-Based Encoding

We discuss two schemes here to achieve this.

**Scheme 1.** Since the actions from the reusable plan may have to be removed or reordered to allow the incorporation of new actions at arbitrary places, we make $K$ copies of the reusable plan and reserve $(K + 1)$ blocks, each of $K'$ steps for accomodating new actions (as shown in Figure 2). This means that to synthesize a $(K + K')$ step plan by reusing a $K$ step plan, we have $(K^2 + (K + 1) * K')$ steps in the encoding. Due to the representation of state at each time step, the possibility of occurrence of any of the $| O |$ actions at the $(K + 1) * K'$ steps reserved for new actions and the explanatory frame axioms, the encoding has $O((K^2 + (K + 1) * K')* | U | +(K + 1) * K'* | O |)$ variables and clauses each. To ensure that the extra $(K^2 + K * (K' - 1))$ steps are mapped to no-ops, $O(K^3 + (K + 1)^2 * K')$ additional clauses are needed.

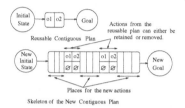

**Fig. 2.** Scheme 1 - Reusing a contiguous plan by making multiple copies of it and reserving multiple blocks of steps for the new actions, $K = 2, K' = 3$.

**Scheme 2.** In this scheme, we specify that each action from the reusable plan will occur at some time step or will never occur. As in the naive state-based encoding, this encoding will have $O((K + K') * (| O | + | U |))$ variables and $O((K + K') * (| U | + | O |))$ clauses. Note that if an action $o_i$ occurs in the reusable plan $s_i$ times, then since $K'$ new steps are added, this action should certainly not occur more than $(K' + s_i)$ times in the final plan. The specification

$$((\vee_{j=0}^{K+K'-1} o_i(j)) \vee \phi_i')$$

does not guarantee this ($\phi_i'$ is introduced to be able to make the clause true if $o_i$ is not required). In this scheme, we can restrict the maximum number of occurrences of $o_i$ in the final plan to $(K' + s_i)$, but only by adding $O((K + K')^{(K'+s_i+1)})$ new clauses which state that no choice of $(K' + s_i + 1)$ out of $(K + K')$ steps should have $o_i$ occurring at them. Thus though this scheme yields an encoding that has fewer variables and fewer clauses than the encoding in scheme 1, it requires more clauses to guarantee a control over the number of occurrences of an action.

### 4.3 Ensuring the Reuse of a Plan

We discuss here various ways of constraining or solving the encodings from sections 4.1 and 4.2 for ensuring that a plan will be maximally reused.

**1. Weighted Satisfiability** - One can assign weights to the clauses that represent the constraints from the reusable plan and maximize the sum of the

| Plan Type | Type of Encoding | |
|---|---|---|
| | State-based | Causal |
| Contiguous | $O(K^2 * \mid U \mid)$ vars | $O(K^2 * \mid U \mid)$ vars |
| | $O(K^2 * \mid U \mid + K^3 * \alpha)$ clauses | $O(K^3 * \mid U \mid)$ clauses |
| Causal | $O(C * K^2)$ vars | $O(K^2 * \mid U \mid)$ vars |
| | $O(C * K^3)$ clauses | $O(K^3 * \mid U \mid)$ clauses |

**Fig. 3.** Sizes of the encodings for plan merging. $C$ denotes the sum of the number of causal links in the individual plans.

weights of the satisfied clauses. MAX-SAT solvers can be used to find models that contain as much part of the reusable plan that can be recycled as possible.
**2. Iterative Solving -** One can create a $K$ step encoding such that it contains only the constraints from the reusable plan and try to solve it. If a solution can be found by removing actions from the reusable plan or if the reusable plan also solves the new problem, one can terminate. If there is no model of this encoding, one can increase the number of steps in the encoding by 1 and iteratively solve the resulting encodings. This ensures that the plan is maximally reused.

[Nebel & Koehler 95] prove that deriving the maximally reusable subplan is not easier than planning. Also, as they point out, maximal reuse makes sense only in the replanning scenario, where costs are charged for not executing the steps already planned. The conventional reuse strategies like those in [Kambhampati & Hendler 92] are heuristic and do not guarantee maximal recycling of the reusable plan. In our empirical evaluation, the plans were maximally recycled because the values of $K'$ chosen were same as the minimum number of new actions required to be included in the reusable plans.

## 5   Discussion

A comparison of the asymptotic sizes of various encodings for the plan merging and plan reuse cases is shown in Figures 3, 4. Note that the main attraction of the state-based encodings, in the non-incremental planning is their lowest size [Ernst et al 97] and the causal encodings are not used because they are hard to solve [Mali & Kambhampati 99]. We have shown that in the plan merging scenario, in the presence of the order preservation restriction, the size of the state-based encoding approaches the size of the causal encoding.

| Plan Type | Type of Encoding | |
| --- | --- | --- |
| | State-based | Causal |
| Contiguous | $O((K + K')*$ $(\mid O \mid + \mid U \mid))$ vars $O((K + K')*$ $(\mid U \mid + \mid O \mid))$ clauses | $O((K + K')^2 * \mid U \mid)$ vars $O((K + K')^3 * \mid U \mid)$ clauses |
| Causal | $O(c * (K + K')^2)$ vars $O(c * (K + K')^3)$ clauses | $O((K'^2 + K * K')$ $* \mid U \mid + \quad c)$ vars $O((K'^3 + K'^2 * K)$ $* \mid U \mid + K' * c)$ clauses |

**Fig. 4.** The sizes of the encodings for plan reuse. The size of the encoding for the reuse of a contiguous plan using the state-based encoding assumes that scheme 2 in section 4.2 is used. $c$ denotes the number of causal links in the reusable plan.

Our empirical results are shown in Figures 5 and 6[2] The number of steps in the encodings were same as the number of actions in the plans. The encodings were generated and solved on a Sun Ultra with 128 M RAM. As argued in the sections 3 and 4, the causal encodings for the reuse or merging of causal plans are far smaller than the naive causal encodings, since the domains of bindings of the steps from the reusable or mergeable plans are much smaller than $\mid O \mid$ and this shrinkage can be propagated to trim the portions of the encodings dependent on these steps. The causal encodings for causal plan merging and causal plan reuse were far smaller and also faster to solve than the naive causal encodings. Though the state-based encodings are the smallest in the generative planning, the state-based encodings for merging contiguous plans were sometimes harder to solve than the causal encodings for merging the causal plans (as shown in Figure 6), due to the blowup in their size (because of multiple copies of the plans) needed to respect the order preservation restriction and because they could not be significantly simplified like the causal encodings, by propagating the knowledge of the step⟶action mapping from the mergeable plans. We also found that neither the causal encodings for merging contiguous plans nor the state-based encodings for merging causal plans were easier to solve, possibly due to the introduction of intermediate variables, the multiple copies of the plans and the choice of encodings that are based on different type of planning than the representation of the mergeable plans, which made the explicit variable dependency in the plans less visible to the systematic solver.

We found that plan reuse as satisfiability is not necessarily faster than generative planning as satisfiability, especially when a contiguous plan is reused using the state-based encoding. This is however not a surprise, since the sizes of these

---

[2] The   encodings   were   solved   with   the   "satz"   solver   at
http://aida.intellektik.informatik.th-darmstadt.de/~hoos/SATLIB

encodings are almost same. We also found that the causal encodings for reusing causal plans are harder to solve than the state-based encodings for reusing contiguous plans, because despite the reduction in the sizes of the causal encodings achieved by constraint propagation, $O((K + K')^2)$ variables like $p_i \prec p_j$ are still needed to represent the all possible $\prec$ relations between the steps.

Our work ignores the costs of goal decomposition and generation of subplans that fulfill subgoals (in plan merging) and the plan retrieval cost in plan reuse. In future, we intend to augment the declarative approach of satisfiability with procedural control, to efficiently handle the problems of goal decomposition, subplan generation and plan retrieval.

| Domain, Plan Actions | CPSB V, C, T | S V, C, T | CPC V, C, T | CU V, C, T |
|---|---|---|---|---|
| Tsp, 14 $K = 7, K' = 7$ k = 14 | 638 1744 0.06 | 631 1639 0.07 | 1562 10003 3.59 | 7785 88873 2.48 |
| Tsp, 29 $K = 20, K' = 9$ k = 29 | 2631 7474 0.37 | 2611 6874 0.3 | 5098 55766 1.42 | 58320 1509973 * |
| Ferry, 79 $K = 52, K' = 27$ k = 79 | 8410 65231 1.3 | 8358 61071 1.27 | 33164 1198060 * | * * * |
| Logistics, 35 $K = 19, K' = 16$ k = 35 | 3240 10060 0.94 | 3132 9063 0.61 | 9118 138402 - | 73637 2343912 * |
| Blocks, 12 $K = 9, K' = 3$ k = 12 | 543 1788 0.12 | 534 1563 0.11 | 675 3156 0.34 | 6123 59715 1.6 |

**Fig. 5. Empirical results for plan reuse and generative planning.** The state-based encoding in case **CPSB** was generated using scheme 2 in section 4.2. $K, K'$ denote the number of steps in the reusable plan and those newly added to it respectively. In this figure as well as figure 6, $V, C, T$ denote the number of variables and clauses in and the times needed to solve the encodings respectively. Times are in CPU seconds. $k$ denotes the number of steps in the naive encodings. **CPSB** denotes the contiguous plan, state-based encoding case, **S** denote the naive state-based encoding, **CPC** denotes the causal plans, causal encoding case and **CU** denotes the naive causal encoding. A "-" indicates that the encoding was not solved within 10 minutes of CPU time. A "*" denotes that the encoding was too large to store. Tsp is the traveling salesperson domain.

| Domain, | CPSB | S | CPC | CU |
|---|---|---|---|---|
| Plan Actions | V, C, T | V, C, T | V, C, T | V, C, T |
| Tsp, 14 | 5909 | 631 | 379 | 7785 |
| $K = k = 14$ | 16017 | 1639 | 2616 | 88873 |
| $m = 5$ | 0.29 | 0.07 | 0.43 | 2.48 |
| Tsp, 29 | 50519 | 2611 | 1219 | 58320 |
| $K = k = 29$ | 161560 | 6874 | 23022 | 1509973 |
| $m = 5$ | - | 0.3 | 10.05 | * |
| Ferry, 19 | 6877 | 588 | 682 | 8535 |
| $K = k = 19$ | 27744 | 2436 | 7231 | 138172 |
| $m = 4$ | 0.5 | 4.05 | 57.75 | - |
| Logistics, 35 | 68740 | 3132 | 1726 | 73637 |
| $K = k = 35$ | 256370 | 9063 | 44326 | 2343912 |
| $m = 5$ | - | 0.61 | 22.69 | * |
| Blocks, 12 | 4494 | 534 | 324 | 6123 |
| $K = k = 12$ | 11775 | 1563 | 1737 | 59715 |
| $m = 4$ | 7.81 | 0.11 | 0.22 | 1.6 |

**Fig. 6.** Empirical results for plan merging and generative planning. $m$ denotes the number of plans merged and $K$ denotes the sum of the number of steps in plans merged.

# 6 Conclusion

We developed a framework for casting the plan reuse and plan merging problems as satisfiability. We reported the complexities of converting the reuse and merging problems into encodings in propositional logic. This analysis and the empirical evaluation lead to several new insights. We showed that the causal encoding of [Kautz et al 96], which has been shown to be hard to solve in generative planning [Mali & Kambhampati 99], can be used to solve several problems (from benchmark domains) of plan merging and plan reuse, due to the reduction in its size achieved by the propagation of the constraints from the plans, though it was still not always the fastest to solve. We also showed that the size of the state-based encoding with explanatory frame axioms [Kautz et al 96] which has been empirically shown to be the smallest and also generally the fastest to solve in the generative planning scenario [Ernst et al 97][Giunchiglia et al 98][Mali & Kambhampati 99], approaches the size of the naive causal encoding, due to its adaptation required to handle the order preservation restriction of plan merging. We also showed that though the satisfiability paradigm is scalable to reusing and merging plans, plan reuse is not always faster than generative planning (however as pointed out in [Nebel & Koehler 95], plan reuse may still be indispensable, if the users are charged for the planning solution provided).

# References

[**Britanik & Marefat 95**] J. Britanik and M. Marefat, Hierarchical plan merging with application to process planning, Procs. of the International Joint Conference on Artificial Intelligence (IJCAI), Vol. 2, 1995, 1677-1683.

[**Ernst et al 97**] Michael Ernst, Todd Millstein and Daniel Weld, Automatic SAT compilation of planning problems, Proccedings of the International Joint Conference on Artificial Intelligence (IJCAI), 1997.

[**Foulser et al 92**] David E. Foulser, Ming Li and Qiang Yang, Theory and algorithms for plan merging, Artificial Intelligence 57, 1992, 143-181.

[**Giunchiglia et al 98**] Enrico Giunchiglia, Alessandro Massarotto and Roberto Sebastiani, Act and the rest will follow: Exploiting determinism in planning as satisfiability, Proceedings of the National Conference on Artificial Intelligence (AAAI), 1998.

[**Hanks & Weld 95**] Steve Hanks and Daniel S. Weld, A domain-independent algorithm for plan adaptation, Journal of Artificial Intelligence Research 2, 1995, 319-360.

[**Kambhampati & Hendler 92**] Subbarao Kambhampati and James A. Hendler, A validation-structure-based theory of plan modification and reuse, Artificial Intelligence 55, 1992, 193-258.

[**Kambhampati et al 96**] S. Kambhampati, L. Ihrig and B. Srivastava, A candidate set-based analysis of subgoal interactions in conjunctive goal planning, Proceedings of the International conference on Artificial Intelligence Planning Systems (AIPS), 1996.

[**Kautz et al 96**] Henry Kautz, David McAllester and Bart Selman, Encoding plans in propositional logic, Proc. of Knowledge Representation and Reasoning Conference, (KRR), 1996.

[**Kautz & Selman 96**] Henry Kautz and Bart Selman, Pushing the envelope: Planning, Propositional logic and Stochastic search, Proc. of the National Conference on Artificial Intelligence (AAAI), 1996.

[**Mali & Kambhampati 99**] Amol D. Mali and Subbarao Kambhampati, On the utility of causal encodings, Proceedings of the National Conference on Artificial Intelligence (AAAI), 1999.

[**Nebel & Koehler 95**] Bernhard Nebel and Jana Koehler, Plan reuse versus plan generation: a theoretical and empirical analysis, Artificial Intelligence 76, 1995, 427-454.

[**Veloso 94**] Manuela M. Veloso, Planning and learning by analogical reasoning, Lecture notes in artificial intelligence 886, Springer-Verlag, 1994.

# SAT-Based Procedures for Temporal Reasoning

Alessandro Armando[1,2], Claudio Castellini[3], and Enrico Giunchiglia[1]

[1] DIST – Università di Genova, Viale Causa 13 – 16145 Genova – Italia
[2] LORIA-INRIA, 615, rue du Jardin Botanique – 54602 Villers les Nancy – France
[3] Div. of Informatics, U. of Edinburgh, 80 South Bridge, Edinburgh EH1 1HN, UK

**Abstract.** In this paper we study the consistency problem for a set of disjunctive temporal constraints [Stergiou and Koubarakis, 1998]. We propose two SAT-based procedures, and show that—on sets of binary randomly generated disjunctive constraints—they perform up to 2 orders of magnitude less consistency checks than the best procedure presented in [Stergiou and Koubarakis, 1998]. On these tests, our experimental analysis confirms Stergiou and Koubarakis's result about the existence of an easy-hard-easy pattern whose peak corresponds to a value in between 6 and 7 of the ratio of clauses to variables.

## 1   Introduction

Temporal reasoning is a traditional area of research, involved in planning (see, e.g., [Allen *et al.*, 1991]) scheduling ([Cheng and Smith, 1994]) and other application areas such as temporal databases ([Brusoni *et al.*, 1996]). One of the most studied problems in temporal reasoning (see, e.g., [Dechter *et al.*, 1991]) involves reasoning about sets of formulas of the form

$$l_1 \leq x - y \leq u_1 \vee \ldots \vee l_n \leq x - y \leq u_n, \tag{1}$$

$(n \geq 1)$ where $l_i, u_i$ are real numbers and $x, y$ are variables ranging over the reals. The *consistency problem* is to determine whether there exists an assignment to variables satisfying all the given formulas. More recently, Stergiou and Koubarakis [Stergiou and Koubarakis, 1998]:

- Extended the framework studied by, e.g., [Dechter *et al.*, 1991], so to allow also *disjunctive temporal constraints*, i.e. formulas of the form

$$x_1 - y_1 \leq u_1 \vee \ldots \vee x_n - y_n \leq u_n, \tag{2}$$

$(n \geq 1)$ where $x_1, y_1, x_2, y_2, \ldots, x_n, y_n$ are variables ranging over the reals, and $u_1, u_2, \ldots, u_n$ are real numbers. Notice that (1) involves only two variables, while this no longer holds for (2). This added generality may be useful in fields such as scheduling, management of temporal databases and natural language processing ([Koubarakis, 1997,Stergiou and Koubarakis, 1998]).

S. Biundo and M. Fox (Eds.): ECP-99, LNAI 1809, pp. 97–108, 2000.
© Springer-Verlag Berlin Heidelberg 2000

- Proposed four progressively more efficient algorithms for solving the consistency problem for sets of disjunctive temporal constraints. Essentially, they use ($i$) (variations of) Smullyan's tableau procedure for propositional logic [Smullyan, 1968] to incrementally generate sets of formulas of the form $x - y \leq u$, each set propositionally satisfying the given set of disjunctive temporal constraints, and ($ii$) Chleq's procedure [Chleq, 1995] to check the consistency of the assignments generated. The proposed algorithms differ in the heuristics adopted for performing the first step.
- On the ground of a theoretical analysis and experimental evaluations conducted using variously generated random tests they concluded that forward checking [Haralick and Elliott, 1980] gives the best performance. The forward checking heuristics amounts to valuating to false (and hence eliminating from the problem at hand) each disjunct whose negation is entailed by the assignment generated so far. In the following, we use SK to denote Stergiou and Koubarakis's procedure implementing forward checking.
- Finally, by plotting the number of consistency checks performed against the ratio of clauses to variables, they observed an easy-hard-easy pattern centered in between 6 and 7. Furthermore they noticed that the hard region does not coincide with the transition from soluble to insoluble problems.

In this paper we propose a SAT-based procedure for solving the consistency problem for a set of disjunctive temporal constraints. According to the SAT-based methodology [Giunchiglia and Sebastiani, 1996, Giunchiglia *et al.*, 2000], we still first

- *generate* a (possibly partial) valuation for the disjuncts which propositionally satisfies the input clauses, and then
- *test* that the generated valuation is indeed consistent.

However, given that the generation step involves propositional reasoning only, we adopt a state-of-the-art SAT decider for generating valuations. Because of this, we inherit the many optimizations and heuristic strategies (improving the average case behavior) which are implemented in current SAT solvers. Thus, we first propose a procedure, $\text{TSAT}^{FC}$, which differs from SK in that we use a state-of-the-art SAT solver in place of the tableau-like procedure implemented in SK. A comparison with SK reveals that $\text{TSAT}^{FC}$ performs up to 1 order of magnitude less consistency checks than SK. Furthermore, by looking at the trace of $\text{TSAT}^{FC}$, we discovered that many of the failed consistency checks are caused by the presence of pairwise mutually exclusive constraints, and observed that such consistency checks (and consequent possible failures) are performed over and over in different parts of the search tree. To avoid such redundant computations, we have implemented a preprocessing routine (called IS(2)) which checks the consistency of each pair of disjuncts, say $c_1$ and $c_2$, involving the same pair of variables: whenever $c_1$ and $c_2$ are mutually inconsistent, the clause $(\neg c_1 \vee \neg c_2)$ is added to the problem. The resulting procedure, $\text{TSAT}^{FC}_{IS(2)}$, performs up to 1 order of magnitude less consistency checks than $\text{TSAT}^{FC}$, and thus up to 2 orders less than SK.

```
function NAIVETSAT(φ : DTP) : boolean
  loop do
    choose a valuation μ P-satisfying φ
      if there is no such valuation then return False;
      if (μ is T-consistent) then return True;
  end
```

**Fig. 1.** NAIVETSAT

The paper is structured as follows. In Section 2, we introduce the basic concepts and terminology that will be used in the paper, and present our SAT-based procedures. In Section 3 we describe our experimental results: we first review the testing methodology presented in [Stergiou and Koubarakis, 1998] and used to evaluate SK; we then carry out a comparative analysis of our procedures against SK using sets of variously generated binary constraints, and show the 2 orders of magnitude improvement. In Section 4 we highlight the differences of the tested procedures, and explain their different behaviors. We end the paper with some concluding remarks in Section 5.

## 2   SAT-Based Procedures for Temporal Reasoning

A *temporal constraint* is a linear inequality of the form $x - y \leq r$, where $x$ and $y$ are variables ranging over the real numbers and $r$ is a real constant. A *disjunctive temporal constraint* is a disjunction of the form $c_1 \vee \cdots \vee c_n$ where $c_1, \ldots, c_n$ are temporal constraints, and $n \geq 1$. A *disjunctive temporal problem* (DTP) is a finite set of disjunctive temporal constraints to be intended conjunctively. An *assignment* is a function which maps each variable into a real number. An assignment $\sigma$ *T-satisfies*

- a temporal constraint $x - y \leq r$ if it is indeed the case that $\sigma(x) - \sigma(y) \leq r$;
- a disjunctive temporal constraint if it T-satisfies at least one of the disjuncts occurring in it;
- a DTP if it T-satisfies each of its elements.

We say that a DTP is *T-consistent* if there exists an assignment which T-satisfies it, and that it is *T-inconsistent* if no such an assignment exists. A *literal* is either a temporal constraint, e.g. $x_1 - x_2 \leq 3$, or the negation of a temporal constraint, e.g. $\neg(x_2 - x_1 \leq 5)$. A *clause* is a disjunction of literals. Thus disjunctive temporal constraints are clauses without occurrences of the negation symbol. A *valuation* is a set of literals, e.g., $\{x_1 - x_2 \leq 3, \neg(x_2 - x_1 \leq 5)\}$. If $c$ is a temporal constraint, then $\bar{c}$ abbreviates $\neg c$ and $\bar{\bar{c}}$ abbreviates $c$. We say that a valuation $\mu$ *P-satisfies* a DTP $\varphi$ if $\mu$ entails $\varphi$ by propositional reasoning only.

The simplest and most general formulation of a SAT-based decision procedure for checking the consistency of a DTP $\varphi$ is given by the non-deterministic algorithm NAIVETSAT given in Figure 1. The soundness of the algorithm relies

---

**function** TSAT($\varphi$ : DTP, $\mu$ : VALUATION): **boolean**
1: **if** $\varphi = \{\}$ **then return** TCONSIST($\mu$);
2: **if** $\{\} \in \varphi$ **then return** False;                /* backtrack */
3: **if** *unit_clause?*($\varphi, l$) **then**                /* unit */
      **return** TSAT(*simplify*($\varphi, l$), $\mu \cup \{l\}$);
4: $l := choose\_literal(\varphi)$;
5: **return** TSAT(*simplify*($\varphi, l$), $\mu \cup \{l\}$) **or**     /* split */
      TSAT(*simplify*($\varphi, \bar{l}$), $\mu \cup \{\bar{l}\}$);

---

**Fig. 2.** TSAT: a sketch of the algorithm

on the observation that if a valuation P-satisfying $\varphi$ is T-consistent, then $\varphi$ is T-consistent. As far as the completeness of the algorithm is concerned, it suffices to notice that if none of the valuations P-satisfying $\varphi$ is T-consistent, then $\varphi$ is not T-consistent. The efficiency of the algorithm rests on the way the valuations P-satisfying $\varphi$ are chosen. Notice that this is essentially the propositional satisfiability problem, for which very efficient procedures are available.

We developed a SAT-based decision procedure for checking the T-satisfiability of DTPs, TSAT, on top of Böhm's procedure, the winner of a 1992 competition involving various SAT-solvers [Buro and Buning, 1992][1]. An abstract presentation of TSAT is given in Figure 2. Given a DTP $\varphi$ and a valuation $\mu$ (initially set to the empty set), TSAT generates the valuations P-satisfying the input DTP $\varphi$, whereas the call to TCONSIST checks the T-consistency of the valuations generated by TSAT. *unit_clause?*($\varphi, l$) checks the existence of a temporal constraint $\{lit\} \in \varphi$ (a "unit clause" in SAT terminology), and assigns the literal *lit* to the variable $l$ occurring as its second argument. If a temporal constraint $l$ belongs to $\varphi$, then any valuation P-satisfying $\varphi$ must necessarily contain $l$. *simplify*($\varphi, l$) simplifies $\varphi$ by

1. first removing all the disjunctive temporal constraint containing $l$ as a disjunct, and
2. then replacing all the disjunctive temporal constraints of the form $cl \cup \{\bar{l}\}$ with $cl$.

Finally *choose-literal*($\varphi$) selects the literal (from those occurring in $\varphi$) which is the best candidate for splitting according to some heuristic criterion.

Forward Checking (FC) can be incorporated into TSAT in various ways. At the abstract level that we have used so far, it is enough to replace the two calls of the form *simplify*($\varphi, l$) in the last two lines, with *fc-simplify*($\varphi, \mu \cup \{l\}$). *fc-simplify*($\varphi, \mu \cup \{l\}$) accomplishes a simplification activity similar to *simplify*'s one, the only difference being that it performs an additional step:

---

[1] Böhm's procedure is an efficient implementation of the Davis & Putnam procedure [Davis and Putnam, 1960] which is available at the URL http://www.informatik.uni-koeln.de/ls_juenger/boehm.html.

3. replacing all the disjunctive temporal constraints of the form $cl \cup \{l'\}$ with $cl$ if $\mu \cup \{l\} \cup \{l'\}$ is not T-consistent.

The check for T-consistency can be readily carried out by invoking TCONSIST. *fc-simplify* is computationally more expensive than *simplify* since it requires a number of T-consistency checks roughly equal to the number of literals occurring in $\varphi$. However in many cases the stronger form of simplification carried out by *fc-simplify* has the positive effect of pruning the search space in a significant way. We call TSAT$^{FC}$ the TSAT algorithm modified in this way.

A common feature of TSAT and TSAT$^{FC}$ is that all the consistency checks are carried on-line. An alternative is to preprocess the DTP and look for sets of temporal constraints in the input DTP which are not T-consistent. If $S$ is one of such sets, the clause $\vee_{c \in S} \neg c$ can be added to the DTP at hand without affecting its consistency. If $n$ is a positive integer, we call IS($n$) the routine which

1. checks for T-consistency all the sets containing up to $n$ temporal constraints taken form those occurring in the input DTP, and
2. extends the input DTP with the clauses encoding the T-inconsistent sets.

Of course, such a preprocessing step makes sense only for small values of $n$. In our experience, we got benefits for $n = 2$. In this case, the IS(2) routine can be restricted to check only pairs of temporal constraints which involve the same pair of variables with opposite signs. (Pairs of temporal constraints involving more than two variables are trivially T-consistent.) Also notice that each performed consistency check is very simple, and can be performed in constant time. We call TSAT$^{FC}_{IS(2)}$ the procedure obtained by adding the IS(2) preprocessing routine to TSAT$^{FC}$.

# 3    Experimental Evaluation

In order to compare experimentally our algorithms with SK, we adopted the same random generation model employed in [Stergiou and Koubarakis, 1998]. Sets of DTPs are generated in terms of the tuple of parameters $\langle k, n, m, L \rangle$, where $k$ is the number of disjuncts per clause, $n$ is the number of arithmetic variables occurring in the problem, $m$ is the number of clauses, and finally $L$ is a positive integer number such that all the constants are taken from the interval $[-L, L]$. For a given $\langle k, n, m, L \rangle$, a DTP is produced by randomly generating $m$ clauses of length $k$. Each clause, $x_1 - y_1 \leq r_1 \vee \cdots \vee x_k - y_k \leq r_k$, is constructed by generating each disjunct $x_i - y_i \leq r_i$ by randomly choosing $x_i$ and $y_i$ with probability $1/n$ out of $n$ variables (but discharging pairs of identical variables) and taking $r_i$ to be a randomly selected integer in the interval $[-L, L]$. Furthermore, clauses with two (or more) identical disjuncts are discharged.

As in [Stergiou and Koubarakis, 1998] we generated our set of tests by taking $k = 2$ and $L = 100$. The class of binary disjunctive constraints is of particular interest since many problems from planning and scheduling can be mapped in such a class. The results of such experiments are plotted in the graphs of Figure 3 for $n = 10, 12, 15$, and 20. Each curve represents the total number of consistency checks versus the ratio of clauses to variables, $r = m/n$, which ranges from 2 to 14. All points are the median value among 100 randomly generated samples.[2] Each graph has curves for $\text{SK}^3$, $\text{TSAT}^{FC}$, and $\text{TSAT}^{FC}_{IS(2)}$ in it. For $\text{TSAT}^{FC}_{IS(2)}$ we also consider the number of consistency checks performed by IS(2) which grows quadratically as $r$ increases.[4]

As shown by the graphs, the algorithms show qualitatively similar behaviors (which are in turn similar to those given in [Stergiou and Koubarakis, 1998]). However $\text{TSAT}^{FC}$ performs uniformly better than SK, and $\text{TSAT}^{FC}_{IS(2)}$ performs uniformly better than $\text{TSAT}^{FC}$. Moreover, the gap in performance between the algorithms increases with the difficulty of the problems (i.e. as the values of $n$ increases). Considering the median number of consistency checks performed by the procedures,

- for $n = 10$ and $r > 6$, $\text{TSAT}^{FC}$ performs roughly one third of the consistency checks performed by SK, and $\text{TSAT}^{FC}_{IS(2)}$ performs roughly half of the consistency checks performed by $\text{TSAT}^{FC}$.
- for $n = 20$ and $r > 6$, $\text{TSAT}^{FC}$ performs almost 1 order of magnitude less consistency checks than SK, and $\text{TSAT}^{FC}_{IS(2)}$ performs almost 1 order of magnitude less consistency checks than $\text{TSAT}^{FC}$, and hence almost 2 orders less than SK.

Our experimental results show that our procedures (on these tests) present an easy-hard-easy pattern. In most cases, the curves representing our procedures reach their peak when $r = 6$, while for SK the peak is usually obtained when $r = 7$. However, considering the plots corresponding to $n = 20$ (which are much

---

[2] For practical reasons, we have used a timeout mechanism to stop the execution of a system on a DTP after 1000 seconds of CPU time. Because of this, we cannot plot the value representing the media of the consistency checks performed by the systems.

[3] We have used an implementation of SK which has been kindly made available to us by Kostas Stergiou.

[4] More in detail, assume that the number of temporal constraints $N$, be equal to $2m$ (which is in turn equal to $2nr$, since $r = m/n$). This assumption is reasonable since the probability of having two occurrences of the same temporal constraint in a DTP is negligible. For instance, when $n = 20$ and $r = 14$ the probability is about 0.015. There are $c = \frac{N^2 - N}{2} = 2r^2n^2 - rn$ possible pairs of temporal constraints, and the probability that two of them involve the same two variables with opposite signs is

$$p_c = \frac{1}{2} \frac{1}{\binom{n}{2}} = \frac{1}{n^2 - n}.$$

Therefore the average number of performed consistency checks is given by

$$c \cdot p_c = \frac{2r^2n - r}{n - 1}.$$

sharper than the others) we agree with Stergiou and Koubarakis's conjecture about the existence of an easy-hard-easy pattern whose peak is obtained for a value of $r$ in between 6 and 7. Furthermore, considering all the plots, it seems that

- the 50% of T-satisfiable DTPs is obtained when $5 \leq r \leq 6$, and
- the transition from solvable to unsolvable problems becomes steeper as $n$ increases.

To better highlight these facts, we have run $\text{TSAT}^{\text{FC}}_{\text{IS}(2)}$ on the two additional set of tests corresponding to $n = 25$ and $n = 30$. Results of these test sets are shown in Figure 4.

# 4    Explaining SK, Tsat$^{\text{FC}}$ and Tsat$^{\text{FC}}_{\text{IS}(2)}$ Behaviors

## 4.1    SK vs. Tsat$^{\text{FC}}$

To illustrate why $\text{TSAT}^{\text{FC}}$ performs better than SK, consider the following DTP $\varphi$:

$$x_1 - x_2 \leq 3$$
$$x_1 - x_3 \leq 4 \vee x_4 - x_3 \leq -2$$
$$x_2 - x_4 \leq 2 \vee x_3 - x_2 \leq 1$$
$$\vdots$$

(in the following, for any DTP $\psi$, $\psi(i)$ denotes the $i$-th disjunction displayed, and $\psi(i,j)$ the $j$-th disjunct of the $i$-th disjunction displayed in $\psi$. Thus, for example, $\varphi(1)$ is $x_1 - x_2 \leq 3$ and $\varphi(2,2)$ is $x_4 - x_3 \leq -2$) where the dots stand for further (possibly many) unspecified clauses such that no T-consistent extension of the evaluation $\{\varphi(1,1), \varphi(2,1)\}$ P-satisfying the whole DTP exists. Let us consider the behavior of SK and $\text{TSAT}$ ($\text{TSAT}^{\text{FC}}$) when $\{\varphi(1,1), \varphi(2,1)\}$ is the evaluation built so far. Since no T-consistent extension of $\{\varphi(1,1), \varphi(2,1)\}$ exists, after some search, failure is necessarily detected and both procedures backtrack and remove $\varphi(2,1)$ from the current valuation.

The main difference between SK and $\text{TSAT}$ ($\text{TSAT}^{\text{FC}}$) is that SK goes on with the valuation $\{\varphi(1,1), \varphi(2,2)\}$, whereas $\text{TSAT}$ ($\text{TSAT}^{\text{FC}}$) proceeds with the valuation $\{\varphi(1,1), \neg\varphi(2,1)\}$ which is immediately extended (via simplification and unit propagation) to $\{\varphi(1,1), \neg\varphi(2,1), \varphi(2,2)\}$. As we are going to see, working with $\{\varphi(1,1), \neg\varphi(2,1), \varphi(2,2)\}$ in place of $\{\varphi(1,1), \varphi(2,2)\}$ may lead to considerable savings. Let us now assume that both procedures extend their own current valuation with $\varphi(3,1)$. Since $\{\varphi(1,1), \neg\varphi(2,1), \varphi(2,2), \varphi(3,1)\}$ is T-inconsistent, $\text{TSAT}$ ($\text{TSAT}^{\text{FC}}$) stops immediately and goes on considering the next disjunct, i.e. $\varphi(3,2)$. On the other hand SK may waste a big amount of resources in the vain attempt of finding a T-consistent extension of $\{\varphi(1,1), \varphi(2,2), \varphi(3,1)\}$ P-satisfying $\varphi$.

The key observation is that $\{\varphi(1,1), \varphi(2,2), \varphi(3,1)\}$ entails $x_1 - x_3 \leq 3$ and therefore implies $\varphi(2,1)$. Thus,

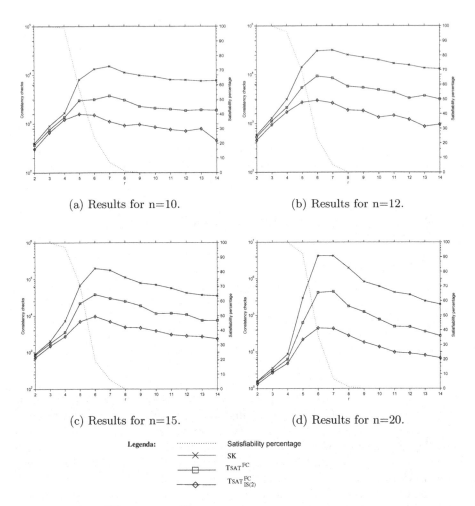

(a) Results for n=10.

(b) Results for n=12.

(c) Results for n=15.

(d) Results for n=20.

**Legenda:**  ················   Satisfiability percentage
—✕—   SK
—▭—   $\text{TSAT}^{\text{FC}}$
—◇—   $\text{TSAT}^{\text{FC}}_{\text{IS}(2)}$

**Fig. 3.** SK, $\text{TSAT}^{\text{FC}}$ and $\text{TSAT}^{\text{FC}}_{\text{IS}(2)}$ median number of consistency checks and percentage of T-satisfiable problems as a function of the ratio of clauses to variables $(r = m/n)$.

- while SK has to (re-)perform some computation in order to determine the T-inconsistency of any valuation $(i)$ extending $\{\varphi(1,1), \varphi(2,2), \varphi(3,1)\}$, and $(ii)$ P-satisfying the whole DTP,
- $\text{TSAT}^{\text{FC}}$ does not perform such computation since it considers the valuation $\{\varphi(1,1), \neg\varphi(2,1), \varphi(2,2), \varphi(3,1)\}$.

At a more abstract level, the fundamental difference between the two search procedures lies in the different form of branching they implement. While SK uses "syntactic branching":

$$\frac{c_1 \lor c_2}{c_1 \quad c_2}$$

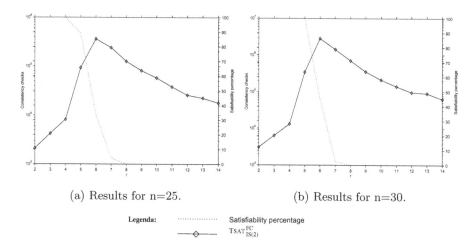

(a) Results for n=25.                    (b) Results for n=30.

Legenda:    ·················  Satisfiability percentage
            ——◇——  $\mathrm{TSAT}^{FC}_{IS(2)}$

**Fig. 4.** $\mathrm{TSAT}^{FC}_{IS(2)}$ median number of consistency checks and percentage of T-satisfiable problems as a function of the ratio of clauses to variables ($r = m/n$).

(given a disjunction $(c_1 \vee c_2)$ it first considers the case in which $c_1$ is true and then the case in which $c_2$ is true), $\mathrm{TSAT}$ (and thus $\mathrm{TSAT}^{FC}$ and $\mathrm{TSAT}^{FC}_{IS(2)}$) performs "semantic branching" [Freeman, 1995]:

$$\frac{\phantom{c_1 \qquad \neg c_1}}{c_1 \qquad \neg c_1}$$

i.e., it selects a not yet evaluated constraint $c_1$, and first considers the case in which it is true and then the case in which it is false. Notice that in the second case, the conjunction of $\neg c_1$ with $(c_1 \vee c_2)$ yields $c_2$ as a unit. As already observed in [D'Agostino, 1992,Giunchiglia and Sebastiani, 1996], syntactic branching may lead to redundant exploration of parts of the search space. That this is indeed the case also in the context of temporal reasoning is shown by the above example.

## 4.2   Tsat$^{FC}$ vs. Tsat$^{FC}_{IS(2)}$

To illustrate the advantages of using the IS(2) heuristics, consider the following disjunctive temporal problem $\varphi$:

$$x_3 - x_2 \leq 6 \vee x_1 - x_4 \leq 4$$
$$x_3 - x_1 \leq 1 \vee x_4 - x_1 \leq -5 \qquad (3)$$
$$x_1 - x_3 \leq -2 \vee x_2 - x_3 \leq -7$$

By checking the T-consistency of the 3 pairs of temporal constraints involving the same variables with opposite signs, we discover that $\{x_3 - x_2 \leq 6, x_2 - x_3 \leq -7\}$, $\{x_1 - x_4 \leq 4, x_4 - x_1 \leq -5\}$, and $\{x_3 - x_1 \leq 1, x_1 - x_3 \leq -2\}$ are T-inconsistent.

Therefore the IS(2) heuristics adds the following clauses to the initial DTP:

$$\neg(x_3 - x_2 \leq 6) \vee \neg(x_2 - x_3 \leq -7)$$
$$\neg(x_1 - x_4 \leq 4) \vee \neg(x_4 - x_1 \leq -5)$$
$$\neg(x_3 - x_1 \leq 1) \vee \neg(x_1 - x_3 \leq -2)$$

Three are the advantages of considering the extended DTP:

- First, if two mutually inconsistent temporal constraints occur in the input DTP, $\text{TSAT}^{FC}$ can check and discover such mutual inconsistency over and over again in different part of the search tree. By adding the clauses encoding such mutual exclusion, $\text{TSAT}^{FC}_{IS(2)}$ performs such checks only once: adding one of the temporal constraints to the current assignment causes the automatic addition of the negation of the other without performing any other consistency check.
- Second, IS(2) adds binary clauses possibly leading to extensive applications of the unit rule. In the case of the example, it is easy to check that after the first initial splitting step (assigning a truth value to a literal in the extended DTP), all the other literals get assigned by the unit step.
- Third, in (almost) all the DTPs that we randomly generate and use as our tests, there are no two occurrences of a same temporal constraint. As a consequence the heuristic implemented by Böhm's (and inherited by $\text{TSAT}^{FC}$ and $\text{TSAT}^{FC}_{IS(2)}$) provides no guidance in the choice of the best literal to use during the splitting step. We recall that Böhm's procedure chooses one among the literals which have the greatest number of occurrences in the clauses of minimal length [Buro and Buning, 1992] (in SAT terminology, Böhm's heuristic is a MOMS [Freeman, 1995]). Adding the clauses introduced by IS(2) has the positive effect that $\text{TSAT}^{FC}_{IS(2)}$ will choose the literal with the greatest number of occurrences in the newly added clauses, and hence possibly leading to the greatest number of applications of the unit rule.

Notice however that IS(2) is indeed a "greedy" heuristic: it performs some computation without any information about its usefulness. For sure, there are DTPs in which IS(2) may lead to a computationally worse behavior.

## 5    Conclusions

In this paper we have proposed a SAT-based approach to disjunctive temporal reasoning. We have presented two decision procedures ($\text{TSAT}^{FC}$ and $\text{TSAT}^{FC}_{IS(2)}$), the second differing from the first in that it uses IS(2), a simple preprocessing routine which greatly improves performances. Our procedures, when applied to sets of randomly generated binary disjunctive temporal constraints, perform less consistency checks than the fastest of the procedures proposed by Stergiou and Koubarakis. We have explained the different computational behaviors of SK and $\text{TSAT}^{FC}$ as due to the different types of propositional search performed. In fact, while SK performs syntactic branching, $\text{TSAT}^{FC}$ performs semantic branching.

We have argued the superiority of semantic branching w.r.t. syntactic branching, using a paradigmatic example as a case study. As for the IS(2) heuristics, we have shown that it is easy to perform, and that it greatly improves performances of $\text{T}_{\text{SAT}}^{FC}$.

Our experimental analysis confirms the existence of an easy-hard-easy pattern of the number of consistency checks, if plotted against the ratio of clauses to variables. We agree with Stergiou and Koubarakis's conjecture that the peak is obtained for a value of $r$ in between 6 and 7. Furthermore, our tests show that the 50% of T-satisfiable DTPs is obtained when $5 \leq r \leq 6$. Also this fact agrees with the experiments presented in [Stergiou and Koubarakis, 1998].

In the future, we plan to implement other forms of pruning strategy, like those described in [Prosser, 1993]. An extensive comparative analysis will reveal how these strategies behave and interact in a SAT-based framework.

Moreover, we plan to extend the SAT-based approach to the class of Disjunctive Linear Relation Sets [Jonsson and Bäckström, 1996] which subsumes the DTP framework, but also many other temporal formalisms.

The implementation of the procedures described in this paper can be obtained on request from the second author. The complete test sets we have used to evaluate the procedures (including those presented in this paper) are publicly available on the Internet at the URL: `http://www.mrg.dist.unige.it/~drwho/Tsat`.

**Acknowledgments.** We are grateful to Silvio Ranise, Roberto Sebastiani and Armando Tacchella for useful discussions related to the subject of this paper. Thanks to Kostas Stergiou for making the SK procedure available to us.

The last author has been supported by the Italian Space Agency under the project "Un sistema Intelligente per la supervisione di robot autonomi nello spazio".

# References

[Allen et al., 1991] J. Allen, H. Kautz, and R. Pelavin (Eds.). *Reasoning about Plans*. Morgan Kaufmann, 1991.

[Brusoni et al., 1996] V. Brusoni, L. Console, B. Pernici, and P. Terenziani. LaTeR: An Efficient, General Purpose Manager of Temporal Information. *IEEE Software*, 1996.

[Buro and Buning, 1992] M. Buro and H. Buning. Report on a SAT competition. Technical Report 110, University of Paderborn, Germany, November 1992.

[Cheng and Smith, 1994] Cheng-Chung Cheng and Stephen F. Smith. Generating feasible schedules under complex metric constraints. In *Proceedings of the Twelfth National Conference on Artificial Intelligence (AAAI-94)*, volume 2, pages 1086–1091, Seattle, Washington, USA, August 1994. AAAI Press/MIT Press.

[Chleq, 1995] N. Chleq. Efficient algorithms for networks of quantitative temporal constraints. In *Proceedings of CONSTRAINTS95*, pages 40–45, April 1995.

[D'Agostino, 1992] M. D'Agostino. Are Tableaux an Improvement on Truth-Tables? *Journal of Logic, Language and Information*, 1:235–252, 1992.

[Davis and Putnam, 1960] M. Davis and H. Putnam. A computing procedure for quantification theory. *Journal of the ACM*, 7:201–215, 1960.

[Dechter *et al.*, 1991] R. Dechter, I. Meiri, and J. Pearl. Temporal constraint networks. *Artificial Intelligence*, 49(1-3):61–95, January 1991.

[Freeman, 1995] Jon W. Freeman. *Improvements to propositional satisfiability search algorithms*. PhD thesis, University of Pennsylvania, 1995.

[Giunchiglia and Sebastiani, 1996] F. Giunchiglia and R. Sebastiani. Building decision procedures for modal logics from propositional decision procedures - the case study of modal K. In *Proc. CADE-96*, Lecture Notes in Artificial Intelligence, New Brunswick, NJ, USA, August 1996. Springer Verlag.

[Giunchiglia *et al.*, 2000] E. Giunchiglia, F. Giunchiglia, and A. Tacchella. SAT-Based Decision Procedures for Classical Modal Logics, 2000. Accepted for publication in Journal of Automated Reasoning. Available at http://www.mrg.dist.unige.it/~enrico.

[Haralick and Elliott, 1980] R. M. Haralick and G. L. Elliott. Increasing Tree Search Efficiency for Constraint Satisfaction Problems. *Acta Informatica*, 14:263–313, 1980.

[Jonsson and Bäckström, 1996] Peter Jonsson and Christer Bäckström. A linear-programming approach to temporal reasoning. In *Proceedings of the Thirteenth National Conference on Artificial Intelligence and the Eighth Innovative Applications of Artificial Intelligence Conference*, pages 1235–1241, Menlo Park, August 4–8 1996. AAAI Press / MIT Press.

[Koubarakis, 1997] Manolis Koubarakis. The complexity of query evaluation in indefinite temporal constraint databases. *Theoretical Computer Science*, 171(1–2):25–60, 15 January 1997.

[Prosser, 1993] Patrick Prosser. Hybrid algorithms for the constraint satisfaction problem. *Computational Intelligence*, 9(3):268–299, 1993.

[Smullyan, 1968] R. M. Smullyan. *First-Order Logic*. Springer-Verlag, NY, 1968.

[Stergiou and Koubarakis, 1998] Kostas Stergiou and Manolis Koubarakis. Backtracking algorithms for disjunctions of temporal constraints. In *Proc. AAAI*, 1998.

# Numeric State Variables in Constraint-Based Planning

Jussi Rintanen[1] and Hartmut Jungholt[2]

[1] Albert-Ludwigs-Universität Freiburg
Institut für Informatik
Am Flughafen 17, 79110 Freiburg im Breisgau
Germany
[2] Universität Ulm
Fakultät für Informatik
Albert-Einstein-Allee, 89069 Ulm
Germany

**Abstract.** We extend a planning algorithm to cover simple forms of arithmetics. The operator preconditions can refer to the values of numeric variables and the operator postconditions can modify the values of numeric variables. The basis planning algorithm is based on techniques from propositional satisfiability testing and does not restrict to forward or backward chaining. When several operations affect a numeric variable by increasing and decreasing its value in parallel, the effects have to be combined in a meaningful way. This problem is especially acute in planning algorithms that maintain an incomplete state description of every time point of a plan execution. The approach we take requires that for operators that are executed in parallel, all linearizations of the operations to total orders behave equivalently. We provide an efficient and general solution to the problem.

## 1 Introduction

In automated planning and for example in generating counterexamples in verification of safety properties of transition systems, the goal is to find a sequence of state transitions that lead from a given initial state to a state that satisfies certain properties. There are many possible ways of describing transition systems. The most basic description used in automated planning is based on operators that are applicable when a number of Boolean state variables are true or false and that make a number of Boolean state variables true or false. Many transition systems can be described by using this kind of operators. However, the descriptions would be more concise if no restriction to two-valued Boolean state variables were made. State variables with $n$ possible values can always be described with $\lceil \log_2 n \rceil$ Boolean state variables. However, replacing an operator description that refers to many-valued variables by ones that refer to Boolean variables only leads to a big increase in the number of operators. This reduces the efficiency of planning algorithms, and is therefore often not feasible.

S. Biundo and M. Fox (Eds.): ECP-99, LNAI 1809, pp. 109–121, 2000.
© Springer-Verlag Berlin Heidelberg 2000

Apart from the practical problems in reducing multi-valued state variables to Boolean state variables, there are semantic problems related to parallelism that are not addressed by that reduction. It is often the case that parallel operations affect the same numeric variables, for example the amount of money. The reductive approach to multi-valued variables in this case is not applicable because operations affecting the same variables that other parallel operations refer to in their preconditions, are not allowed. Therefore parallel operations have to be given a semantics that is aware of numeric variables and operations on them. For example, it should be possible to simultaneously execute a number of operations that increase and decrease the amount of money so that under the semantics the result is well-defined.

In this paper we investigate implementing parallel operations in a general planning framework based on constraint propagation. The operators may increase or decrease state variables with a fixed amount, or assign a fixed value to a state variable. Despite the restricted form of changes we consider, the work gives a solution to the problem of parallelism also for more general cases, for example when state variables may be assigned new values that are complex expressions that refer to constants and other numeric state variables.

## 2   Numeric State Variables in Operator Definitions

We extend the notion of operators with state variables that may take numeric values. Numeric state variables may be referred to in operator preconditions, and operators' postconditions may change the values of the state variables. The basic problems inherent in handling numeric state variables can be demonstrated with integers and constant increments and decrements. Preconditions of operators may include expressions $R = [n..m]$ which requires that $R \geq n$ and $R \leq m$. Postconditions of operators may be assignments $R := n$, $R := R + n$ and $R := R - n$ that respectively assign a specific value to $R$ or increase or decrease the value of $R$ by $n$. We do not allow disjunctions of preconditions $R = [n..m]$, as this would complicate the reasoning we have to perform. Conjunctions of preconditions $R = [n..m]$ are equivalent to intervals that are intersections of the conjunct intervals. Unbounded intervals can be formed with the constant infinite $\infty$ as $[-\infty..n]$ and $[n..\infty]$. The sum $\infty + n$ for an integer $n$ is $\infty$. Notice that ordinary Boolean state variables can be understood as numeric variables, for example by interpreting the interval $[0..0]$ as false and the interval $[1..1]$ as true.

## 3   Parallelism

Most conventional planners that allow parallelism, including planners of the partial-order planning paradigm [7] and more recent planning algorithms [1], restrict to cases where two parallel operations may not have variables in common in their postconditions, and variables occurring in the postconditions of one may not occur in the preconditions of the other. Two operators related in this way are said to be *dependent* because executing one may disable or enable the execution of the other, or falsify the effects of the other. The purpose of these restrictions is to guarantee a well-defined meaning for parallelism: parallelism means the possibility of interleaving the operators to a sequence in all possible ways, and the effect of executing the operators has to be the same for all interleavings.

In many applications, for example when the variables represent money or some physical resources, the requirements on parallel operators stated earlier are too strict, and in some applications may prevent parallelism altogether. For example, if all operators are associated with a cost and no two operators increasing the cost may be executed in parallel, no parallelism is possible.

Therefore we decided to try to relax these requirements. It turns out that interesting forms of parallelism, not considered in earlier work in planning, are possible. A set of operators executed in parallel have to fulfill the requirement used in earlier work on planning: parallelism is allowed as far as the execution of a set of operators in all possible interleavings produces the same result. Instead of using the syntactic condition used in connection with Boolean state variables, we develop a more abstract criterion parallel operations have to fulfill, and show that it can be implemented efficiently. It is not obvious how this can be done.

*Example 1.* Consider the operators $R = [1..2] \Rightarrow R := R + 2$ and $R = [1..5] \Rightarrow R := R + 2$. Initially $R = 1$. The operators can be interleaved in two ways. In the first interleaving the first operator is executed first, the value of $R$ is increased by 2 to 3, and then the second operator is executed, $R$ getting the value 5. However, the second interleaving does not lead to a legal sequence of operations. First the second operator is applied, increasing the value of $R$ to 3. Now the precondition of the first operator is not true, and the interleaving does not represent a valid execution. Hence the operators cannot be executed in parallel.

## 4    The Basis Algorithm

To investigate the problem of parallel updates of numeric state variables in practise, we decided to implement our techniques in an existing automated planner. The reason for extending the algorithm presented by Rintanen [8] to handle numeric state variables is that the algorithm subsumes both forward-chaining and backward-chaining, as the algorithm can simulate them by making branching decisions in a temporally directed manner starting from the goal state and proceeding towards the initial state, or vice versa. The algorithm by Rintanen is motivated by the planning as satisfiability approach to classical planning [3].

Each run of the main procedure of the algorithm is parameterized by the plan length. The main procedure is called with increased plan lengths 1, 2, 3, 4 and so on, until a plan is found. The algorithm maintains a state description for each time point of the plan execution. This description is incomplete in the sense that not all truth values of state variables need to be known. Initially, it includes the unique initial state and a description of the goal states.

For a given plan length and initial and goal states, planning proceeds by binary search. The algorithm chooses an operator $o$ and a time point $t$, and records in its data structures that operator $o$ is applied at $t$ (or respectively that is not applied at $t$.) Each decision to apply or not apply is followed by reasoning concerning the new situation. For example for an applied operator its preconditions are true at $t$ and postconditions are true at $t+1$ and dependent operators cannot be applied at $t$. The decision to not apply a certain operator at $t$ may indicate that a certain state variable retains its truth-value between $t$ and $t+1$, thereby allowing to infer the value of the variable at $t$ if its value at $t+1$ was known, and vice versa. A more detailed description of the constraint reasoning involved has been published elsewhere [8]. Upon detecting the inexistence of plans in the current subtree, the algorithm backtracks.

Figure 1 shows a part of a search tree produced by the algorithm. Oi represents a particular operator application (an operator and a time point), and N-Oi that the operator application is not performed. Each arrow corresponds to a decision to (not) apply a certain operator at a certain point of time, and inferred applications and inapplications of other operations. For example, in the first subtree of the first child node of the root, choosing to not apply O2 leads to inferring the operation O3.

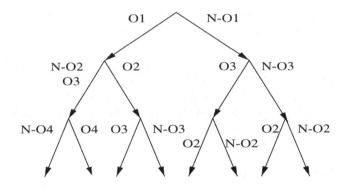

**Fig. 1.** Part of a search tree of the planning algorithm

When reaching a leaf node in the search tree without detecting any inconsistencies, that is, we have committed to the application and non-application of all operators at all time points, a plan has been found.

An important technique discovered in the context of algorithms for propositional satisfiability is detection of failed literals by unit resolution [6]. This technique in the setting of our algorithm proceeds as follows. An operation is labeled APPLIED (respectively NOT-APPLIED), and the inferences directly following from this are made. If an inconsistency is found, the operator must be labeled NOT-APPLIED instead (respectively APPLIED.) This technique is very useful because it allows to infer operator application and inapplications by inexpensive (low polynomial time) computation, in many cases leading to a big reduction in the depth of the search tree. A linear reduction in the search tree depth corresponds to an exponential reduction in the number of nodes in the search tree and consequently in the computation time.

# 5   Extension to the Basis Algorithm

The initial state of a problem instance specifies the values of all numeric state variables uniquely, and the goal specifies values of some numeric state variables as an interval. When the main program of the planner is run, all the processing of numeric state variables goes through a single function call that indicates that the main program has decided to apply or not to apply a certain operator. In addition, there are functions for registering a backtracking point and for undoing all the changes done since the previous backtracking point was registered.

From the point of view of numeric state variables, the algorithmic question that has to be solved is the following. Given a sequence of operators, labeled APPLIED or NOT-APPLIED (corresponding to a directed path starting from the root of the tree in Figure 1), decide whether this information is consistent with the initial values of the variables, with the values determined by the goal, and with the intermediate values inferred for the variables. Consistency here means that there is a plan execution (and hence also a plan) that executes operations labeled APPLIED and does not execute operations labeled NOT-APPLIED. Executing an operator means that its preconditions are true and the changes that take place exactly correspond to the changes caused by the operations that are executed.

For the correctness of the planning algorithm, the constraint reasoner has to satisfy the following two requirements.

1. When all operator applications have been labeled, an inconsistency must be reported if there are no executions where exactly the APPLIED operations are performed.
2. If an inconsistency is reported, there may not be executions that apply operations labeled APPLIED and do not apply operations labeled NOT-APPLIED.

These requirements for example allow refraining from doing any constraint reasoning as long as some operation applications are not labeled. This strategy, even though correct, would not be very efficient as there are usually a lot of possibilities of detecting inconsistencies much earlier and to avoid traversing parts of the search tree that do not contain plans.

For simplicity of presentation, in the following we discuss the case with only one numeric state variable $R$. Inferences performed for different numeric variables do not interact and are done in complete separation. The value of $R$ at a time point $t$ is represented by $R_t$.

The two kinds of changes in state variables are handled differently. Assignments $R := n$ at $t$ are simply performed by assigning $R_t := [n..n]$. Assignments obviously cannot be performed in parallel with other assignments or increments or decrements because different linearizations would give different results. The more complicated case is that of increments and decrements.

For reasoning about the values of $R$ efficiently, we introduce the following auxiliary variables.

$M_t^I$ Maximum possible increase at time $t$. This is the sum of the increases of operators that are not labeled NOT-APPLIED at $t$, minus the decreases of operators that are labeled APPLIED at $t$.

$M_t^D$ Maximum possible decrease at time $t$. This is symmetrically the sum of the decreases of operators that are not labeled NOT-APPLIED at $t$, minus the increases of operators that are labeled APPLIED at $t$.

$I_t^I$ Maximum intermediate increase at time $t$. This is the sum of increases of operators labeled APPLIED at $t$. The variable is needed for the test that the preconditions are satisfied under all interleavings.

$I_t^D$ Maximum intermediate decrease at time $t$. This is the sum of decreases of operators labeled APPLIED at $t$.

When all operator applications have been labeled, $M_t^I = -M_t^D = I_t^I - I_t^D$ for all $t$. The values of these auxiliary variables can be recomputed every time they are needed. Alternatively, and more efficiently, they can be incrementally maintained when operations get labels. Initially, $M_t^I$ is assigned the sum of increments of $R$ in postconditions of all operators for all $t$, $M_t^D$ similarly the sum of decrements of $R$, and $I_t^I$ and $I_t^D$ are assigned 0.

When an operator that increases $R$ by $n$ is labeled APPLIED at $t$, set $M_t^D := M_t^D - n$ and $I_t^I := I_t^I + n$.

When an operator that decreases $R$ by $n$ is labeled APPLIED at $t$, set $M_t^I := M_t^I - n$ and $I_t^D := I_t^D + n$.

When an operator that increases $R$ by $n$ is labeled NOT-APPLIED at $t$, set $M_t^I := M_t^I - n$.

When an operator that decreases $R$ by $n$ is labeled NOT-APPLIED at $t$, set $M_t^D := M_t^D - n$.

The following rules for propagating constraints on neighboring values of $R$ are needed. The first two are applied only when all operators making constant-value assignments to $R$ at $t$ are labeled NOT-APPLIED. The rules are applied whenever some of the variables involved change.

1. If $R_t = [a..b]$, then $R_{t+1} := R_{t+1} \cap [a - M_t^D..b + M_t^I]$.
2. If $R_{t+1} = [a..b]$, then $R_t := R_t \cap [a - M_t^I..b + M_t^D]$.
3. If an operator with precondition $R = [a..b]$ is applied at $t$, then $R_t := R_t \cap [a..b]$.

When a number of operators are applied in parallel at $t$, it is not sufficient to test that the preconditions of the operators are satisfied at $t$. It is also necessary to test that the preconditions are satisfied in all intermediate states of executions of all linearizations of the operators. The standard interpretation of parallelism requires this: all interleavings/linearizations have to be possible and the result of executing the operations has to be the same in all cases.

The test that the executions of all linearizations are possible is based on the auxiliary variables $I_t^I$ and $I_t^D$ that indicate how high and how low the value of the state variable can get when performing the execution of the operators between $t$ and $t+1$.

Assume that $R = [a..b]$ at $t$. Now the value of $R$ can get as high as $a + I_t^I$ under some linearization of the operators executed between $t$ and $t+1$ (and higher if the value turns out to be higher than $a$, for example $b$), and as low as $b - I_t^D$. We have to test that the preconditions of none of the operators are violated under any of the linearizations.

Consider an operator $o$ with the precondition $R = [p..q]$ that does not affect the value of $R$. Now if $b - I_t^D < p$, then there is a linearization of the operators during the execution of which the precondition of $o$ is not fulfilled: execute first all operators that decrease the value of $R$. Similarly, if $a + I_t^I > q$, we have detected an inconsistency.

In the general case the operator $o$ may have a precondition $R = [p..q]$ and also affect the value of $R$. Because the change caused by $o$ cannot affect the fulfillment of its own precondition, the effect of $o$ on the value of $R$ has to be ignored when testing the precondition of $o$ with respect to all interleavings. If $o$ increases $R$ by $n$, define $n^+ = n$ and $n^- = 0$. If $o$ decreases $R$ by $n$, define $n^+ = 0$ and $n^- = n$. Now we can state the test for operator preconditions under parallelism in the general case. Assume that the value of $R$ at $t$ is $[a..b]$. If $b - (I_t^D - n^-) < p$, we have detected an inconsistency. Similarly, if $a + (I_t^I - n^+) > q$, we have detected an inconsistency. These checks are subsumed by the following constraint on the value of $R$.

$$R_t := R_t \cap [p + (I_t^D - n^-)..q - (I_t^I - n^+)]$$

This constraint stems from the possibility of executing all other operators that increase (or decrease) $R$ before executing $o$. This way the intermediate value of $R$ might violate the precondition of $o$. To avoid this violation, the initial value of $R$ has to satisfy the constraint.

The correctness of the linearization test may not be obvious, for example whether the test allows all those sets of parallel operations that can be executed in any order with the same final result. Assume that no execution violates the precondition of a particular operator. Then we have to be able to first execute all decrementing (incrementing) operators and the particular operator immediately afterwards so that the lower-bound (upper-bound) of the precondition is not violated. This is exactly the test we perform.

*Example 2.* Consider the operators $R = [1..5] \Rightarrow R := R + 2$ and $R = [4..6] \Rightarrow R := R - 2$ that are applied in parallel at $t$. From the preconditions we directly get the constraint $R_t = [4..5]$. From the precondition of the first operator under the interleaving consideration we get $R_t := R_t \cap [1 + (2 - 0)..5 - (2 - 2)]$. From the second operator we get $R_t := R_t \cap [4 + (2 - 2)..6 - (2 - 0)]$. Hence $R_t := [4..5] \cap [3..5] \cap [4..4] = [4..4]$, and 4 is the only possible value for $R$ at $t$: otherwise there would be an interleaving in which one of the preconditions is violated.

For example for the initial value $R = 5$, executing the first operator makes the precondition of the second false.

We illustrate the propagation of constraints and the treatment of parallel operators with two further examples.

*Example 3.* Consider two operators that may respectively increase the value of $R$ by 5 and decrease the value of $R$ by 2. Initially the value of $R$ is 0, and at time point 3 it is 1. We construct a sequence of operations that take us from the initial value 0 to the goal value 1.

The computation of the values of $R$ at $t \in \{1, 2\}$ is based on the values of $R$ at 0 and 3 and the values of $M_t^I$ and $M_t^D$. For example, the value of $R_1$ is the intersection of $[0 - M_0^D..0 + M_0^I]$ and $[1 - M_2^I - M_1^I..1 + M_2^D + M_1^D]$, that is $[-2..5] \cap [-9..5] = [-2..5]$. (See the leftmost table below.) Now we apply op-2 at 1 and commit to not applying op5 at 1. (See the rightmost table below.)

| $t$ | 0 | 1 | 2 | 3 |
|---|---|---|---|---|
| $R_t$ | $[0..0]$ | $[-2..5]$ | $[-4..3]$ | $[1..1]$ |
| $M_t^I$ | 5 | 5 | 5 | |
| $M_t^D$ | 2 | 2 | 2 | |
| $I_t^I$ | 0 | 0 | 0 | |
| $I_t^D$ | 0 | 0 | 0 | |
| op5 | | | | |
| op-2 | | | | |

| $t$ | 0 | 1 | 2 | 3 |
|---|---|---|---|---|
| $R_t$ | $[0..0]$ | $[-2..5]$ | $[-4..3]$ | $[1..1]$ |
| $M_t^I$ | 5 | -2 | 5 | |
| $M_t^D$ | 2 | 2 | 2 | |
| $I_t^I$ | 0 | 0 | 0 | |
| $I_t^D$ | 0 | 2 | 0 | |
| op5 | | F | | |
| op-2 | | T | | |

And we apply op5 at 2 and commit to not applying op-2 at 1. (See the leftmost table below.) And finally, the only possibility left is to apply op-2 at 0. (See the rightmost table below.)

| $t$ | 0 | 1 | 2 | 3 |
|---|---|---|---|---|
| $R_t$ | $[0..0]$ | $[-2.. - 2]$ | $[-4.. - 4]$ | $[1..1]$ |
| $M_t^I$ | 5 | -2 | 5 | |
| $M_t^D$ | 2 | 2 | -5 | |
| $I_t^I$ | 0 | 0 | 5 | |
| $I_t^D$ | 0 | 2 | 0 | |
| op5 | | F | T | |
| op-2 | | T | F | |

| $t$ | 0 | 1 | 2 | 3 |
|---|---|---|---|---|
| $R_t$ | $[0..0]$ | $[-2.. - 2]$ | $[-4.. - 4]$ | $[1..1]$ |
| $M_t^I$ | -2 | -2 | 5 | |
| $M_t^D$ | 2 | 2 | -5 | |
| $I_t^I$ | 0 | 0 | 5 | |
| $I_t^D$ | 2 | 2 | 0 | |
| op5 | F | F | T | |
| op-2 | T | T | F | |

Next we consider an example in which it is essential to check the interleavings.

*Example 4.* There are three operations, get5DM, lose2DM and lose4DM that respectively change the value of $R$ by 5, -2 and -4. The initial value of $R$ is 5. The decrementing operations have as their precondition that the value of $R$ is respectively at least 2 and at least 4.

This scenario could be viewed as transactions on somebody's bank account.

We now try to schedule one instance of each of these operations on two points of time that we call Monday and Tuesday (there may be Boolean variables and a goal – invisible to the constraint reasoner – that enforces this task.) The initial situation is as shown below on the left. Then we try to see whether it is possible to apply both lose2DM and lose4DM on Monday (and not on Tuesday.) Before applying the linearization constraints the situation is as shown below on the right.

| $t$ | Mon | Tue | Wed |
|---|---|---|---|
| $R_t$ | $[5..5]$ | $[-1..10]$ | $[-7..15]$ |
| $M_t^I$ | 5 | 5 | |
| $M_t^D$ | 6 | 6 | |
| $I_t^I$ | 0 | 0 | |
| $I_t^D$ | 0 | 0 | |
| get5DM | | | |
| lose2DM | | | |
| lose4DM | | | |

| $t$ | Mon | Tue | Wed |
|---|---|---|---|
| $R_t$ | $[5..5]$ | $[-1..4]$ | $[-1..9]$ |
| $M_t^I$ | -1 | 5 | |
| $M_t^D$ | 6 | 0 | |
| $I_t^I$ | 0 | 0 | |
| $I_t^D$ | 6 | 0 | |
| get5DM | | | |
| lose2DM | T | F | |
| lose4DM | T | F | |

The preconditions of both lose2DM and lose4DM are true on Monday and it still seems possible to achieve a positive balance for Tuesday as the interval upper bound is 4. However, we have to apply the rules $R_{\mathrm{Mon}} := R_{\mathrm{Mon}} \cap [2 + (I_{\mathrm{Mon}}^D - 2)..\infty - (I_{\mathrm{Mon}}^I - 0)] = R_{\mathrm{Mon}} \cap [6..\infty]$ (imposed by lose2DM), and $R_{\mathrm{Mon}} := R_{\mathrm{Mon}} \cap [4 + (I_{\mathrm{Mon}}^D - 4)..\infty - (I_{\mathrm{Mon}}^I - 0)] = R_{\mathrm{Mon}} \cap [6..\infty]$ (imposed by lose4DM). These both violate $R_{\mathrm{Mon}} = [5..5]$ and therefore indicate that whichever operator is applied first, the precondition of the other would not be fulfilled.

Therefore we lose2DM and get5DM on Monday and lose4DM on Tuesday.

| $t$ | Mon | Tue | Wed |
|---|---|---|---|
| $R_t$ | $[5..5]$ | $[8..8]$ | $[4..4]$ |
| $M_t^I$ | 3 | -4 | |
| $M_t^D$ | -3 | 4 | |
| $I_t^I$ | 5 | 0 | |
| $I_t^D$ | 2 | 4 | |
| get5DM | T | F | |
| lose2DM | T | F | |
| lose4DM | F | T | |

## 6    Experiments

To evaluate the efficiency of the planner, we ran it on a number of benchmarks earlier used in connection with another planner that handles numeric state variables [11] and extensions of some benchmarks used in connection with classical planners [2]. Our current implementation handles only increments and decrements in operator effects. Therefore in the benchmarks of Wolfman and Weld [11] we replaced constant assignments (fill the tank of an airplane or a truck) by increments (if the tank is less than half full, add fuel half the capacity.) We do not expect this modification to affect the runtimes significantly.

In Table 1 we give runtimes for the logistics benchmarks. These are numeric versions of the well-known logistics benchmarks for classical planners. The runtimes inside parentheses are for the computation after the plan length has been determined. The total runtimes are given outside parentheses; in the LPSAT case it is the sum of the runtimes of finding the plan (of length $n$) and showing that plans of length $n-1$ do not exist [10]. Wolfman and Weld ran their program LPSAT on a 450 MHz Pentium II that is probably slightly faster than the 296 MHz Sun Ultra we used. The number of non-leaf search tree nodes is for our planner.

**Table 1.** Runtimes of four benchmarks on LPSAT and on our planner (in seconds)

| problem | LPSAT | | we | | nodes |
|---------|-------|--|----|--|-------|
| log-a | 20.35 | (12.1) | 11.8 | (6.0) | 47 |
| log-b | 591.2 | (576) | 65.7 | (23.1) | 80 |
| log-c | 849.7 | (830) | 66.3 | (23.7) | 73 |
| log-d | > 3600 (> 3600) | | 227.6 (140.5) | | 477 |

The logistics domain is very sensitive to the criteria according to which branching variables are chosen. The benchmarks are solved efficiently because the branching heuristic reliably guides the search to a plan needing no or very little backtracking. An earlier version of our planner that did not perform all possible inferences when evaluating branching variables and did not consider numeric state variables in the branching heuristic was not able to solve these benchmarks. The heuristic made early a wrong branching decision and this was discovered only after going 20 or 30 nodes deep in the search tree. Finding the way out by exhaustive search was not possible in a reasonable amount of time.

We also ran benchmarks from the trains domain of Dimopoulos et al. [2]. The numeric versions of these benchmarks add constrains on the total number of trips between cities. We determined the smallest possible numbers of trips with which these benchmarks remain solvable, and forced the planner to find such solutions. Both planners were run with post-serializable operations [2] which means that parallelism for Boolean variables was allowed as long as there is one linearization of the operators.

We ran three variants of this benchmark and give the results in Table 2. The first runtimes are with numeric state variables and the second runtimes for the original non-numeric versions. The runtimes in parentheses are for finding the plan when the length is known. The total runtimes are given outside parentheses.

**Table 2.** Runtimes of three benchmarks on our planners (in seconds)

| problem | numeric | | nodes | non-numeric | nodes |
|---------|---------|--|-------|-------------|-------|
| train-a10 | 13.5 | (9.7) | 17 | 13.0 (9.0) | 15 |
| train-b12 | 69.7 | (58.7) | 14 | 81.0 (69.5) | 17 |
| train-c13 | 153.8 (103.8) | | 27 | 180.4 (124.7) | 31 |

# 7  Related Work

Koehler [5] extends the Graphplan algorithm [1] to handle numeric state variables. The numeric preconditions and effects Koehler considers are more general than the ones considered by us. She, for example, allows multiplication and division of resource values by constants and linear equations involving two resource values. The main difference to our work is that Koehler allows only a restricted amount of parallelism: two operators, one of which increases and the other decreases a numeric state variable, cannot be executed in parallel. Parallel operations are often crucial in achieving efficient planning because for $n$ independent operators not all $n!$ linearizations have to be considered separately. The Graphplan framework restricts to backward-chaining and does not have the generality satisfiability planning or constraint-based planning have.

Kautz and Walser [4] show how integer programming can be a basis for efficient domain-independent planning with numeric state variables. Their framework makes it possible to directly minimize – for a fixed plan length – certain numeric values, for example the number of operators in the plan or the maximum value of a numeric state variable during a plan execution. Kautz and Walser also consider parallel operations like we do. Their main results are based on a local search algorithm that is not capable of determining the inexistence of plans satisfying certain properties; for example that there are no plans of certain length. Our planning algorithm systematically goes through the search space representing all possible plans and is therefore capable of determining the inexistence of plans having a certain property.

Vossen et al. [9] do classical planning by translating problem instances to integer programming. They use a standard integer programming solver and still are able to show that integer programming provides a solution method for classical planning that approaches the efficiency of general-purpose satisfiability algorithms on the same problems. There is the obvious possibility to extend the translations to cover numeric state variables. This is however not discussed further by Vossen et al.

Wolfman and Weld [11] present LPSAT, a combination of decision procedures for propositional satisfiability and linear programming. For satisfiability they use a variant of the Davis-Putnam procedure, and for linear programming an implementation of the Simplex algorithm. The algorithm used by Wolfman and Weld is systematic and is therefore capable of reporting the inexistence of solutions.

# 8  Conclusions

Obvious extensions to our framework of numeric variables are more complex operator preconditions and more complex effects of operators. In this work we have restricted to very simple operator preconditions to make it possible to represent incompletely known values of numeric variables as intervals. For example disjunctive preconditions would require unions of intervals, and this leads to more complex constraint reasoning. Also more complex updates make the constraint reasoning more complicated. However, the problems of parallel updates are present in the current framework in their full extent, and we have presented solutions to these problems. For incorporating more complex updates in the current framework, techniques from reasoning with more complex arithmetic expressions could be directly applied.

**Acknowledgements.** This research was funded by the Deutsche Forschungsgemeinschaft SFB 527 and carried out while the first author was at the University of Ulm.

# References

1. Avrim L. Blum and Merrick L. Furst. Fast planning through planning graph analysis. *Artificial Intelligence*, 90(1-2):281–300, 1997.
2. Yannis Dimopoulos, Bernhard Nebel, and Jana Koehler. Encoding planning problems in nonmonotonic logic programs. In *Proceedings of the Fourth European Conference on Planning (ECP'97)*, pages 169–181. Springer-Verlag, September 1997.
3. Henry Kautz and Bart Selman. Pushing the envelope: planning, propositional logic, and stochastic search. In *Proceedings of the Thirteenth National Conference on Artificial Intelligence and the Eighth Innovative Applications of Artificial Intelligence Conference*, pages 1194–1201, Menlo Park, California, August 1996. AAAI Press / The MIT Press.
4. Henry Kautz and Joachim Walser. State-space planning by integer optimization. In *Proceedings of the Sixteenth National Conference on Artificial Intelligence*, pages 526–533, 1999.
5. Jana Koehler. Planning under resource constraints. In *Proceedings of the 13th European Conference on Artificial Intelligence*, pages 489–493. John Wiley & Sons, 1998.
6. Chu Min Li and Anbulagan. Heuristics based on unit propagation for satisfiability problems. In *Proceedings of the 15th International Joint Conference on Artificial Intelligence*, pages 366–371, Nagoya, Japan, August 1997.
7. David A. McAllester and David Rosenblitt. Systematic nonlinear planning. In T. L. Dean and K. McKeown, editors, *Proceedings of the 9th National Conference on Artificial Intelligence*, pages 634–639, Anaheim, California, 1991. The MIT Press.
8. Jussi Rintanen. A planning algorithm not based on directional search. In A. G. Cohn, L. K. Schubert, and S. C. Shapiro, editors, *Principles of Knowledge Representation and Reasoning: Proceedings of the Sixth International Conference (KR '98)*, pages 617–624, Trento, Italy, June 1998. Morgan Kaufmann Publishers.
9. Thomas Vossen, Michael Ball, Amnon Lotem, and Dana Nau. On the use of integer programming models in AI planning. In Thomas Dean, editor, *Proceedings of the 16th International Joint Conference on Artificial Intelligence*, volume I, pages 304–309, Stockholm, 1999. Morgan Kaufmann Publishers.
10. Steven A. Wolfman, August 1999. email correspondence.
11. Steven A. Wolfman and Daniel S. Weld. The LPSAT engine & its application to resource planning. In Thomas Dean, editor, *Proceedings of the 16th International Joint Conference on Artificial Intelligence*, volume I, pages 310–315, Stockholm, 1999. Morgan Kaufmann Publishers.

# Appendix: Sample Output from the Planner

The following is output from the planner on a simplified logistics problem. The output is after the planner has reached plan length 7 and has not yet performed any search.

```
                                              01234567
                  at(bostruck,bosairpo)  F   TTFF
                    at(bostruck,bospo)   T   FFTT
                     at(p1,bosairpo)     FFFTFFFF
                       at(p1,bospo)      FFFFFFFT
                      at(p1,laairpo)     TTFFFFFF
                        at(p1,lapo)      FFFFFFFF
                   at(plane1,bosairpo)   TFFTT
                    at(plane1,laairpo)   FTTFF
                      in(p1,bostruck)    FFFFFTTF
                       in(p1,plane1)     FFTTFFFF
       drive-truck(bostruck,bosairpo,bospo,bos)  .  ..T.
       drive-truck(bostruck,bospo,bosairpo,bos)     ...
              fly-plane-bos-la(plane1)   T...
              fly-plane-la-bos(plane1)   ..T..
          load-plane(plane1,p1,bosairpo) .......
          load-plane(plane1,p1,laairpo)  .T.....
         load-truck(bostruck,p1,bosairpo) ....T..
          load-truck(bostruck,p1,bospo)  .......
                refuel-plane(bosairpo)   ...
                 refuel-plane(laairpo)   .....
         refuel-truck(bostruck,bosairpo) ...  ..
           refuel-truck(bostruck,bospo)  .......
        unload-plane(plane1,p1,bosairpo) ...T...
        unload-plane(plane1,p1,laairpo)  .......
       unload-truck(bostruck,p1,bosairpo) .......
        unload-truck(bostruck,p1,bospo)  ......T
```

| Variable | 0 | 1 | 2 | 3 | 4 | 5 | 6 |
|---|---|---|---|---|---|---|---|
| bostruck.fuel | 100_ 100 | 80_ 100 | 40_ 100 | 0_ 100 | -20_ 160 | 20_ 220 | 0_ 200 |
| mpi: | 0 | 0 | 0 | 60 | 60 | -20 | 0 |
| mpd: | 20 | 40 | 40 | 20 | 0 | 20 | 0 |
| ami: | 0 | 0 | 0 | 0 | 0 | 0 | 0 |
| amd: | 0 | 0 | 0 | 0 | 0 | 20 | 0 |
| plane1.fuel | 400_ 400 | 250_ 250 | 250_ 250 | 100_ 100 | 100_ 300 | -50_ 500 | -350_ 900 |
| mpi: | -150 | 0 | -150 | 200 | 200 | 400 | 400 |
| mpd: | 150 | 0 | 150 | 0 | 150 | 300 | 300 |
| ami: | 0 | 0 | 0 | 0 | 0 | 0 | 0 |
| amd: | 150 | 0 | 150 | 0 | 0 | 0 | 0 |

# Hierarchical Task Network Planning as Satisfiability*

Amol Dattatraya Mali

Dept. of Elect. Engg. & Computer Science, P.O.Box 784,
University of Wisconsin, Milwaukee, WI 53201, USA
mali@miller.cs.uwm.edu

**Abstract.** The satisfiability paradigm has been hitherto applied to planning with only primitive actions. On the other hand, hierarchical task networks have been successfully used in many real world planning applications. Adapting the satisfiability paradigm to hierarchical task network planning, we show how the guidance from the task networks can be used to significantly reduce the sizes of the propositional encodings. We report promising empirical results on various encodings that demonstrate an orders of magnitude reduction in the solving times.

## 1 Introduction

Given a planning problem $\langle I, G, O \rangle$, where $I$ is the initial state of the world, $G$ is the goal state and $O$ is the set of executable actions, [Kautz & Selman 96] showed that finding a $k$-step solution to this problem can be cast as propositional satisfiability. The basic idea is to generate a propositional formula (called an *encoding*) such that any model of this formula will correspond to a $k$ step solution to the original problem. The clauses in the encoding thus capture various constraints required for proving that a $k$-length action sequence is a solution to the planning problem. Promising results were obtained by using this paradigm, as shown by [Kautz & Selman 96],[Ernst et al 97] and [Giunchiglia et al 98].

However this paradigm has been applied to planning with only primitive actions. On the other hand, HTN planners have been used in several fielded applications including space platform construction, satellite planning and control [Tate 77], beer factory production line scheduling, military operations planning [Wilkins 88], image processing for science data analysis and deep space network antenna operations [Estlin *et al.* 97]. Casting HTN planning as satisfiability is also one of the challenges proposed recently [Kambhampati 97]. Though the satisfiability paradigm has been adapted to handle HTN planning [Mali & Kambhampati 98], no empirical results on the performance of the encodings of HTN

---

* I thank Subbarao Kambhampati and the anonymous referees of ECP-99 for useful comments on this work. This work was performed while the author was a graduate student at Arizona State University. The college of engineering and applied sciences at Univ. of Wisconsin, Milwaukee provided financial support for attending the conference and presenting the work.

planning are available. [Mali & Kambhampati 98] describe three encodings of HTN planning and report the number of clauses and variables that these encodings contain. They show that the HTN encodings contain more clauses and more variables than the action-based encodings of [Kautz et al 96]. It has thus remained unclear whether the HTN encodings are easier to solve. This is the question that has motivated our work. Our work makes the following contributions.

- We provide a procedure for pre-processing the causal HTN encodings of [Mali & Kambhampati 98] and show that it reduces the sizes of the encodings significantly and that these smaller encodings are also faster to solve.
- We modify the HTN encodings of [Mali & Kambhampati 98], so that their models obey the *criterion of parsing* (the models of the HTN encodings must be same as the the plans that can be parsed by the task reduction schemas).
- We provide an empirical evaluation of the HTN encodings of [Mali & Kambhampati 98] to illustrate the effectiveness of our pre-processing procedure and show that the pre-processed causal HTN encodings can be smaller than the smallest of the action-based encodings and also the fastest to solve, on some problems.

This paper is organized as follows. In section 2, we discuss the basics of the HTN encodings and explain some notation used to represent the constraints from the task reduction schemas, in the encodings. In section 3, we discuss procedures of pre-processing the HTN encodings and discuss how they reduce the sizes of the encodings. In section 4, we show how the HTN encodings can be augmented with constraints to obey the criterion of parsing, without significantly increasing their size. We also show how these constraints avoid the generation of non-minimal plans. In section 5, we describe an empirical evaluation of the HTN encodings and the action-based encodings and discuss the insights obtained. We report conclusions in section 6.

## 2    Basics of HTN Encodings

The propositional encodings for HTN planning can be developed by constraining the action-based encodings of [Kautz et al 96], such that their satisfying models also conform to the grammar of solutions of interest to a user, specified by the task reduction schemas. Since most implemented HTN planners [Wilkins 88][Tate 77] share a lot of structure of the partial order planners, the "causal encodings" in [Kautz *et al.* 96] provide a natural starting place for developing encodings for HTN planning. Since HTN planning can be viewed as just an augmentation of action-based planning with a grammar of legal solutions, there is no reason to constrain ourselves to causal encodings. Indeed, an HTN encoding based on the state-based encoding of [Kautz *et al.* 96] also exists, as shown by [Mali & Kambhampati 98]. The representation we use for the task reduction schemas is consistent with that used in [Erol 95].

In what follows, we assume that $o_i$ denotes a STRIPS style ground primitive action. $p_i$ denotes a primitive step that is mapped to a ground primitive action or the null action (no-op) $\phi$ that has no pre-conditions and no effects. $O$ denotes the set of all ground primitive actions from the domain.

$N_i$ denotes a ground non-primitive task. $d_i$ denotes the number of reduction schemas of $N_i$. $s_i$ denotes a non-primitive step that is mapped to one and only one non-primitive task. $r_{ij}$ denotes $j$th reduction schema of $N_i$, since a task may be reduced in multiple ways. Each reduction schema $r_{ij}$ may contain the following types of constraints (that capture the causal links and orderings between non-primitive tasks and/or primitive actions) - (i) mapping from an action symbol to a primitive action, for example, $o_1 : load(x, l_1, R)$ (ii) mapping from a non-primitive task symbol to non-primitive task name, for example, $N_2 : achieve(at(R, l_2))$ (iii) $o_s \prec N_p$ (where $\prec$ denotes temporal precedence) (iv) $N_p \prec o_s$ (v) $o_p \xrightarrow{f} o_q$, which denotes the causal link where $o_p$ is the contributor and $o_q$ is the consumer. (vi) $? \xrightarrow{f} o_p$ (which denotes that the contributor is not known a priori, and will be introduced in the partial plan as planning progresses, possibly by the reduction of some other task) (vii) $o_q \xrightarrow{f} ?$ (viii) $N_q \xrightarrow{f} o_p$ (ix) $o_p \xrightarrow{f} N_q$ (x) $N_p \prec N_q$ (xi) $N_p \xrightarrow{f} N_q$ (xii) $o_p \prec o_q$ (xiii) $? \xrightarrow{f} N_q$ and (xiv) $N_p \xrightarrow{f} ?$. The semantics of these constraints is discussed in [Mali & Kambhampati 98].

Mapping a step $s_k$ to $N_i$ (denoted by $s_k = N_i$) means choosing $N_i$ to solve the planning problem. In this case, it is necessary to carry out $N_i$ by satisfying the constraints in some reduction schema of $N_i$. $t(r_{ij})_k$ then represents a conjunction of the disjunction of all potential ways of satisfying each constraint in $r_{ij}$. $m$ denotes the number of non-primitive tasks used in an HTN encoding. $M_i$ denotes the maximum number of primitive actions in reduction schema of $N_i$ and the reduction schema containing $M_i$ primitive actions is denoted by $r_{iJ}$. $M = max(\{M_i \mid i \in [1, m]\})$.

When $s_k$ is mapped to $N_i$ and the reduction schema $r_{ij}$ is used to reduce $N_i$, primitive steps ranging from $p_{a'_{ijk}}$ to $p_{a''_{ijk}}$ are mapped to the primitive actions from $r_{ij}$, where,

$$a'_{ijk} = (k - 1) * M + 1, a''_{ijk} = a'_{ijk} + b_{ij} - 1$$

where $b_{ij}$ is the number of primitive actions in $r_{ij}$.

The total number of primitive steps in an HTN encoding are $T$, where $T = M * K$, $K$ being the number of non-primitive steps chosen. This can be viewed as allocating $M$ primitive steps ranging from $p_{i*M}$ to $p_{(i+1)*M-1}$ to each non-primitive step $s_i$ ($M$ is computed automatically by examining the reduction schemas and is not supplied by a user). Such an allocation done a priori does not affect the soundness and completeness of an HTN encoding, since all potential orderings between the primitive steps are represented and the primitive steps can be interleaved as per need. A plan may contain less than $T$ primitive actions and less than $K$ reduction schemas may be required to synthesize the plan. Such plans can be found by mapping the excess steps to $\phi$.

We refer to the encodings that are not pre-processed or simplified as "naive" encodings. All potential ways of satisfying the constraints in the reduction schemas and all potential choices of the reduction schemas are represented in the encodings, to achieve soundness and completeness. Note that we do not handle recursive task reduction schemas.

## 2.1    Causal HTN Encodings

The task reduction schemas can be used to control planning either in a top-down or a bottom-up way. In the top-down way, which is followed in most implemented HTN planners, planning starts with non-primitive tasks, and the reduction schemas are used to gradually reduce them into more concrete actions (while taking care of ensuing interactions), as in [Erol 95]. In the bottom-up way [Barrett & Weld 94], the (partial) solutions generated by an action-based planner are incrementally parsed with the help of reduction schemas and the branches leading to solutions that cannot be parsed are pruned. Two causal HTN encodings that are motivated by these notions of decomposition and parsing check are proposed in [Mali & Kambhampati 98]. We review them next. Note that our top-down and bottom-up HTN encodings differ only in the presence of certain variables and clauses. The encodings are not directional and they are not necessarily solved in the top-down or bottom-up styles.

**Top-Down Causal HTN Encoding -** The notion of task decomposition is captured in this encoding with the constraint $\wedge_{i=1}^{m} \wedge_{k=1}^{K} ((s_k = N_i) \Rightarrow (\vee_{j=1}^{d_i} t(r_{ij})_k))$. Consider the constraint $N_p \prec N_q$ from the reduction schema of a non-primitive task to which the step $s_k$ is mapped. To represent this, one of the constraints we specify is,

$$\vee_{w_1=1, w_1 \neq k}^{K} \vee_{w_2=1, w_2 \neq w_1, w_2 \neq k}^{K} ((s_{w_1} = N_p) \wedge (s_{w_2} = N_q) \wedge (\wedge_{u_1 = a'_{pJw_1}}^{a''_{pJw_1}}$$

$$\wedge_{u_2 = a'_{qJw_2}}^{a''_{qJw_2}} (p_{u_1} \prec p_{u_2})))$$

(since there must exist non-primitive steps that are mapped to $N_p$ and $N_q$ respectively and each primitive step mapped to an action in the reduction of $N_p$ must precede each primitive step mapped to an action in the reduction of $N_q$). By introducing $(K-1)*(K-2)$ new "intermediate" variables, each implying a conjunction in the disjunction above, we can reduce the number of clauses there from exponential to $O(K^2 * M_p * M_q)$.

**Bottom-Up Causal HTN Encoding -** The non-primitive step and non-primitive task symbols are not used in the bottom-up encoding. It uses only primitive steps that are mapped to actions. The notion of the allocation of certain number of primitive steps to the reduction of a task is however present in the bottom up encoding as well. Thus the number of primitive steps in the bottom-up encoding is $T(K*M)$. The bottom up encoding contains the constraint that some $t(r_{ij})_k$, $k \in [1, K], i \in [1, m], j \in [1, d_i]$ must be true, to ensure that the constraints from some reduction schemas are respected by the solution.

Due to the absence of $s_i, N_j$ symbols, different number of clauses and variables are required to represent certain constraints in the bottom-up HTN encodings than the top-down HTN encodings. For example, consider the constraint $N_p \prec N_q$ (contained in $r_{ij}$) that needs to be true when $t(r_{ij})_k$ is true. To represent this in the encoding, one of the constraints we specify is,

$$\vee_{z_1=1}^{d_p} \vee_{z_2=1}^{d_q} \vee_{w_1=1, w_1 \neq k}^{K} \vee_{w_2=1, w_2 \neq k, w_1 \neq w_2}^{K} (t(r_{pu_1})_{w_1} \wedge$$
$$t(r_{qu_2})_{w_2} \wedge (\wedge_{u_1=a'_{pz_1 w_1}}^{a''_{pz_1 w_1}} (\wedge_{u_2=a'_{qz_2 w_2}}^{a''_{qz_2 w_2}} (p_{u_1} \prec p_{u_2}))))$$

This requires $O(K^2 * M_p * M_q * d_p * d_q)$ clauses and $O(K^2 * d_p * d_q)$ intermediate variables, both being higher than those required in the top-down causal encoding by the factor of $O(d_p * d_q)$. This difference in the number of clauses and intermediate variables also applies to the representation of the constraint $N_p \xrightarrow{f} N_q$.

The causal HTN encodings contain $O(K * m * d * Z + T^3 * | \Gamma |)$ clauses and $O(K * m * d * V + T^2 * | \Gamma |)$ variables, $Z, V$ being the number of clauses and variables used to represent all potential ways of satisfying all the constraints in a reduction schema of a non-primitive task. $d = max(\{d_i \mid i \in [1, m]\})$. $\Gamma$ denotes the set of ground pre-conditions of actions in the domain. An action-based causal encoding with $T$ primitive steps, without the constraints the reduction schemas, contains $O(T^3 * | \Gamma |)$ clauses (to resolve all potential threats to all potential causal links) and $O(T^2 * | \Gamma |)$ variables (all potential causal links), showing that the causal HTN encodings contain more clauses and more variables than the action-based causal encoding of [Kautz et al 96].

## 2.2   State-Based HTN Encoding

To synthesize this encoding, [Mali & Kambhampati 98] constrain the state-based encoding of [Kautz et al 96] with the constraints from the task reduction schemas. This encoding contains $T(K*M)$ contiguous steps. In this encoding, $o_i(t)$ denotes that the primitive action $o_i$ occurs at time $t$. $f_i(t)$ denotes that the proposition $f_i$ is true at time t.

Consider the causal link $o_p \xrightarrow{f} o_q$. We convert this causal link into an ordering and an interval preservation constraint, as shown below.

$$\vee_{i=0}^{T-2} \vee_{j=i+1}^{T-1} (o_p(i) \wedge o_q(j) \wedge (\wedge_{t=i+1}^{j} f(t)))$$

$U$ denotes the set of ground pre-condition and effect propositions in the domain. The state-based HTN encoding contains $O(b' * d^2 * m * T^3 * | O | + T * (| O | + | U |))$ clauses and $O(b' * d^2 * m * T^2 * | O | + T * (| O | + | U |))$ variables respectively, where $b' = max(\{b_{ij} \mid i \in [1, m], j \in [1, d_i]\})$, higher than the number of clauses $(O(T * (| O | + | U |)))$ and variables $(O(T * (| O | + | U |)))$ in the action-based state-based encoding of [Kautz et al 96] (these expressions of [Kautz et al 96] follow from the fact that the world state is represented at each time step (in the state-based encodings) and an action occurring at time $t$ implies the truth of its pre-conditions at $t$ and the truth of its effects at $(t + 1)$. Explanatory frame axioms are required to insist that if $(f(t) \wedge \neg f(t+1))$ is true, some action deleting $f$ must occur at $t$).

# 3   Pre-processing

In this section, we show how the HTN encodings can be pre-processed to reduce their sizes.

## 3.1   Causal HTN Encodings

Consider the allocation of primitive steps to the non-primitive steps in Fig. 1. Each non-primitive step may be mapped to either $N_1$ or $N_2$ which have 3 and 2 reductions respectively. The primitive actions from the reduction schemas of these tasks are also shown there. These actions may have ordering constraints between them and there may be additional constraints in the reduction schemas (not shown in Fig. 1). Since the primitive steps in the causal encodings are partially ordered and all such orderings are represented, one can even fix the potential step→action mapping ($p_i = o_j$) a priori, in the form of a disjunction of step→action mappings e.g. One can infer that $((p_1 = o_1) \vee (p_1 = o_3) \vee (p_1 = o_6) \vee (p_1 = \phi) \vee (p_1 = o_9) \vee (p_1 = o_{13}))$. In particular, an $i$ th primitive step allocated to a non-primitive step $s_j$ will be mapped to either $\phi$ or only the $i$ th primitive action in some reduction schema of some non-primitive task, rather than any of the $\mid O \mid$ actions in the domain (which is the case for the naive encodings). For each $p_i, i \in [1, T]$, a naive encoding represents the mapping $((\vee_{j=1}^{13}(p_i = o_j)) \vee (p_i = \phi))$. We refer to the reduction in the domain of $p_i$ as the reduction in the disjunction in the step→action mapping. This reduction can be propagated to significantly reduce the sizes of the causal HTN encodings. Hence the sizes reported by [Mali & Kambhampati 98] should not be interpreted to conclude that in practice, the HTN encodings will always be larger than the corresponding action-based encodings.

The naive causal encodings contain all $O(T^2 * \mid \Gamma \mid)$ potential causal links. However, with the reduced disjunction in the step→action mapping, before generating any causal link $p_i \xrightarrow{f} p_j$, one can check (a) if $p_i$ may be mapped to an action that adds $f$ and (b) if $p_j$ may be mapped to an action that needs $f$. If either (a) or (b) is false, this causal link variable need not be created. If this link is not created, no clauses need to be generated to resolve threats to this link. Even if (a) and (b) are both true, before generating the threat resolution clause $((p_i \xrightarrow{f} p_j) \wedge Dels(p_s, f)) \Rightarrow ((p_s \prec p_i) \vee (p_j \prec p_s))$, one can check if $p_s$ may be mapped to an action that deletes $f$. If this is not the case, this clause need not be generated. Action-based causal encodings contain $O(T * \mid O \mid^2)$ clauses to specify that a step cannot be bound to more than one action. With the reduced disjunction in the step→action mapping, many of these $O(T * \mid O \mid^2)$ clauses will not have to be generated. Though we have discussed here how only the dominant terms in the sizes of the causal HTN encodings are lowered by the propagation of the reduced disjunction in the step→action mapping, other less dominant terms are lowered as well. Though the bottom-up causal HTN encoding does not contain the $s_i$ and $N_i$ symbols, it does contain the $t(r_{ij})_k$ symbols and such a reduction in the disjunction in the step→action mapping is doable there as well. We have integrated this pre-processing with the code that generates the causal HTN encodings. Most current SAT-based planners generate encodings and then simplify them. However many encodings are too large to store (as shown in Fig. 2), making it impossible to simplify them later. We generate a simplified encoding rather than simplifying a generated encoding, using our pre-processing procedures.

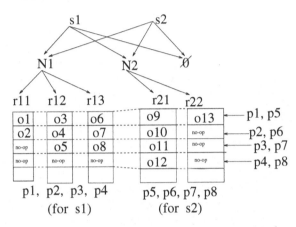

**Fig. 1.** The step allocation in the top-down causal HTN encoding, K = 2, m = 2, M = 4, T = K * M = 8

Is the reduction in the disjunction in the step→action mapping always guaranteed? A closer examination shows that when there are at least $\theta$ reduction schemas, such that $\theta \geq |O|$, such that $i$ th reduction schema $(1 \leq i \leq |O|)$ contains $\gamma$ occurrences of $o_i$, $\gamma = M$, any primitive step in the encoding may be mapped to any of the $|O|$ actions in the domain, like the naive action-based encodings, without yielding a reduction in the disjunction in the step→action mapping. In all other cases, there will be a reduction in the disjunction in the step→action mapping.

## 3.2   State-Based HTN Encoding

We use the planning graph [Blum & Furst 95] to frugally represent the potential ways of satisfying the constraints from the reduction schemas. The planning graph is grown starting from the initial state (0th proposition level), by applying all actions whose pre-conditions are satisfied in $I$, computing the next proposition level (which is a union of the propositions in the previous proposition level and the add and delete effects of the actions applicable in the previous proposition level) and repeating this process. The constraint $o_2 \prec o_4$ is represented in the naive encoding as $\vee_{i=0}^{T-2}(o_2(i) \wedge (\vee_{j=i+1}^{T-1} o_4(j)))$. However by examining the actions in the planning graph, one can know the time steps at which $o_2$ and $o_4$ may occur and use this information to represent $o_2 \prec o_4$ with fewer variables and fewer clauses. Similarly, the causal links can be represented with fewer variables and fewer clauses.

# 4    Properties of the Models

In this section, we discuss how the HTN encodings of [Mali & Kambhampati 98] can be constrained, so that their models obey the parsing criterion. In the causal HTN encodings, in case the total number of reduction schemas chosen to solve the problem is less than $K$ (and since the number of primitive steps allocated is intended for the use of $K$ reduction schemas), the extra primitive steps may be mapped to non-null actions, yielding non-minimal plans (plans from which some actions can be removed, preserving the plan correctness). To prevent this, we include clauses in these encodings which state that such excess primitive steps must be mapped to $\phi$. That is, (1) if all the $t(r_{ij})_k$ variables, $i \in [1, m], j \in [1, d_i]$ are false, the corresponding $M$ primitive steps allocated to $s_k$ must be mapped to $\phi$. (2) Also, if $t(r_{ij})_k$ is true, the excess $(M - b_{ij})$ primitive steps must also be mapped to $\phi$. Specifying these restrictions requires $O(K * M * m * d)$ clauses. These prevent non-minimal plans. By mapping the redundant steps to $\phi$, we also prevent them from being mapped to actions outside the reduction schemas used, enforcing the criterion of parsing.

Though the steps that will be mapped to the primitive actions from a reduction schema are not known a priori, an allocation of $M$ steps to a reduction schema exists in the state-based HTN encoding as well. Thus the state-based HTN encodings too can be augmented with extra constraints to prevent non-minimality. However, since the explanatory frame axioms allow parallel actions, the models of the state-based encodings can still be non-minimal plans and violate the parsing criterion, unless they are constrained with the only one action at a step restriction.

# 5    Discussion

Reviewing the HTN encodings of [Mali & Kambhampati 98], we proposed preprocessing methods to reduce their sizes. To test the effectiveness of our methods, we conducted an empirical evaluation of the encodings on several benchmark domains. [1] Only those task reduction schemas that were required for solving the problems were used in generating the encodings. The encodings were generated and solved on a Sun Ultra with 128 M RAM.

The descriptions of the house building and tire world domains were respectively obtained from http://www.aiai.ed.ac.uk/ $\sim$ oplan/ and http://pecan.srv. cs.
cmu.edu/afs/cs.cmu.edu /local/mosaic/common/omega/Web/ People/avrim/ graphplan.html. The encodings were solved with the "satz" solver, except those in the Tsp (traveling sales person domain) which were solved with version 35 of Walksat. [2] All the encodings were simplified by unit propagation and pro-

---

[1]  Available at http://www.cs.yale.edu/HTML/YALE/CS/HyPlans/ mcdermott.html
[2]  The solvers are available at http://aida.intellektik.informatik.th-darmstadt.de/ $\sim$ hoos/SATLIB.

cessed using procedures from section 3, before being solved.[3] Like [Ernst et al 97], we do not report the times required to simplify the encodings, since these times were small. A comparison of the sizes of the causal HTN encodings in Fig. 2 (pre-processed and naive) shows that our pre-processing procedure (which is based on the propagation of the reduction in the disjunction in the step→action mapping) reduces both the number of clauses and the variables by orders of magnitude. This also yields an orders of magnitude improvement in the times required to solve the encodings. The sizes of the top-down and bottom-up causal HTN encodings (whether they are pre-processed or naive does not matter) and even their solving times are very close, since the bottom-up causal HTN encodings differed from the top-down causal HTN encodings only in the presence of non-primitive step and non-primitive task symbols.

The HTN encodings contained $T$ ($K * M$) primitive steps and this was generally larger than the number of actions in the plans, as can be seen from Fig. 2. The action-based encodings however contained only $k$ primitive steps, $k$ being the number of actions in the plans, $k \ll T$ (thus the HTN encodings had $(T - k)$ redundant primitive steps, this also shows that the HTN encodings will have no redundant primitive steps, if the domain knowledge is partitioned among the reduction schemas such that $M = \frac{k}{K}$). The action-based causal encodings were significantly larger than the pre-processed causal HTN encodings, despite the fewer primitive steps used, illustrating the power of the pre-processing method from section 3.1. Also, we used only those actions that appeared in the reduction schemas, in the action-based encodings. Since this will generally not be the case in practice (the number of primitive actions in the reduction schemas will be far lower than the number of primitive actions in the domain), the action-based encodings will be far larger.

We found that the sizes of the state-based HTN encodings pre-processed using the planning graph information and the naive state-based HTN encodings, as well as their solving times did not differ significantly, on most of the problems. This is because any action that occurs at action level $i$ in the plan graph also occurs at all $j$ th action levels, $j > i$ and the sizes of the encodings do not reduce significantly, due to this redundancy. On most of the problems, the state-based HTN encodings still had fewer clauses and fewer variables than both the naive and the pre-processed causal HTN encodings. However the number of variables in the state-based HTN encodings was higher than those in both the naive and pre-processed causal HTN encodings, on problems from the house building and tire world domains. This is because the reduction schemas in these domains contained lot of causal links and precedence constraints and the representation of these constraints on the contiguous steps required more clauses and more variables. Because of this, the state-based HTN encodings were larger and also harder to solve, than the state-based action-based encodings and the causal pre-processed HTN encodings, on these problems.

---

[3] The unit propagation code used is from the Feb. 96 version of the Satplan system (/satplan/system/simplify/unitprop) available at ftp://ftp.research.att.com/dist/ai.

The state-based action-based encodings were the smallest on most of the problems, however the plans found by solving them, as well as those found by solving the state-based HTN encodings were highly non-minimal, both because the number of steps in the encodings (action-based) were higher (in some cases) than the length of the shortest plan and because the explanatory frame axioms allow parallel actions. For example, the plan found using the state-based action-based encoding, in the tire world had 12 redundant actions and the plan found using the state-based HTN encoding (for the same problem) had 9 redundant actions, because of the cyclicity possible, e.g. $open(container)$, $close(container)$, $do\_up(nut, hub)$, $undo(nut, hub)$, $jack\_up(hub)$, $jack\_down(hub)$, $tighten(nut, hub)$ and $loosen(nut, hub)$ etc. The models of the state-based HTN encoding and the state-based action-based encodings for the logistics problem (for which a 21 action plan exists), contained 49 and 42 redundant flights respectively, e.g. $fly(R_1, London, Paris)$ and $fly(R_1, Paris, London)$ etc. Similarly, the model of the state-based HTN encoding for the problem from the Ferry domain (for which a 19 action plan exists) contained 8 redundant actions, e.g. $sail(a, b)$, $sail(b, a)$. The causal encodings did not suffer from this problem of non-minimality of the plans, showing that their models conform to the user intent better than the models of the state-based encodings. To guarantee minimal plans using the state-based HTN encodings, the restriction that only one action should occur at a time step needs to be added to them. The number of non-primitive steps $K$ we used in the encodings was same as the number of reduction schemas required to solve the problems. If the value of $K$ chosen is larger than the necessary, or the reduction schemas supplied themselves contain redundancy, the models of the HTN encodings may still be non-minimal plans.

The causal encodings have been shown to be hard to solve in the action-based planning scenario [Mali & Kambhampati 99]. Our pre-processing method was so effective, that in some domains like the tire world, the pre-processed causal HTN encodings were the smallest (smaller than even the state-based encodings with explanatory frame axioms which have been found to be the smallest in the action-based planning scenario [Ernst et al 97] and [Giunchiglia et al 98]). The structure in Fig. 1 that shows the potential step $\longrightarrow$ action mappings can be modified using some soundness and completeness preserving heuristics (without changing the task reduction schemas and without violating the user intent), to yield a larger reduction in the encoding size and we intend to perform an empirical study of the performance of such heuristics in future.

We want to stress that though we compared the performance and sizes of the action-based encodings and the HTN encodings to see if the information from the HTNs could be used to make the encodings easier to solve, efficiency is not always the primary reason for using the HTNs. Since the HTNs provide a grammar of the solutions desired by the users, it is some times necessary to use them to respect the user intent, even if plan synthesis is not faster.

[Kautz & Selman 98] have shown that adding domain specific knowledge to the action-based encodings (though it increases the size of an encoding) can be used to solve them faster. This is because the domain specific knowledge can be

| Domain, Plan Actions | TDCP V, C, T | TDCN V, C, T | BUCP V, C, T | BUCN V, C, T | AC V, C, T | SBH V, C, T | AS V, C, T |
|---|---|---|---|---|---|---|---|
| Ferry, 19 | 1715 | 19753 | 1699 | 19737 | 8535 | 1242 | 588 |
| K = 2, T = 30 | 31162 | 526697 | 30687 | 526657 | 138172 | 4856 | 2436 |
| k = 19 | 11.31 | * | 11.42 | * | * | 60.81 | 3.96 |
| Ferry, 39 | 6141 | 131893 | 6115 | 131867 | 57880 | 4577 | 2178 |
| K = 2, T = 60 | 249153 | 7438877 | 247318 | 7438812 | 2067472 | 21406 | 11431 |
| k = 39 | * | - | * | - | * | * | * |
| Ferry, 27 | 3153 | 48985 | 3133 | 48965 | 21361 | 2324 | 1104 |
| K = 2, T = 42 | 85220 | 1886741 | 84309 | 1886691 | 511168 | 9820 | 5080 |
| k = 27 | 45.96 | - | 45.63 | - | * | * | * |
| Ferry, 31 | 4051 | 70765 | 4029 | 70743 | 30942 | 2991 | 1422 |
| K = 2, T = 48 | 127531 | 3146849 | 126348 | 3146794 | 861556 | 13094 | 6855 |
| k = 31 | * | - | * | - | * | * | * |
| Tsp, 14 | 607 | 9889 | 573 | 9855 | 7785 | 1097 | 631 |
| K = 2, T = 16 | 4445 | 130021 | 4240 | 129936 | 88873 | 2884 | 1639 |
| k = 14 | 7.15 | 26.47 | 7.24 | 24.09 | X | X | 0.07 |
| Tsp, 19 | 1291 | 27448 | 1222 | 27379 | 17880 | 2568 | 1141 |
| K = 3, T = 24 | 14765 | 572087 | 14327 | 571925 | 289048 | 6566 | 2984 |
| k = 19 | 22.83 | X | X | X | X | X | 0.13 |
| Blocks, 12 | 348 | 6135 | 337 | 6124 | 6123 | 569 | 534 |
| K = 1, T = 12 | 1886 | 58969 | 1789 | 58938 | 59715 | 1886 | 1671 |
| k = 12 | 0.1 | 2.18 | 0.1 | 2.2 | 44.44 | 0.12 | 0.09 |
| Logistics, 18 | 1856 | 11162 | 1778 | 11084 | 11048 | 2114 | 836 |
| K = 6, T = 18 | 13154 | 168263 | 12864 | 168126 | 168764 | 5577 | 2408 |
| k = 18 | 10.77 | 260.9 | 8.51 | 265.32 | * | 3.71 | 0.21 |
| Logistics, 21 | 2787 | 16360 | 2682 | 16255 | 16206 | 3109 | 1079 |
| K = 7, T = 21 | 23003 | 293490 | 22633 | 293330 | 294078 | 8066 | 3144 |
| k = 21 | 24.38 | * | 18.06 | * | * | 9.53 | 0.29 |
| Build House, 23 | 1267 | 17377 | 1252 | 17362 | 16056 | 29451 | 1105 |
| K = 3, T = 24 | 14645 | 353735 | 14333 | 353699 | 313239 | 293189 | 2532 |
| k = 23 | 3.78 | * | 3.88 | * | * | * | 0.28 |
| Tire World, 11 | 284 | 4476 | 280 | 4472 | 4471 | 1994 | 465 |
| K = 1, T = 11 | 1480 | 38657 | 1415 | 38647 | 39429 | 10605 | 1567 |
| k = 11 | 0.04 | 1.09 | 0.05 | 1.07 | 31.46 | 0.68 | 0.05 |

**Fig. 2.** Empirical results on various encodings. $V, C, T$ denote the number of variables, clauses in and the times needed to solve the encodings respectively. Times are in CPU seconds. A "*" indicates that the encoding could not solved within 10 minutes. A "-" denotes that the encoding was too large to store. X denotes that the encoding could not be solved with Walksat with default parameters. $k$ denotes the number of steps in the action-based encodings. TDCP and TDCN denote top-down causal pre-processed and top-down causal naive HTN encodings respectively. BUCP and BUCN denote bottom-up causal pre-processed and bottom-up causal naive HTN encodings respectively. SBH denotes the state-based HTN encoding and AC and AS denote action-based causal and action-based state-based encodings respectively.

efficiently propagated to reduce the size of an encoding. Our work differs from this work in at least two aspects - **(1) Nature of Knowledge -** [Kautz & Selman 98] use state constraints, e.g. once a vehicle is loaded (at time $t$), it should move immediately (at time $(t + 1)$). The reduction schemas we used contain different types of constraints like the causal links and precedences. Certain encodings are naturally compatible with certain constraints, e.g. the state constraints and the state-based encodings, the causal links, partial order and the causal encodings. Just like adding state constraints may significantly increase the size of a causal encoding, adding causal constraints (like links and $\prec$) significantly increases the size of a state-based encoding, as illustrated by our results in the domains like house building and tire world. **(2) Organization of the Knowledge -** The domain specific knowledge in [Kautz & Selman 98] is not necessarily problem dependent. On the other hand, the reduction schemas generally contain constraints that must be satisfied by the solution of a specific problem in the domain. The commitment to using a reduction schema means a commitment to satisfying all of its constraints. However the constraints in [Kautz & Selman 98] are not grouped together as those in the reduction schemas and thus one can choose to add some constraints to an encoding ignoring others.

# 6    Conclusion

Hitherto, it was unclear if the HTN encodings of [Mali & Kambhampati 98] were easier to solve than the action-based encodings and whether they were solvable at all, since their sizes are higher than the corresponding action-based encodings and the action-based causal encodings have been shown to be hard to solve [Mali & Kambhampati 99]. We showed that the causal HTN encodings are significantly easier to solve than the causal action-based encodings, by developing a pre-processing procedure that reduced their sizes significantly. We also added constraints to the HTN encodings of [Mali & Kambhampati 98] to force their models to be minimal plans and obey the parsing criterion, to respect the user intent. We showed that the models of the causal HTN encodings indeed have these properties.

**References**

**[Barrett & Weld 94]** Anthony Barrett and Daniel Weld, Task-Decomposition via plan parsing, Proceedings of the National Conference on Artificial Intelligence (AAAI), 1994, 1117-1122.

**[Blum & Furst 95]** Avrim Blum and Merrick Furst, Fast planning via planning graph analysis, Proceedings of the International Joint Conference on Artificial Intelligence (IJCAI), 1995.

**[Ernst et al 97]** Michael Ernst, Todd Millstein and Daniel Weld, Automatic SAT compilation of planning problems, Proccedings of the International Joint Conference on Artificial Intelligence (IJCAI), 1997.

[**Erol 95**] Kutluhan Erol, Hierarchical task network planning: Formalization, Analysis and Implementation, Ph.D thesis, Dept. of computer science, Univ. of Maryland, College Park, 1995.

[**Estlin *et al.* 97**] Tara Estlin, Steve Chien and Xuemei Wang, An argument for a hybrid HTN/Operator-based approach to planning, Proceedings of the European Conference on Planning (ECP), 1997, 184-196.

[**Giunchiglia et al 98**] Enrico Giunchiglia, Alessandro Massarotto and Roberto Sebastiani, Act and the rest will follow: Exploiting determinism in planning as satisfiability, Proceedings of the National Conference on Artificial Intelligence (AAAI), 1998.

[**Kambhampati *et al.* 98**] Subbarao Kambhampati, Amol Mali & Biplav Srivastava, Hybrid planning in partially hierarchical domains, Proceedings of the National Conference on Artificial Intelligence (AAAI), 1998.

[**Kambhampati 97**] Subbarao Kambhampati, Challenges in bridging the plan synthesis paradigms, Proceedings of the International Joint Conference on Artificial Intelligence (IJCAI), 1997.

[**Kautz *et al.* 96**] Henry Kautz, David McAllester and Bart Selman, Encoding plans in propositional logic, Proceedings of the conference on Knowledge Representation & Reasoning (KRR), 1996.

[**Kautz & Selman 96**] Henry Kautz and Bart Selman, Pushing the envelope: Planning, Propositional logic and Stochastic search, Proceedings of the National Conference on Artificial Intelligence (AAAI), 1996.

[**Kautz & Selman 98**] Henry Kautz and Bart Selman, The role of domain-specific knowledge in the planning as satisfiability framework, Proceedings of the international conference on Artificial Intelligence Planning Systems (AIPS), 1998.

[**Mali & Kambhampati 98**] Amol D. Mali and Subbarao Kambhampati, Encoding HTN planning in propositional logic, Proceedings of the International Conference on AI Planning Systems (AIPS), 1998.

[**Mali & Kambhampati 99**] Amol D. Mali and Subbarao Kambhampati, On the utility of causal encodings, Proceedings of the National Conference on Artificial Intelligence (AAAI), 1999.

[**Tate 77**] Austin Tate, Generating project networks, Proceedings of the International Joint Conference on Artificial Intelligence (IJCAI), 1977, 888-893.

[**Wilkins 88**] David Wilkins, Practical planning: Extending the classical AI planning paradigm, Morgan Kaufmann, 1988.

# Exhibiting Knowledge in Planning Problems to Minimize State Encoding Length

Stefan Edelkamp and Malte Helmert

Institut für Informatik
Albert-Ludwigs-Universität, Am Flughafen 17, D-79110 Freiburg, Germany,
e-mail: {edelkamp,helmert}@informatik.uni-freiburg.de

**Abstract.** In this paper we present a general-purposed algorithm for transforming a planning problem specified in Strips into a concise state description for single state or symbolic exploration.

The process of finding a state description consists of four phases. In the first phase we symbolically analyze the domain specification to determine constant and one-way predicates, i.e. predicates that remain unchanged by all operators or toggle in only one direction, respectively.

In the second phase we symbolically merge predicates which lead to a drastic reduction of state encoding size, while in the third phase we constrain the domains of the predicates to be considered by enumerating the operators of the planning problem. The fourth phase combines the result of the previous phases.

## 1  Introduction

Single-state space search has a long tradition in AI. We distinguish memory sensitive search algorithms like A* [12] that store the explored subgraph of the search space and approaches like IDA* and DFBnB [15] that consume linear space with respect to the search depth. Especially on current machines memory sensitive algorithms exhaust main memory within a very short time.

On the other hand linear search algorithms explore the search tree of generating paths, which might be exponentially larger than the underlying problem graph. Several techniques such as transposition tables [21], finite state machine pruning [23], and heuristic pattern databases [3] have been proposed. They respectively store a considerable part of the search space, exhibit the regular structure of the search problems, and improve the lower bound of the search by retrieving solutions to problem relaxations. Last but not least, in the last decade several memory restricted algorithms have been proposed [4,22]. All memory restricted search approaches cache states at the limit of main memory.

Since finite state machine pruning is applicable only to a very restricted class of symmetric problems, single-state space search algorithms store millions of fully or partially defined states. Finding a good compression of state space is crucial. The first step is to efficiently encode each state; if we are facing millions of states we are better off with a small state description length.

The encoding length is measured in bits. For example one instance to the well-known Fifteen Puzzle can be compressed to 64 bits, 4 bits for each tile.

S. Biundo and M. Fox (Eds.): ECP-99, LNAI 1809, pp. 135–147, 2000.

Single-state algorithms have been successful in solving "well-informed do-mains", i.e. problems with a fairly elaborated lower bound [13,16], for good esti-mates lead to smaller parts of the search tree to be considered. In solving one specific problem, manually encoding the state space representations can be de-vised to the user. In case of AI planning, however, we are dealing with a family of very different domains, merely sharing the same, very general specification language. Therefore planners have to be general-purpose. Domain-dependent knowledge has either to be omitted or to be inferred by the machine.

Planning domains usually have large branching factors, with the branching factor being defined as the average number of successors of a state within plan-ning space. Due to the resulting huge search spaces planning resists almost all approaches of single-state space search. As indicated above automated finite state pruning is generally not available although there is some promising research on symmetry leading to good results in at least some domains [10].

On the other hand, domain-independent heuristic guidance in form of a lo-wer bound can be devised, e.g. by counting the number of facts missing from the goal state. However, these heuristics are too weak to regain tractability. Moreo-ver, new theoretical results in heuristic single-state search prove that while finite state machine pruning can effectively reduce the branching factor, in the limit heuristics cannot [5,17]. The influence of lower bounds on the solution length can best be thought of as a decrease in search depth. Therefore, even when incor-porated with lower bound information, the problem of large branching factors when applying single-state space searching algorithms to planning domains re-mains unsolved. As a solution we propose a promising symbolic search technique also favoring a small binary encoding length.

## 2  Symbolic Exploration

Driven by the success of model checking in exploring search spaces of $10^{20}$ sta-tes and beyond, the new trend in search is reachability analysis [19]. Symbolic exploration bypasses the typical exponential growth of the search tree in many applications. However, the length of the state description severely influences the execution time of the relevant algorithms. In symbolic exploration the rule of thumb for tractability is to choose encodings of not much more than 100 bits.

Edelkamp and Reffel have shown that and how symbolic exploration leads to promising results in solving current challenges to single-agent search such as the Fifteen Puzzle and Sokoban [6]. Recent results show that these methods contribute substantial improvements to deterministic planning [7].

The idea in symbolically representing a set $S$ is to devise a boolean function $\phi_S$ with input variables corresponding to bits in the state description that eva-luates to true if and only if the input $a$ is the encoding of one element $s$ in $S$. The drawback of choosing boolean formulae to describe $\phi_S$ is that satisfiability checking is NP-complete. The unique representation with binary decision dia-grams (BDDs) can grow exponentially in size, but, fortunately, this characteristic seldom appears in practice [2].

BDDs allow to efficiently encode sets of states. For example let $\{0, 1, 2, 3\}$ be the set of states encoded by their binary value. The characteristic function of a single state is the minterm of the encoding, e.g. $\phi_{\{0\}}(x) = \overline{x_1} \land \overline{x_2}$. The resulting $BDD$ has two inner nodes. The crucial observation is that the $BDD$ representation of $S$ increases by far slower than $|S|$. For example the $BDD$ for $\phi_{\{0,1\}} = \overline{x_1}$ consists of one internal node and the $BDD$ for $\phi_{\{0,1,2,3\}}$ is given by the 1-sink only.

An operator can also been seen as an encoding of a set. In contrast to the previous situation a member of the transition relation corresponds to a pair of states $(s', s)$ if $s'$ is a predecessor of $s$. Subsequently, the transition relation $T$ evaluates to 1 if and only if $s'$ is a predecessor of $s$. Enumerating the cross product of the entire state space is by far too expensive. Fortunately, we can set up $T$ symbolically by defining which variables change due to an operator and which variables do not.

Let $S_i$ be the set of states reachable from the start state in $i$ steps, initialized by $S_0 = \{s\}$. The following equation determines $\phi_{S_i}$ given both $\phi_{S_{i-1}}$ and the transition relation: $\phi_{S_i}(s) = \exists s' (\phi_{S_{i-1}}(s') \land T(s', s))$. In other words we perform breadth first search with $BDDs$. A state s belongs to $S_i$ if it has a predecessor in the set $S_{i-1}$ and there exists an operator which transforms $s'$ into $s$. Note that on the right hand side of the equation $\phi$ depends on $s'$ compared to $s$ on the left hand side. Thus, it is necessary to substitute $s$ with $s'$ in the $BDD$ for $\phi_{S_i}$. Fortunately, this substitution corresponds to a simple renaming of the variables.

Therefore, the key operation in the exploration is the *relational product* $\exists v(f \land g)$ of a variable vector $v$ and two boolean functions $f$ and $g$. Since existential quantification of one boolean variable $x_i$ in the boolean function $f$ is equal to disjunction $f|_{x_i=0} \lor f|_{x_i=1}$, the quantification of $v$ results in a sequence of subproblem disjunctions. Although computing the relational product is NP-hard in general, specialized algorithms have been developed leading to an efficient determination for many practical applications.

In order to terminate the search we test, if a state is contained in the intersection of the symbolic representation of the set $S_i$ and the set of goal states $G$. This is achieved by evaluating the relational product $goalReached = \exists x (\phi_{S_i} \land \phi_G)$. Since we enumerated $S_0, \ldots, S_{i-1}$ in case $goalReached$ evaluates to 1, $i$ is known to be the optimal solution length.

# 3   Parsing

We evaluate our algorithms on the AIPS'98 planning contest problems[1], mostly given in Strips [8]. An operator in Strips consists of pre- and postconditions. The latter, so-called effects, divide into an add list and a delete list.

Extending Strips leads to ADL with first order specification of conditional effects [20] and PDDL, a layered planning description domain language. Although symbolic exploration and the translation process described in this paper are not restricted to Strips, for the ease of presentation we will keep this focus.

---

[1] http://ftp.cs.yale.edu/pub/mcdermott/aipscomp-results.html.

A PDDL-given problem consists of two parts. In the domain specific part, predicates and actions are defined. A predicate is given by its name and its parameters, and actions are given by their names, parameters, preconditions, and effects. One example domain, Logistics, is given as follows[2].

```
(define (domain logistics-strips)
  (:predicates (OBJ ?obj) (TRUCK ?tru) (LOCATION ?loc) (AIRPLANE ?plane) (CITY ?city)
               (AIRPORT ?airport) (at ?obj ?loc)  (in ?obj ?obj) (in-city ?obj ?city))
(:action LOAD-TRUCK
  :parameters (?obj ?tru ?loc)
  :precondition (and (OBJ ?obj) (TRUCK ?tru) (LOCATION ?loc) (at ?tru ?loc) (at ?obj ?loc))
  :effect (and (not (at ?obj ?loc)) (in ?obj ?tru)))
(:action UNLOAD-TRUCK
  :parameters (?obj ?tru ?loc)
  :precondition (and (OBJ ?obj) (TRUCK ?tru) (LOCATION ?loc) (at ?tru ?loc) (in ?obj ?tru))
  :effect (and (not (in ?obj ?tru)) (at ?obj ?loc)))
(:action DRIVE-TRUCK
  :parameters (?tru ?loc-from ?loc-to ?city)
  :precondition  (and (TRUCK ?tru) (LOCATION ?loc-from) (LOCATION ?loc-to) (CITY ?city)
                      (at ?tru ?loc-from) (in-city ?loc-from ?city) (in-city ?loc-to ?city))
  :effect (and (not (at ?tru ?loc-from)) (at ?tru ?loc-to))) ... )
```

The problem specific part defines the objects to be dealt with and describes initial and goal states, consisting of a list of facts (instantiations to predicates).

```
(define (problem strips-log-x-1)
  (:domain logistics-strips)
  (:objects package6 package5 package4 package3 package2 package1
            city6 city5 city4 city3 city2 city1
            truck6 truck5 truck4 truck3 truck2 truck1 plane2 plane1
            city6-1 city5-1 city4-1 city3-1 city2-1 city1-1
            city6-2 city5-2 city4-2 city3-2 city2-2 city1-2)
  (:init (and (OBJ package1) (OBJ package2) ... (at package6 city3-1) .. ))
  (:goal (and (at package6 city1-2) ... )))
```

Using current software development tools, parsing a PDDL specification is easy. In our case we applied the Unix programs *flex* and *bison* for lexically analyzing the input and parsing the result into data structures. We used the standard template library, STL for short, for handling the different structures conveniently. The information is parsed into vectors of predicates, actions and objects. All of them can be addressed by their name or a unique numeric identifier, with STL maps allowing conversions from the former ones to the latter ones. Having set up the data structures, we are ready to start analyzing the problem.

# 4   Constant and One-Way Predicates

A *constant predicate* is defined as a predicate whose instantiations are not affected by any operator in the domain. Since Strips does not support types, constant predicates are often used for labeling different kinds of objects, as is the case for the TRUCK predicate in the Logistics domain. Another use of constant predicates is to provide persistent links between objects, e.g. the in-city predicate in the Logistics domain which associates locations with cities. Obviously, constant predicates can be omitted in any state encoding.

---

[2] Dots (...) are printed if source fragments are omitted.

Instantiations of *one-way* predicates do change over time, but only in one direction. There are no one-way predicates in the Logistics domain; for an example consider the Grid domain, where doors can be opened with a key and not be closed again. Thus `locked` and `open` are both one-way predicates. Those predicates need to be encoded only for those objects that are not listed as open in the initial state. In PDDL neither constant nor one-way predicates are marked and thus both have to be inferred by an algorithm. We iterate on all actions, keeping track of all predicates appearing in any effect lists. Constant predicates are those that appear in none of those lists, one-way predicates are those that appear either as add effects or as delete effects, but not both.

## 5   Merging Predicates

Consider the Logistics problem given above, which serves as an example for the remaining phases. There are 32 objects, six packages, six trucks, two airplanes, six cities, six airports, and six other locations. A naive state encoding would use a single bit for each possible fact, leading to a space requirement of $6 \cdot 32 + 3 \cdot 32^2 = 3264$ bits per state, since we have to encode six unary and three binary predicates. Having detected constant predicates, we only need to encode the `at` and `in` predicates, thus using only $2 \cdot 32^2 = 2048$ bits per state. Although this value is obviously better, it is far from being satisfying.

A human reader will certainly notice that it is not necessary to consider all instantiations of the `at` predicate independently. If a given package $p$ is at location $a$, it cannot be at another location $b$ at the same time. Thus it is sufficient to encode *where* $p$ is located, i.e. we only need to store an object number which takes only $\lceil \log 32 \rceil = 5$ bits per package. How can such information be deduced from the domain specification? To tell the truth, this is not possible, since the information does not only depend on the operators themselves but also on the initial state of the problem. If the initial state included the facts (`at` $p$ $a$) as well as (`at` $p$ $b$), then $p$ could be at multiple locations at the same time.

However, we can try to prove that the number of locations a given object is at cannot increase over time. For a given state, we define $\#at_2(p)$ as the number of objects $q$ for which the fact (`at` $p$ $q$) is true. If there is no operator that can increase this value, then $\#at_2(p)$ is limited by its initial value, i.e. by the number of corresponding facts in the initial state. In this case we say that `at` is *balanced* in the second parameter. Note that this definition can be generalized for $n$-ary predicates, defining $\#pred_i(p_1, \ldots, p_{i-1}, p_{i+1}, \ldots, p_n)$ as the number of objects $p_i$ for which the fact (`pred` $p_1$ ... $p_n$) is true. If we knew that `at` was balanced in the second parameter, we would be facing one of the following situations:

- $\#at_2(p) = 0$: We have to store no information about the location of $p$.
- $\#at_2(p) = 1$: The location of $p$ can be encoded by using an object index, i.e. we need $\lceil \log o \rceil$ bits, where $o$ denotes the number of objects $p$ can be assigned to in our problem.
- $\#at_2(p) > 1$: In this case, we stick to naive encoding.

So can we prove that the balance requirement for at is fulfilled? Unfortunately we cannot, since there are some operators that increase $\#at_2$, namely the UNLOAD-TRUCK operator. However, we note that whenever $\#at_2(p)$ increases, $\#in_2(p)$ decreases, and vice versa. If we were to merge at and in into a new predicate at+in, this predicate would be balanced, since $\#(at+in)_2 = \#at_2 + \#in_2$ remains invariant no matter what operator is applied.

We now want to outline the algorithm for checking the balance of $\#pred_i$ for a given predicate $pred$ and parameter $i$: For each action $a$ and each of its add effects $e$, we check if $e$ is referring to predicate pred. If so, we look for a corresponding delete effect, i.e. a delete effect with predicate pred and the same argument list as $e$, except for the $i$-th argument which is allowed to be (and normally will be) different. If we find such a delete effect, it balances the add effect, and there is no need to worry.

If there is no corresponding delete effect, we search the delete effect list for any effect with a matching argument list (again, we ignore parameter $i$), no matter what predicate it is referring to. If we do not find such an effect, our balance check fails. If we do find one, referring to predicate other, then we recursively call our algorithm with the merged predicate pred+other. Note that "matching argument list" does not necessarily mean that other takes its arguments in the same order as pred, which makes the actual implementation somewhat more complicated.

It is even possible to match other if that predicate takes one parameter less than pred, since parameter $i$ does not need to be matched. This is a special case in which $\#other_i(p_1, \ldots, p_{i-1}, p_{i+1}, \ldots, p_n)$ can either be 1 or 0, depending on whether the corresponding fact is true or not, since there is no parameter $p_i$ here. Examples of this situation can be found in the Gripper domain, where carry ?ball ?gripper can be merged with free ?gripper.

If there are several candidates for other, all of them are checked, maybe proving balance of different sets of merged predicates. In this case, all of them are returned by the algorithm. It is of course possible that more than two predicates are merged in order to satisfy a balance requirement since there can be multiple levels of recursion. This algorithm checks the $i$-th parameter of predicate pred. Executing it for all predicates in our domain and all possible values of $i$ and collecting the results yields an exhaustive list of balanced merged predicates.

In the case of the Logistics domain, our algorithm exhibits that merging at and in gives us the predicate at+in which is balanced in the second parameter. Looking at the initial facts stated in the problem specification, we see that we can store the locations of trucks, airplanes and packages by using six bits each, since $\#(at+in)_2$ evaluates to one for those objects, and that we do not need to encode anything else, since the other objects start off with a count of zero.

Note that $\lceil \log 32 \rceil = 5$ bits are not sufficient for encoding locations at our current level of information, since we not only have to store the index of the object we are referring to, but also which of the two predicates at or in is actually meant. Thus our encoding size can be reduced to $(6+6+2) \cdot 6 = 84$ bits, which is already a reasonable result and sufficient for many purposes.

# 6   Exploring Predicate Space

However, we can do better. In most cases it is not necessary to allow the full range of objects for the balanced predicates we have detected, since e.g. a package can only be at a location or in a vehicle (truck or airplane), but never at another package, in a location, and so on.

If a fact is present in the initial state or can be instantiated by any valid sequence of operators, we call it *reachable*, otherwise it is called *unreachable*.

Many facts can be proven to be unreachable directly from the operators themselves, since actions like LOAD-TRUCK require the object the package is put into to be a truck. However, there are some kinds of unreachable facts we do not want to miss that cannot be spotted that way.

For example, DRIVE-TRUCK can only move a truck between locations in the same city, since for a truck to move from $a$ to $b$, there must be a city $c$, so that (in-city $a$ $c$) and (in-city $b$ $c$) are true. Belonging to the same city is no concept that is modeled directly in our Strips definition.

For those reasons, we do not restrict our analysis to the domain specification and instead take the entire problem specification into account. What we want to do is an exploration of predicate space, i.e. we try to enumerate all instantiations of predicates that are reachable by beginning with the initial set of facts and extending it in a kind of breadth-first search.

Note that we are exploring predicate space, not search space. We do not store any kind of state information, and only keep track of which facts we consider reachable. Thus, our algorithm can do one-side errors, i.e. consider a fact reachable although it is not, because we do not pay attention to mutual exclusion of preconditions. If a fact $f$ can be reached by an operator with preconditions $g$ and $h$, and we already consider $g$ and $h$ reachable, then $f$ is considered reachable, although it might be the case that $g$ and $h$ can never be instantiated at the same time. This is a price we have to pay and are willing to pay for reasons of efficiency. Anyway, if we were able to decide reliably if a given combination of facts could be instantiated at the same time, there would hardly remain any planning problem to be solved. We tested two different algorithms for exploring predicate space, *Action-Based Exploration* and *Fact-Based Exploration*.

## 6.1   Action-Based Exploration

In the action-centered approach, the set of reachable facts is initialized with the facts denoted by the initial state. We then instantiate all operators whose preconditions can be satisfied by only using facts that we have marked as reachable, marking new facts as reachable according to the add effect lists of the instantiated operators. We then again instantiate all operators according to the extended set of reachable facts. This process is iterated until no further facts are marked, at which time we know that there are no more reachable facts.

Our implementation of the algorithm is somewhat more tricky than it might seem, since we do not want to enumerate all possible argument lists for the operators we are instantiating, which might take far too long for difficult problems (there are e.g. $84^7 \approx 3 \cdot 10^{13}$ different possible instantiations for the drink operator in problem Mprime-14 from the AIPS'98 competition).

To overcome this problem, we apply a backtracking technique, extending the list of arguments one at a time and immediately checking if there is an unsatisfied precondition, in which case we do not try to add another argument. E.g., considering the LOAD-TRUCK operator, it is no use to go on searching for valid instantiations if the ?obj parameter has been assigned an object $o$ for which (OBJ $o$) has not been marked as reachable.

There is a second important optimization to be applied here: Due to the knowledge we already have accumulated, we know that OBJ is a constant predicate and thus there is no need to dynamically check if a given object satisfies this predicate. This can be calculated beforehand, as well as other preconditions referring to constant predicates.

So what we do is to statically constrain the domains of the operator parameters by using our knowledge about constant and one-way predicates. For each parameter, we pre-compute which objects possibly could make the corresponding preconditions true. When instantiating operators later, we only pick parameters from those sets. Note that due to this pre-computation we do not have to check preconditions concerning constant unary predicates at all during the actual instantiation phase.

For one-way predicates, we are also able to constrain the domains of the corresponding parameters, although we cannot be as restrictive as in the case of constant predicates. E.g., in the Grid example mentioned above, there is no use in trying to open doors that are already open in the initial state. However, we cannot make any assumption about doors that are closed initially.

There are two drawbacks of the action-based algorithm. Firstly, the same instantiations of actions tend to be checked multiple times. If an operator is being instantiated with a given parameter list, it will be instantiated again in all further iterations. Secondly, after a few iterations, changes to the set of reachable facts tend to become smaller and less frequent, but even if only a single fact is added to the set during an iteration, we have to evaluate all actions again, which is bad. Small changes should have less drastic consequences.

## 6.2   Fact-Based Exploration

Therefore we shift the focus of the exploration phase from actions to facts. Our second algorithm makes use of a queue in which all facts that are scheduled to be inserted into the set of reachable facts are stored. Initially, this queue consists of the facts in the initial state, while the set of reachable facts is empty. We then repeatedly remove the first fact $f$ from the queue, add it to the set of reachable facts and instantiate all operators whose preconditions are a subset of our set of reachable facts *and include* $f$. Add effects of these operators that are not yet stored in either the set of reachable facts or the fact queue are added to the back of the fact queue. This process is iterated until the fact queue is empty.

**Table 1.** Encoding the Logistics problem 1-01 with 42 bits.

```
(5 bits) package6
    at  city6-1 city5-1 city4-1 city3-1 city2-1 city1-1
        city6-2 city5-2 city4-2 city3-2 city2-2 city1-2
    in  truck6 truck5 truck4 truck3 truck2 truck1 plane2 plane1
 ...
(5 bits) package1
    at  city6-1 city5-1 city4-1 city3-1 city2-1 city1-1
        city6-2 city5-2 city4-2 city3-2 city2-2 city1-2
    in  truck6 truck5 truck4 truck3 truck2 truck1 plane2 plane1
(1 bit) truck6
    at  city6-1 city6-2
 ...
(1 bit) truck1
    at  city1-1 city1-2
(3 bits) plane2
    at  city6-2 city5-2 city4-2 city3-2 city2-2 city1-2
(3 bits) plane2
    at  city6-2 city5-2 city4-2 city3-2 city2-2 city1-2
```

Although it does not look as if much was gained at first glance, this algorithm is a big improvement to the first one.

The key difference is that when instantiating actions, only those instantiations need to be checked for which $f$ is one of the preconditions, which means that we can *bind* all parameters appearing in that precondition to the corresponding values of $f$, thus reducing the number of degrees of freedom of the argument lists. Of course, the backtracking and constraining techniques mentioned above apply here as well. The problem of multiple operator instantiations does not arise. We never instantiate an operator that has been instantiated with the same parameter list before, since we require $f$ to be one of the preconditions, and in previous iterations of the loop, $f$ was not regarded a reachable fact.

Returning to our Logistics problem, we now know that a package can only be at a location, in a truck or in an airplane. An airplane can only be at an airport, and a truck can only be at a location which must be in the same city as the location the truck started at.

## 7   Combining Balancing and Predicate Space Exploration

All we need to do in order to receive the encoding we are aiming at is to combine the results of the previous two phases. Note that these results are very different: While predicate space exploration yields information about the facts themselves, balanced predicates state information about the *relationship* between different facts. Both phases are independent of each other, and to minimize our state encoding, we need to combine the results.

In our example, this is simple. We have but one predicate to encode, the at+in predicate created in the merge phase. This leads to an encoding of 42 bits (cf. Table 1), which is the output of our algorithm. However, there are cases in which it is not obvious how the problem should be encoded. In the Gripper domain (constant predicates omitted) the merge step returns the balanced predicates at-robby, carry+free, and at+carry; at-robby is an original operator, while carry+free and at+carry have been merged.

We do not need to encode each of the merged predicates, since this would mean encoding carry twice. If we had already encoded carry+free and now wanted to encode the at+carry predicate for a given object $x$, with $n$ facts of the type (at $x$ $y$) and $m$ facts of the type (carry $x$ $y$), we would only need $\lceil \log(n+1) \rceil$ bits for storing the information, since we only have to know which of the at-facts is true, or if there is no such fact. In the latter case, we know that some fact of the type (carry $x$ $y$) is involved and can look up which one it is in the encoding of carry+free. However, encoding at+carry first, thus reducing the space needed by carry+free is another possibility for encoding states, and is in fact the better alternative in this case. Since we cannot know which encoding yields the best results, we try them out systematically.

Although there is no need for using heuristics here since the number of conflicting possibilities is generally very small, we want to mention that as a rule of thumb it is generally a good idea to encode predicates that cover the largest number of facts first.

Predicates that are neither constant nor appear in any of the balanced merge predicates are encoded naively, using one bit for each possible fact. Those predicates are rare. In fact, in the considered benchmark set we only encountered them in the Grid domain, and there only for encoding the locked state of doors which obviously cannot further be compressed.

## 8    Experimental Results

In this section we provide data on the achieved compression to the state descriptions of the AIPS'98 planning competition problems. The problem suite consists of six different Strips domains, namely *Movie*, *Gripper*, *Logistics*, *Mystery*, *Mprime*, and *Grid*. In Table 1 we have exemplarily given the full state description for the first problem in the Logistics suite. The exhibited knowledge in the encoding can be easily extracted in form of *state invariants*, e.g. a package is either a location in a truck or in an airplane, each truck is restricted to exactly one city, and airplanes operate on airports only.

Table 2 depicts the state description length of all problems in the competition. Manually encoding some of the domains and comparing the results we often failed to devise a smaller state description length by hand.

Almost all of the execution time is spent on exploring predicate space. All the other phases added together never took more than a second of execution time. The time spent on exploring predicate space is not necessarily lost. When symbolically exploring planning space using BDDs, the operators need to be instantiated anyway for building the transition function, and if we keep track of all operator instantiations in the exploration phase this process can be sped up greatly.

**Table 2.** Length of state encodings and elapsed time of the AIPS'98 benchmark set. The data was generated on a Sun Ultra Sparc Station.

| | Movie | | Gripper | | Logistics | | Mystery | | Mprime | | Grid | |
|---|---|---|---|---|---|---|---|---|---|---|---|---|
| problem | bits | sec | bits | sec | bits | sec | bits | sec | bits | sec | bits | sec |
| 1-01 | 6 | <1 | 11 | <1 | 42 | <1 | 28 | <1 | 32 | <1 | | |
| 1-02 | 6 | <1 | 15 | <1 | 56 | <1 | 117 | <1 | 121 | <1 | | |
| 1-03 | 6 | <1 | 19 | <1 | 98 | <1 | 77 | <1 | 89 | <1 | | |
| 1-04 | 6 | <1 | 23 | <1 | 115 | <1 | 50 | <1 | 63 | <1 | | |
| 1-05 | 6 | <1 | 27 | <1 | 35 | <1 | 86 | <1 | 96 | <1 | | |
| 1-06 | 6 | <1 | 31 | <1 | 174 | <1 | 148 | <1 | 179 | <1 | | |
| 1-07 | 6 | <1 | 35 | <1 | 95 | <1 | 82 | <1 | 126 | 1 | | |
| 1-08 | 6 | <1 | 39 | <1 | 254 | 1 | 90 | <1 | 142 | <1 | | |
| 1-09 | 6 | <1 | 43 | <1 | 184 | <1 | 83 | <1 | 93 | <1 | | |
| 1-10 | 6 | <1 | 47 | <1 | 162 | <1 | 291 | <1 | 315 | <1 | | |
| 1-11 | 6 | <1 | 51 | <1 | 104 | 1 | 52 | 1 | 61 | 1 | | |
| 1-12 | 6 | <1 | 55 | <1 | 195 | <1 | 42 | <1 | 56 | <1 | | |
| 1-13 | 6 | <1 | 59 | <1 | 287 | 1 | 291 | <1 | 323 | 1 | | |
| 1-14 | 6 | <1 | 63 | <1 | 282 | 1 | 320 | 1 | 346 | 1 | | |
| 1-15 | 6 | <1 | 67 | <1 | 144 | <1 | 184 | <1 | 210 | 1 | | |
| 1-16 | 6 | <1 | 71 | <1 | 205 | <1 | 90 | <1 | 120 | <1 | | |
| 1-17 | 6 | <1 | 75 | <1 | 190 | <1 | 188 | <1 | 202 | <1 | | |
| 1-18 | 6 | <1 | 79 | <1 | 270 | 1 | 112 | 1 | 160 | 1 | | |
| 1-19 | 6 | <1 | 83 | <1 | 256 | <1 | 129 | <1 | 163 | <1 | | |
| 1-20 | 6 | <1 | 87 | <1 | 264 | 2 | 144 | <1 | 169 | 2 | | |
| 1-21 | 6 | <1 | | | 300 | 1 | 205 | <1 | 230 | <1 | | |
| 1-22 | 6 | <1 | | | 530 | 3 | 234 | 1 | 283 | 1 | | |
| 1-23 | 6 | <1 | | | 166 | <1 | 157 | <1 | 186 | <1 | | |
| 1-24 | 6 | <1 | | | 336 | <1 | 229 | <1 | 263 | 1 | | |
| 1-25 | 6 | <1 | | | 343 | 5 | 23 | <1 | 26 | <1 | | |
| 1-26 | 6 | <1 | | | 382 | 3 | 67 | <1 | 86 | <1 | | |
| 1-27 | 6 | <1 | | | 604 | 4 | 63 | <1 | 67 | <1 | | |
| 1-28 | 6 | <1 | | | 818 | <1 | 38 | <1 | 41 | <1 | | |
| 1-29 | 6 | <1 | | | 566 | 1 | 74 | <1 | 86 | <1 | | |
| 1-30 | 6 | <1 | | | 470 | 8 | 109 | <1 | 117 | 1 | | |
| 2-01 | | | | | 26 | <1 | | | 120 | <1 | 67 | <1 |
| 2-02 | | | | | 28 | <1 | | | 84 | <1 | 83 | <1 |
| 2-03 | | | | | 39 | <1 | | | 269 | <1 | 93 | <1 |
| 2-04 | | | | | 64 | <1 | | | 129 | <1 | 107 | <1 |
| 2-05 | | | | | 63 | <1 | | | 46 | <1 | 139 | 1 |

# 9   Related Work and Conclusion

There is some work in literature dealing with reformulation of planning problems. However, research mainly concentrates on inferring state invariants instead of minimizing the state description length.

Fox and Long, for example, have contributed several suggestions that have been implemented in the planner *Stan* (for STate ANalysis) [18]. The project is based on Graphplan [1] and uses a variety of techniques to exhibit domain-dependent information. In this context the automatic inference of state invariants is important. The pre-processor *Tim* (Type Inference Module) explores planning domains in order to find typings of untyped parameters [9]. The information is found by an algorithm starting with a projection of actions to their parameters establishing so-called *properties*, i.e. predicates together with the argument position filled by the objects. Given the properties and operators, transition rules are inferred (e.g. $on_1 \rightarrow on_1$ in Logistics) which constitute a finites state machine corresponding to the property exchanges. Types are found by exploration of membership patterns starting with the initial set.

Given the inferred type specification, in an additional analysis step three major state invariants can be found: *identity invariants, membership invariants* and *unique state invariants*. E.g in *Blocks World* we have $\forall x, y, z$ on$(y, x) \wedge$ on$(z, x) \rightarrow y = z$, $\forall x \; \exists y$ on$(y, x) \vee$ clear$(x)$, and $\forall x \; \neg(\exists y$ on$(y, x) \wedge$ clear$(x))$ Furthermore, *Tim* has been extended to infer *cardinality constraints* such as $|\{x|\text{at-robot}(x)\}| = 1$ in *Gripper*. Tim is sound but not complete, i.e., it will find correct invariants but not all of them. Very recent unpublished work by Fox and Long focus *Mobile Analysis*, which constructs maps of locations that can be navigated by a mobile through an operator schema that gives the mobility.

The problem of finding state invariants is also addressed by Gerevini and Schubert [11]. Their planner *Discoplan* discovers two kinds of invariance rules, *single-valued* and *implicative constraints*. For example we have on$(x, y) \wedge y \neq z \Rightarrow \neg$on$(x, z)$ and on$(x, y) \wedge y \neq$ table $\Rightarrow \neg$clear$(y)$ in *Blocks World*. While *Tim* improves explorations in *Graphplan*, in case of *Discoplan* the invariants improve satisfiability planning such as in *Satplan* [14]. Satplan itself is closely related to our approach of symbolically exploring planning space with BDDs, since both algorithms rely on a specification of the problem with boolean formulae.

The information gathered by *Tim* and *Discoplan* can be compared to our approach of balanced predicates and constraining the domains of predicates, since the presented algorithms exhibit domain-dependent knowledge leading to problem invariants as shown in the given example. As highlighted above the encoding in *Logistics* apparently give *identity invariants membership invariants* and *unique state invariants* as well as some *single-valued* and *implicative constraints*. Even *cardinality constraints* can be extracted from the encodings.

We conjecture that it is possible to extract the same invariants as in Tim and Discoplan from our encodings and that the knowledge inferred by our algorithms is more detailed, but there is an extraction process required to obtain the invariants from the encodings and to prove the assertion. On the other hand we think that the binary encoding length is probably the best performance measure to compare the inferred knowledge of different precompilers.

Literature reveals that an information gathering phase prior to search takes time. Through the efficiency of our approaches the time spent on these efforts is by far shorter than the time needed for constructing the transition function and the symbolic search phase itself. Automatically inferring problem-dependent knowledge in planning problems is challenging but an inevitable necessity for current state space search engines. The paper contributes efficient new algorithms based on symbolical manipulation and search. The promising results of BDD-based exploration according to the achieved encodings are given in [7]. We conclude that our approach to automatically infer compressed state descriptions mainly tailored to symbolic exploration reflects current research and could have a strong impact on current planning systems.

**Acknowledgments.** Thanks to F. Reffel for helpful discussions concerning the symbolic exploration phase. S. Edelkamp and M. Helmert are partially supported by DFG in project Ot-11/3 entitled *Heuristic Search and Its Application in Protocol Verification*.

# References

1. A. Blum and M. Furst. Fast planning through planning graph analysis. In *IJ-CAI'95*, pages 1636–1642, 1995.
2. R. E. Bryant. Symbolic manipulation of boolean functions using a graphical representation. In *ACM/IEEE Design Automation*, pages 688–694, 1985.
3. J. C. Culberson and J. Schaeffer. Searching with pattern databases. In *Proceedings of the Biennial Conference of the Canadian Society for Computational Studies of Intelligence on Advances in AI*, volume 1081 of *LNAI*, pages 402–416, 1996.
4. J. Eckerle and S. Schuierer. Efficient memory-limited graph search. *Lecture Notes in Computer Science*, 981:101–112, 1995.
5. S. Edelkamp and R. E. Korf. The branching factor of regular search spaces. In *AAAI-98*, pages 299–304, 1998.
6. S. Edelkamp and F. Reffel. OBDDs in heuristic search. In *KI-98*, volume 1504 of *LNAI*, pages 81–92, Berlin, 1998. Springer.
7. S. Edelkamp and F. Reffel. Deterministic state space planning with BDDs. Technical Report 120, Computer Science Departement, University of Freiburg, 1999.
8. R. Fikes and N. Nilsson. Strips: A new approach to the application of theorem proving to problem solving.
9. M. Fox and D. Long. The automatic inference of state invariants in TIM. *JAIR*, 9:367–421, 1998.
10. M. Fox and D. Long. The detection and exploration of symmetry in planning problems. Technical Report 1/99, Department of Computer Science, University of Durham, 1999.
11. A. Gerevini and L. Schubert. Inferring state constraints for domain-independent planning. In *AAAI-98*, pages 905–912. AAAI Press, 1998.
12. P. E. Hart, N. J. Nilsson, and B. Raphael. A formal basis for heuristic determination of minimum path cost. *IEEE Trans. on SSC*, 4:100, 1968.
13. A. Junghanns and J. Schaeffer. Single agent search in the presence of deadlocks. In *AAAI-98*, pages 228–233, 1998.
14. H. Kautz and B. Selman. Pushing the envelope: Planning propositional logic, and stochastic search. In *AAAI-96*, pages 1194–1201, 1996.
15. R. E. Korf. Depth-first iterative-deepening: An optimal admissible tree search. *Artificial Intelligence*, 27(1):97–109, 1985.
16. R. E. Korf. Finding optimal solutions to Rubik's Cube using pattern databases. In *AAAI-97*, pages 700–705, 1997.
17. R. E. Korf. Complexity analysis of admissible heuristic search. In *AAAI-98*, pages 305–310, 1998.
18. D. Long and M. Fox. Efficient implementation of the plan graph in STAN. *JAIR*, 10:87–115, 1998.
19. K. McMillan. *Symbolic Model Checking*. Kluwer Academic Press, 1993.
20. E. Pednould. ADL: Exploring the middleground between Strips and situation calculus. *KR-89*, pages 324–332, 1989.
21. A. Reinefeld and T. Marsland. Enhanced iterative-deepening search. *IEEE Transactions on Pattern Analysis and Machine Intelligence*, 16(7):701–710, 1994.
22. S. Russell. Efficient memory-bounded search methods. In *ECAI-92*, pages 1–5. Wiley, 1992.
23. L. A. Taylor and R. E. Korf. Pruning duplicate nodes in depth-first search. In *AAAI-93*, pages 756–761, 1993.

# Action Constraints for Planning

Ulrich Scholz

Darmstadt University of Technology
Alexanderstraße 10, 64283 Darmstadt, Germany
scholz@informatik.tu-darmstadt.de
http://aida.intellektik.informatik.tu-darmstadt.de/~scholz/

**Abstract.** Recent progress in the applications of propositional planning
systems has led to an impressive speed-up of solution time and an increase in tractable problem size. In part, this improvement stems from
the use of domain-dependent knowledge in form of state constraints.
In this paper we introduce a different class of constraints: *action constraints*. They express domain-dependent knowledge about the use of
actions in solution plans and can express strategies which are used by
human planners. The use of action constraints results in a tendency to
better plans. We explain how to calculate and apply action constraints
in the framework of parallel total-order planning, which is the design
of the most powerful planners at the moment. We present two classes
of action constraints and demonstrate their capabilities in the planner
PROBAPLA.

## 1 Introduction

Recent progress in the applications of propositional planning systems has led
to an impressive speed-up of solution time and an increase in tractable problem size. In part, this improvement stems from the use of domain-dependent
knowledge [BK96,KS96]. This knowledge can be formalized as axioms about properties of plans and many planning formalisms can incorporate them easily. A
domain-independent planning system has to generate these domain-dependent
axioms automatically via preprocessing of the planning problem. By now, these
preprocessing methods have concentrated on the generation of state constraints
[GS98,FL98] which are very successful. Their expressiveness does not cover the
whole space of domain-dependent knowledge, which gives rise to the hope that
other types of constraints can advance planning in a similar way.

Human planners take a different approach for using domain-dependent knowledge: They try to avoid actions and action sequences without useful effect or
to try the best combination of actions for solving a subproblem. Similar to state
constraints, this knowledge can be formulated as *action constraints*. These eliminate some preliminary plans and solutions from the search space which allows
to solve larger problems in shorter time. An additional effect is a tendency to
better plans.

S. Biundo and M. Fox (Eds.): ECP-99, LNAI 1809, pp. 148–158, 2000.

Current planner use action constraints as heuristic [KS98] or as post-processing technique for specific domains [AK97] in a domain-dependent way. Our approach is different: We present action constraints for domain-independent planning. This requires to consider several issues which depend on each other: (1) What class of action constraints do we want to use? (2) How does the corresponding constraint look like for the used planning formalism? (3) How can we calculate the domain-dependent knowledge for a given planning problem? (4) How do we use the action constraint once it is calculated, and finally (5) what are the expected improvements of its use? The designer of a planning system has to weigh up these issues. This paper presents some points in her decision space and explains their interconnections.

The paper is organized as follows: In the next section we introduce action constraints and action patterns (1,2). The following section explains how to find and use action constraints based on patterns for a parallel total-order planner and for a given planning problem (2,3,4). Then we present two constraints (1): The rsa constraint deals with the subsumption of action sequences of the length two. The top constraint allows total-order planners to have a similar search space as partial-order ones. The final part of the paper demonstrates these constraints (5) and gives concluding remarks.

## 2    Action Constraints

The quality of a planning method is measured by three criteria: planning speed, the class of planning problems it can handle, and the quality of its solutions. State constraints are a successful method to improve the first two criteria, but unfortunately they are not helpful in regard of the third. Rating the quality of plans and searching for good solutions requires to discern between different solutions and to discard unwanted ones but state constraints hold for *all* plans. Furthermore, the common definition of a good plan is to be short and to contain few actions. These criteria do not involve properties of states. For this reason we introduce constraints about actions which are axioms that hold for all actions in a wanted solution.

Practical planning requires the use of fast and secure techniques. For this reason we introduce a restricted class of action constraints which are based on patterns of actions. Their use is similar to a keyhole technique used by human planners, which examine actions in adjacent time steps. In case that they find a suboptimal pattern of actions they remove it or replace it with a more optimal one. An example is the representation of a door. A human planner does not open and close a door without an action inbetween. Such a sequence of actions is suboptimal in any case and she detects them by examining a small part of a plan.

An *action pattern* is a set of actions together with their mapping to time steps, which is part of a plan. An action constraint is *based* on a set of patterns $\mathcal{P}$ iff it eliminates all plans which contain a pattern of $\mathcal{P}$ and holds for all other plans. *Replacing* an action pattern $A$ with $A'$ in a plan means to remove exactly the actions of $A$ and to insert exactly the actions of $A'$. An action pattern does not depend on other properties of plan, like the activity of facts or the time step it is placed at. This simplifies the calculation and application of the corresponding action constraint.

In this paper we use a compact way to write action patterns: Actions on the same time step are given as a set and adjacent time steps are connected by $\circ$. In case that we assign a single action to a time step, we drop the brackets. For the above example of representing a door, the action constraint can be formulated as $\forall a.\ name(a,\ \text{'open door'}) \rightarrow \nexists a'.\ a \circ a' \wedge name(a', \text{'close door'})$. The replacement operator removes this action pattern by replacing it with the empty pattern.

The elimination of solutions from the plan space can make a planning problem unsolvable for a planning system, as it is the case with bounding the length of the considered plans. On the other hand we expect planning systems to compute a single solution and loosing some solutions can be an advantage. If a technique guarantees to preserve at least one solution of a solvable planning problem, it is safe to apply this technique. We call planning methods with that property *solution-safe*.

Some planning systems, like CNF-based ones, search for arbitrary solutions and stop after they find the first one. Action constraints which eliminate plans other than the optimal ones improve the solution quality of these planners. An action pattern $A$ is called *suboptimal* according to an objective function $o$ iff (1) every solution plan $P$ which contains $A$ can be transformed into a solution $P'$ not containing $A$ by replacing $A$ with another action pattern, and (2) $P'$ is better or at least of equal quality as $P$ according to $o$. Action constraints based on suboptimal patterns are always solution-safe: The better plan does not contain the pattern, so that it complies with the constraint.

The elimination of suboptimal solutions and the general reduction of search space size has an effect on the planning speed and the size of tractable problems. Often it is harder to find an optimal solution than finding an arbitrary one, e. g. it is possible to find a solution to a `blocksworld` problem in polynomial time but finding an optimal solution is NP-complete [Byl94]. On the other hand, reducing the search space can result in a faster search, even if the solution set is not changed.

## 3  Generation and Use

The idea of action constraints is useful for planning in general. For example it seems to be helpful for any planning system and any planning problem to avoid action patterns without effect. Nevertheless, an action constraint for a specific planning problem is domain-dependent knowledge, so domain-independent planners have to calculate them from the problem description. Planning systems use various formalisms, representations, and search methods. Action constraints form one small wheel in this machinery and have to adopt its characteristics.

For example consider the definition of conflict: Parallel planners cannot assign actions to the same time step which add and delete the same fact. With linear

$$a_1: \text{pre } \emptyset \qquad a_2: \text{pre } f_1 \qquad a_3: \text{pre } f_1 \qquad a_4: \text{pre } \emptyset$$
$$\text{add } f_1 \qquad \text{add } f_2 \qquad \text{add } \emptyset \qquad \text{add } f_2$$
$$\text{del } \emptyset \qquad \text{del } f_1 \qquad \text{del } f_1 \qquad \text{del } f_1$$

**Fig. 1.** Example of simple action patterns. An action $a$ *annuls* an action $a'$ iff del($a$) is a superset of add($a'$) and patterns of annulling actions can be suboptimal. The following three anulling action patterns are different in respect to their optimality and replacement operator: $a_1 \circ a_2$ is not suboptimal as $f_2$ can be necessary for a solution and $a_2$ depends on $a_1$. The patterns $a_1 \circ a_3$ and $a_1 \circ a_4$ are suboptimal but require different replacement operators. The first one has no combined effect and can be removed. The second pattern has the combined effect of adding $f_2$. Action $a_4$ does not depend on an effect of $a_1$, so that $a_1$ can be removed.

planning methods these *add/del conflicts* cannot occur. Action patterns which are suboptimal for a linear planner can be optimal for a parallel one, if all its replacements result in an add/del conflict. Some action constraints are based completely on a characteristic of a planning formalism, so they are not usable for planners with different design. For example the patterns for the top constraint, explained in Sect. 5, are usable only for total-order planners.

A generic way to find a pattern-based action axiom for a planning problem is to enumerate and examine all patterns of the considered size. For many classes of constraints there are more efficient algorithms: The actions of a pattern are related, so that enumerating the facts required or changed by an action and the actions which require or change a fact cuts the search space. The same holds for finding a replacement pattern. After calculating all patterns it is easy to find the corresponding axiom: It has to be violated for every plan which contains a suboptimal pattern and not violated otherwise.

Even for simple classes of patterns it can be hard to specify a set of replaceable action patterns and their replacements. Figure 1 presents annulling pairs of actions. The patterns of this class are devided into three subclasses which have to be handled differently.

Planning systems use action constraints in various ways. The action axiom can be encoded into a plan space representation so that it does no longer include the corresponding suboptimal plans or it can be used to prune the search tree directly. In case that a planner calculates the corresponding replacement operator, it can perform the replacements during its search phase. If the replacement operator just eliminates an action, the planner can remove this action from the planning domain in advance. Ambite and Knoblock [AK97] perform the replacements as post-processing. They introduce a planning method which attempts to find an initial solution quickly. Then, they rewrite this possibly suboptimal plan by applying the action constraints as rewriting rules.

In this paper we assume a parallel total-order planning formalism. An action $a$ consists of preconditions, added effects, and deleted effects, abbreviated by pre($a$), add($a$), and del($a$), respectively. Preconditions and effects are sets of positive facts without quantification, and actions are not allowed to add one of their preconditions. An action in a plan has a conflict in case that one of its preconditions is inactive. In addition, two actions on the same time step have a conflict if one deletes a precondition or an added effect of the other. Several successful planning systems use this design for their internal representation: GRAPHPLAN as well as BLACKBOX and other planners of the AIPS'98 planning competition. In the following we present two action constraints, rsa and top.

## 4 Replaceable Sequences of Actions

In many planning domains there are sequences of actions which have the same effects than a single action.[1] Examples are planning domains with locations: Instead of moving the same object in subsequent time steps it is often possible to place it in its final location right away. Eliminating these replaceable sequences of actions rsa leads to more optimal plans. Due to the computational complexity we will examine only sequences of the length two which are not suboptimal otherwise.

A sequence $a_1 \circ a_2$ is replaceable by an action $a$ but $a$ cannot replace $a_1$ or $a_2$ directly if both $a_1$ and $a_2$ have weaker preconditions, more added effects, or less deleted effects than $a$. Their sequence is replaceable by $a$, so that each of $a_1$ and $a_2$ has to have the missing preconditions or deleted effects, and delete or require the extra added effects of the other. According to these cases, replaceable action sequences are defined as follows:

**Definition 1.** *The action sequence $a_1 \circ a_2$ is called* replaceable *by an action $a$ in case that $a$ fulfills*

$\mathsf{pre}(a) \subseteq \mathsf{pre}(a_1) \cup \mathsf{pre}(a_2) \setminus \mathsf{add}(a_1) \land$
$\mathsf{add}(a) \supseteq (\mathsf{add}(a_1) \setminus \mathsf{del}(a_2)) \cup (\mathsf{add}(a_2) \setminus \mathsf{pre}(a_1)) \land$
$\mathsf{del}(a) \subseteq (\mathsf{del}(a_1) \setminus \mathsf{add}(a_2)) \cup \mathsf{del}(a_2)$

If no action is in parallel to a replaceable action sequence $a_1 \circ a_2$, the sequence is suboptimal and its replacement reduces the plan length by one time step. This is always the case for linear plans. Actions in parallel to the sequence can conflict with the replacing action. Placing a replacement in a time step is solution-safe if such a conflict cannot occur. Figure 2(a) exemplifies this case. If the replacing action can cause a conflict in both time steps, the replacement operator has to insert it into an additional time step. Figure 2(b) shows that this can result in plans with suboptimal length. Besides the mentioned add/del problem, these conflicts occur if $a_1$ adds a fact which is required at the next time step or $a_2$ deletes a fact which is required at the preceding time step. The replacing action can add these facts too late or delete them too early, respectively.

---

[1] The same is true for single actions. Due to the limited space we will not exemplify this case.

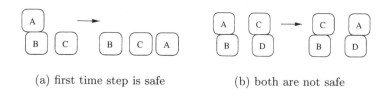

(a) first time step is safe          (b) both are not safe

**Fig. 2.** Examples for replaceable action sequences in the `blocksworld` domain. In Fig. 2(a) the sequence `move(A, B, C)`∘`move(A, C, Table)` can be replaced in a solution-safe way by `move(A, B, Table)` on the time step of the first action. The table is always clear, so that this fact cannot be the cause of a conflict. In Fig. 2(b) we cannot replace any sequence if we want to find a plan with optimal length because it is not possible to insert the replacing action into the first or the second time step: Blocks $A$ and $C$ mutually occupy the target location of each other.

## 5    Searching a Total-Order Representation in Partial-Order Style

Beginning with the view of planning as satisfiability problem [KS92] and the introduction of planning graphs [BF95], the most powerful planners are based on a total-order plan representation. Despite this fact, it is well known that total-order planners can have an exponentially larger search space than partial-order ones [MBD94]. Moreover, the search space of a parallel total-order representation can be even larger.

The reason for the large number of totally ordered plans compared to partially ordered ones is their commitment to a specific order of unrelated actions. Parallel total-order representations have even more possible orderings than linear ones: They have 'parallel' as a third ordering relation. The search space of a partial-order and a total-order planner are related. A *totalization* of a partially ordered plan is a total order over the plan's actions that is consistent with the existing partial order.[2]

In order to construct a totally ordered search space which is similar in size than the corresponding partial one, we partition the set of totally ordered plans such that its representative set is similar to the set of partially ordered plans. The **top** action constraint allows a total-order planner to eliminate all plans besides the representative set. It is related to the orderings introduced by a partial-order planning algorithm.

According to [BW94], a simple partial-order planner inserts an action $a_1$ together with the ordering constraint $a_1 \prec a_2$ into a plan $P$ if $a_1$ adds an unsupported precondition $f$ of $a_2$. In addition, one of $a_3 \prec a_1$ or $a_2 \prec a_3$ is added if an action $a_3$ of $P$ deletes $f$ and could be ordered between $a_1$ and $a_2$. All other action remain unordered. The planner starts with a plan containing $a_0$, $a_\infty$ and $a_0 \prec a_\infty$, where $a_0$ adds all initial facts and $a_\infty$ has the goal facts in

---

[2] This definition is similar to *linearization* in [MBD94], but the word linear in combination with parallel actions seems to be awkward.

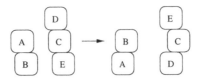

**Fig. 3.** Example for the relation between a total-order and a partial-order search space. Flipping the right tower requires at least three time steps. The two actions for the left tower are $a_1 = \text{move}(A, B, Table)$ and $a_2 = \text{move}(B, Table, A)$, and there are three different ways to assign them to the three time steps. Action $a_1$ adds the fact $\text{on}(A, Table)$ which is required by $a_2$, so that a partial-order planner orders them with $a_1 \prec a_2$. The corresponding **top** constraints require $a_1$ in the first time step and $a_2$ in the second, so that they eliminate the other two solutions.

its preconditions. A plan is a solution if it supports all preconditions of all of its actions.

We expect actions which are unordered in the partially ordered plan to be in parallel in its totally ordered counterpart. Unfortunately, actions which have an add/del conflict caused by unnecessary facts can remain unordered. For this reason we have to extend the above algorithm by the following operator: As part of the introduction of actions, the planner adds one of the ordering constraint $a \prec a'$ or $a' \prec a$ if $a$ and $a'$ are not ordered otherwise and they have an add/del conflict. A total-order planner using **top** action constraints has a solution space similar to the one of this extended planner. Figure 3 gives an example.

**Theorem 1.** *For every partially ordered plan $P_{po}$ found by the extended algorithm there is exactly one totally ordered plan $P_{to}$ which fulfills the following conditions:*

1. *In case that two actions $a, a'$ in $P_{po}$ are ordered with a constraint $a \prec a'$, $a$ is placed in an earlier time step of $P_{to}$ than $a'$.*
2. *For all actions $a'$ of $P_{to}$ (except the ones in the first time step) we have: If $a'$ is in time step $t$, there is an action $a$ in the time step directly preceding $t$ such that the constraint $a \prec a'$ is in $P_{po}$.*

It is obvious that every partially ordered plan has a totalization which fulfills these conditions and vice versa. Also there can be no partially ordered plan which has two different totalizations of this kind. This would require an action $a$ to have two possible time steps but $a$ would have to observe the same ordering relations for both totalizations. This would violate condition two. The corresponding action constraint is $\forall a' \in P. (\nexists a.\ a \text{ is before } a' \text{ in } P) \vee (\exists a.\ a \circ a' \in P \wedge a \xrightarrow{\text{top}} a')$, where $a \xrightarrow{\text{top}} a'$ is defined as $\text{pre}(a) \cap \text{del}(a') \neq \emptyset \vee \text{del}(a) \cap \text{add}(a') \neq \emptyset \vee \text{add}(a) \cap \text{del}(a') \neq \emptyset \vee \text{add}(a) \cap \text{pre}(a') \neq \emptyset$. The relation $\xrightarrow{\text{top}}$ holds for all pairs of actions which are ordered by the partial-order planner.

Planners can use a subset of **top**-relations for an action constraint. An example is the domain-dependent heuristic 'simplifying' presented by Kauts and

**Table 1.** Problem domains and instances used in this paper. The `blocksworld` domain does not have a robot arm. The instances bw_large.a and bw_large.c are well known from the literature. logistics.d is an instance of the logistics-typed-length domain, taken from the BLACKBOX distribution and logistics.e is an extension of it. The $D^mS^{2*}$ domain is designed by Barret and Weld. The table lists the number of actions and facts of these domains, the minimal length of a parallel solution, and the minimal number of actions of that solution. Note that numbers marked with a '*' are the best of our knowledge.

| domain | instance | nb actions | nb facts | optimal length | optimal size |
|---|---|---|---|---|---|
| blocksworld | bw_large.a | 648 | 90 | 4 | 10 |
| | bw_large.c | 3150 | 240 | 8 | 18 |
| logistics | logistics.d | 7180 | 378 | 14 | 73* |
| | logistics.e | 14244 | 538 | 14 | 87* |
| $D^mS^{2*}$ | sfacts40 | 81 | 122 | 81 | 81 |

Selman [KS98]. The set of constraints resulting from this heuristic is subsumed by the set of **top** constraints.

Similar to the parallel case, **top**-style constraints can be used to eliminate unnecessary orderings in linear plans. Here, actions have at most one direct predecessor, so that the second condition of Theorem 1 does not apply. Instead, we assign an index to actions and disallow pairs of actions $a, a'$ for which the following holds: (1) $a$ is on an earlier time step than $a'$, (2) there are no actions $a_1, \ldots, a_n$ with $a \prec a_1 \prec \cdots \prec a_n \prec a'$ and $a \xrightarrow{top} a_1 \xrightarrow{top} \ldots \xrightarrow{top} a_n \xrightarrow{top} a'$, and (3) $\text{index}(a) > \text{index}(a')$. A linear planner using forward search, like TLPLAN [BK96], can use linear **top** constraints easily by maintaining a list of actions which can still meet condition (2) according to the definition of the **top** relation.

# 6 Results and Conclusions

We demonstrate the presented action constraints as part of the planner PROBA-PLA on several large instances of the `blocksworld`, `logistics`, and $D^mS^{2*}$ domain, as explained in Tab. 1. PROBAPLA is a parallel total-order planner which analyzes a planning domain to generate state and action constraints, [Sch97] describes an early version. PROBAPLA builds a bounded plan space representation which is searched via an encoding into CNF. The current version of PROBAPLA requires the specification of the desired plan length together with the planning problem. For this reason we cannot exemplify the effect of the action constraints on the solution length. In addition to **top** and **rsa**, PROBAPLA uses the simple action constraint **const** to eliminate inapplicable actions and generates SMF state constraints, as described in [Sch98].

Table 2 gives the time required for the calculation of the constraints and the number of patterns they find in the planning problems. The `blocksworld` domain has a high interconnection: Every block can be on top of each other and each action affects three blocks. This results in a high number of **top** relations and replaceable sequences together with a high calculation time of the patterns.

**Table 2.** Table presenting the time to generate the action constraints for the problem instances and the number of the corresponding patterns. Time is given in seconds, measured on a SUN Ultra 10 with 300 MHz and 196 MB. Columns labeled #top, #rsa, and #safe give the number of top-relations, replaceable sequences, and replaceable sequences which are solution-safe, respectively.

| instance | t | #top | t | #rsa | #safe |
|----------|-----|--------|-------|-------|-------|
| bw_large.a | 0.3 | 128664 | 4.3 | 8568 | 1008 |
| bw_large.c | 9.0 | $1.2 \cdot 10^6$ | 177.9 | 79170 | 5460 |
| logistics.d | 0 | 4620 | 0.3 | 150 | 150 |
| logistics.e | 0 | 7680 | 0.7 | 276 | 276 |
| sfacts40 | 0 | 4840 | 0 | 0 | 0 |

**Table 3.** Performance of PROBAPLA and BLACKBOX version 3.4. Times for PROBAPLA combine the time to generate the CNF formula and the time to solve this formula with satz-rand version 2.0. Both planners have the optimal plan length as input and the numbers are averaged over 100 runs. Columns labeled size give the total number of actions. In case a planner did not find a solution in every attempt, the column succ gives the percentage of successful runs. PROBAPLA is tested three times: without top and rsa, with top, and with both. The largest $D^m S^{2*}$ instance solvable by BLACKBOX is sfacts13 taking 38.1 seconds. PROBAPLA solves this instance with 27 actions and 41 facts in 1.7 seconds.

| | PROBAPLA | | | with top | | and with rsa | | BLACKBOX | | |
|----------|-------|------|------|-------|-------|-------|------|-------|------|------|
| instance | t | size | succ | t | size | t | size | t | size | succ |
| bw_large.a | 1.0 | 10.9 | | 1.3 | 10.6 | 5.1 | 10 | 2.2 | 10 | |
| bw_large.c | 6.4 | 30.5 | | 16.2 | 30.1 | 185.4 | 29.7 | — | | 0% |
| logistics.d | 9.6 | 105.5 | | 7.3 | 102.4 | 7.2 | 77.3 | 44.6 | 98.0 | 95% |
| logistics.e | 27.7 | 125.0 | | 70.1 | 123.2 | 54.0 | 97.6 | 104.7 | 116 | 56% |
| sfacts40 | 218.4 | 81 | 99% | 142.8 | 81 | 142.8 | 81 | — | | 0% |

As explained in Fig. 2(b), the replacement by moves from blocks to blocks is not safe, so that the fraction of solution-safe rsa is small.

The actions of the logistics domain are much less coupled, eg. airplanes can land on airports only but not on trucks. This results in a small number of top relations and replaceable sequences and their calculation takes much less time. Note that all rsa of the logistics domain are solution-safe. For the $D^m S^{2*}$ domain the number of top relations is high compared to the number of actions. It is a hard domain and PROBAPLA can solve only small problem instances, for which the calculation of the top relations is fast. This domain does not have replaceable action sequences and the search for rsa fails quickly.

Table 3 shows the trade-off between the time required to find a plan and the use of the action constraints rsa and top. In addition, it compares PROBAPLA to the planner BLACKBOX. The high interconnection in the blocksworld domain and the small number of solution-safe replaceable action sequences makes these actions constraints a bad choice for its instances: The increase in quality is minimal compared to the increase in planning time. This is different for the logistics domain. For the longer instance both top and rsa reduced the size of the solutions by 22% while the total run time was less than doubled. For problems of the $D^mS^{2*}$ domain, a solution with optimal length has optimal size, too. The top constraint reduces the run time by one third.

These results show that the application of action constraints is not favorable for all planning domains: The designer of a planning system has to trade their potential effects against their time requirements. The rsa constraint shows the difficulties of finding the right choice: It results in little improvement for the blocksworld domain but performs much better for logistics. Nevertheless the replacement of action sequences is favorable if optimal plans are of high value or the computational effort of computing the constraints is small. The latter can be the case after the application of preprocessing techniques which reduce the number of actions, like the RIFO method [NDK97]. Furthermore Ambite and Knoblock [AK97] showed that rsa performs well as post-processing method for blocksworld.

In this paper we presented pattern-based action constraints as a mean to improve the quality to reduce the run time of planning systems. We explained the connections between the use of action constraints and the optimality of plans, showed that action constraints can be based solely on the specifics of a planning formalism, and gave examples how to calculate and use them. We presented two action constraints in detail and demonstrated their potential for improvements in planning speed and plan quality with the parallel total-order planner PROBAPLA. For both constraints we explained their use with linear total-order planners. Finally we showed that the effects of an action constraint varies for different planning domains.

# References

[AK97]   Joseó Luis Ambite and Craig A. Knoblock. Planning by Rewriting: Efficiently Generating High-Quality Plans. In *Proceedings of the National Conference on Artificial Intelligence (AAAI-97)*, pages 706–713, 1997.

[BF95]   Avrim L. Blum and Merrick L. Furst. Fast planning through planning graph analysis. In *Proceedings of the International Joint Conference on Artificial Intelligence (IJCAI-95)*, pages 1636–1642, San Mateo, CA, 1995. Morgan Kaufmann.

[BK96]   Fahiem Bacchus and Froduald Kabanza. Using Temporal Logic to Control Search in a Forward Chaining Planner. In *M. Ghallab and A. Milani, editors, New Directions in Planning*, pages 141–153. IOS Press, 1996.

[BW94]   Anthony Barret and Daniel S. Weld. Partial-Order Planning: Evaluating Possible Efficiency Gains. *AI magazine*, 67:71–112, 1994.

[Byl94]  Tom Bylander. The Computational Complexity of STRIPS Planning. *Artificial Intelligence Journal*, 69:165–204, 1994.

[FL98]   Maria Fox and Derek Long. The Automatic Inference of State Invariants in TIM. *Journal of Artificial Intelligence Research*, 9:367–421, December 1998.

[GS98]   Alfonso Gerevini and Lenhard Schubert. Inferring State Constraints for Domain-Independent Planning. In *Proceedings of the National Conference on Artificial Intelligence (AAAI-98)*, pages 905–912, 1998.

[KS92]     Henry Kautz and Bart Selman. Planning as Satisfiability. In B. Neumann, editor, *Proceedings of the European Conference on Artificial Intelligence (ECAI)*, pages 359–363. John Wiley & Sons, Ltd, 1992.

[KS96]     Henry Kautz and Bart Selman. Pushing the Envelope: Planning, Propositional Logic, and Stochastic Search. In *Proceedings of the National Conference on Artificial Intelligence (AAAI-96)*, pages 1194–1201, Portland, OR, 1996. Morgan Kaufmann, San Mateo, CA.

[KS98]     Henry Kautz and Bart Selman. The Role of Domain-Specific Knowledge in the Planning as Satisfiability Framework. In *Proc. 4th Intl. Conference on Planning in AI*, June 1998.

[MBD94]    Steven Minton, John Bresina, and Mark Drummond. Total-Order and Partial-Order Planning: A Comparative Analysis. *Journal of Artificial Intelligence Research, pages 227–262, December 1994.*

[NDK97]    Bernhard Nebel, Yannis Dimopoulos, and Jana Koehler. Ignoring Irrelevant Facts and Operators in Plan Generation. In *Proceedings of the European Conference on Planning*, pages 338–350, 1997.

[Sch97]    Ulrich Scholz. Planning by Local Search. Diplomarbeit, Technische Universität Darmstadt, Fachbereich Informatik, Alexanderstraße 10, 64283 Darmstadt, Germany, December 1997.

[Sch98]    Ulrich Scholz. Strategien zur Domänenanalyse. In 12. Workschop *"'Planen und Konfigurieren'" (PuK '98), Bericht tr-ri-98-193, Reihe Informatik*, pages 17–22, FB Mathematik/Informatik, Warburgerstraße 10, 33098 Paderborn, Germany, April 1998.

# Least Commitment on Variable Binding in Presence of Incomplete Knowledge

R. Barruffi, E. Lamma, P. Mello, and M. Milano

DEIS, University of Bologna, Viale Risorgimento 2, 40136 Bologna, Italy,
Fax: 0039 051 6443073
Email: {rbarruffi, elamma, pmello, mmilano}@deis.unibo.it

**Abstract.** Constraint Satisfaction techniques have been recognized to be effective tools for increasing the efficiency of least commitment planners. We focus on least commitment on variable binding. A constraint based approach for this issue has been previously proposed by Yang and Chan [21]. In this setting, the planning problem is mapped onto a Constraint Satisfaction Problem. Its variables represent domain objects and are defined on a finite domain of values; constraints remove inconsistent values from variable domains through constraint propagation. In many applications, however, it is not always convenient, if possible at all, to know in advance all objects belonging to variable domains. Thus, domain values should be retrieved during the plan construction only when needed. The interesting point is that data acquisition for each variable can be guided by the constraint (or the constraints) imposed on the variable itself, in order to retrieve only consistent values. For this purpose, we have extended a Partial Order Planner performing least commitment on variable binding. This extension can cope with incomplete knowledge. We use the Interactive Constraint Satisfaction framework defined in [12] in order to exploit the efficiency deriving from constraint propagation and the possibility of acquiring the domain knowledge during the plan construction. Experimental results and comparisons with related approaches show the effectiveness of the proposed technique.

**Keywords:** Partial Order Planning, Least commitment, Variable Binding, Incomplete Knowledge, Interactive Constraint Satisfaction Techniques

## 1 Introduction

*Least commitment* planning [19] is a technique for deferring not essential decisions during the plan construction. It has been proved to be extremely useful since it avoids to prematurely commit to wrong choices, consequent failures and backtracking steps. In this paper, we focus on least commitment on variable binding activity which allows the planner to efficiently work with action schemata without prematurely commit to any particular action instance [9,21].

Constraint Satisfaction (CS) techniques [8] have been recognized as effective tools for performing least commitment since they allow for active decision

S. Biundo and M. Fox (Eds.): ECP-99, LNAI 1809, pp. 159–171, 2000.

postponement [9,10,20]. In fact, not only can constraint consistency be checked once decisions are made, but constraint propagation can be used in order to avoid inconsistent choices. *Active postponement* can be achieved by mapping the planning problem into a Constraint Satisfaction Problem (CSP).

A CSP is defined on a set of variables $\{v_1, v_2, \ldots, v_n\}$ ranging on a domain of possible values $\{d_1, d_2, \ldots, d_n\}$. Variables are linked by constraints $c(v_{1_i}, \ldots, v_{k_i})$ that define combinations of values that can appear in a consistent solution.

Several works have shown that constraint based techniques improve performances of planners, see for example [10,13,18,20]. In [21] Yang and Chan show the efficiency gain that comes from actively postponing action variable binding by means of constraint satisfaction techniques. They map the planning problem onto a CSP whose variables are those appearing in the action schemata introduced into the plan, and represent domain objects; variable domains contain all the alternative choices; variables are linked by no-codesignation and codesignation constraints. Such an approach is developed in the context of a traditional planner, SNLP [14], which is based on the Close World Assumption [19]. SNLP assumes that it has a complete and consistent knowledge of the initial state of the world. Thus, standard CS techniques are considered where all variable domain values are available at the beginning of the plan construction algorithm.

In many application, however, it is not convenient if possible at all, to retrieve all object values before starting the planning algorithm [5,6,17]. Consider for instance a planner working in a distributed system environment and assembling system configuration plans. The knowledge about the underlying system resources is huge and complex. Thus, loading all the information and keep it consistent is difficult. Variable domains would become huge and constraint propagation algorithms would not be particularly effective since their complexity depends on variable domain size. In addition, in order to generate a plan satisfying a particular goal, we only need to know relevant data.

Consider, as an example, a variable representing a file $F$ situated on a directory $D$ of machine $M$. The traditional approach would consider as $F$ domain the set of all files present in all directories (representing the domain of $D$) on all machines (representing the domain of $M$). Thus, before starting the plan construction algorithm, a data acquisition procedure should retrieve **all** files in the system in order to build the domain of variable $F$. Clearly, this is a computational expensive task and, in general, lot of useless data is retrieved.

A more efficient approach only retrieves values for the domain of $F$ once the machine and the directory of interest are selected. This results in extremely smaller domains. In fact, being variables subject to constraints, those constraints can be exploited in order to guide the knowledge acquisition and retrieve only consistent values, when possible. In the previous example, variable $F$ is subject to the constraint imposing that it should be in a given directory, i.e., *directory*$(F, dir)$. The acquisition of domain values for $F$ can be guided by this constraint and limited to consistent values. Constraints embedding such information gathering mechanism are called *Interactive Constraints* (ICs).

For this purpose, we use the Interactive Constraint Satisfaction Problem (ICSP) framework, defined in [12], where variable domains can be partially or completely unknown at the beginning of the computation. We embed the corresponding constraint solver in a planner which becomes able to cope with incomplete knowledge.

We apply this solution for handling variable binding activity in a Partial Order Planner (POP) [19]. We treat action preconditions and final goal conjuncts as Interactive Constraints (ICs). Thus, before looking for an action satisfying them, the planner tests them in the initial state of the system. It does not need to have a complete representation of the world since the ICSP framework allow it to directly query the system. Thus, when the needed information is unavailable, IC propagation retrieves variable domain values consistent with constraints. Otherwise, ICs prune inconsistent values from known domains. Finally, when variables are already instantiated ICs propagation results in a boolean test. Note that when variable domains are known, our approach is equivalent to that of Yang and Chan [21]. It is worth noting that the information gathering process is transparent to the planner which relies on a constraint solver without taking into account declarative sensing actions.

The structure of this paper is as follows: in section 2 some previous approach are recalled. In particular the work of Yang and Chan is introduced as the background of our work. Then we introduce the main motivations of this paper in section 3. Section 4 provides some definitions of the ICSP framework. Section 5 defines our planning architecture. We present an example in the field of network management in section 6. Experimental results described in section 7 show the effectiveness of the proposed approach. Conclusion and future work follow.

## 2   Delaying Variable Binding Commitments: Previous Approaches

Traditional generative planners such as SNLP [14] and UCPOP [19] perform least commitment on variable binding activity by maintaining a set of codesignation and no-codesignation constraints. Codesignation constraints state that two variables $X$ and $Y$ must be bound to be the same constant $(X = Y)$; vice versa no-codesignation constraints enforce that the two involved variables assume different values $(X \neq Y)$. When variables get instantiated, constraints are checked in a "generate and test" fashion. In case of failure, a backtracking step is performed. In general, this approach leads to a *thrashing* problem concerning failures in different part of the search space due to the same wrong choice performed before.

FSNLP [21] is an extension of SNLP exploiting constraint satisfaction techniques for two purposes: ($i$) to perform least commitment on the instantiation of action variables, ($ii$) to drive the planner in the binding process, in order to reduce the problem search space and the thrashing problem. If a variable domain contains $n$ values, an eager commitment approach generates n branches of the search tree, whereas the constraint-based method helps in reducing these alternatives. This method is used in the context of a traditional Partial Order Planner (POP) whose main loop is the following:

1. *Termination.* **If** $Correct(plan)$ **then** return $plan$ **else**
2. *Causal link protection.* Given the set of causal links $L$, solve existing threats.
3. *Goal selection.* Choose a pair $\langle Q, A_k \rangle$ from the *Agenda* of open conditions, with $A_k$ belonging to the set of plan actions $A$ and $Q$ precondition of $A_k$.
4. *Action selection.* Choose an action $A_j$ (already in $A$ or newly instantiated) whose effects contain a conjunct $R$ that unifies with $Q$. When the chosen action is the dummy action $Start$, it means that $Q$ is already satisfied in the initial state[1].
5. *Consistency checking.* **If** $A_j < A_k$ is consistent with the set of ordering constraints $O$ **and if** the unification between $R$ and $Q$ is consistent with the binding constraint set $B$, **then:**
6. Update sets $B$, $O$, $L$, $A$ and *Agenda*.
7. **Else** go back to step 4.

Yang and Chan in [21] propose a solution where the planning problem is mapped into a CSP whose variables are those appearing in the action schemata introduced into the plan. Variable domains are represented in the form $Domain(X) = \langle O_1, \ldots, O_n \rangle$ meaning that the variable $X$ can be bound to any of the $O_1 \ldots O_n$ objects of the domain description. Codesignation and no-codesignation constraints in $B$ are propagated on variable domains in order to actively drive the binding activity, instead than passively check their consistency after any instantiation. The above algorithm is extended in two points:

- In step 1, the boolean function $Correct(plan)$ checks whether every action precondition has a causal link and if every causal link is safe. Here, an additional termination routine $CSP(plan)$ checks if an assignments for all variables in the plan exists consistent with constraints in $B$. If both $Correct(plan)$ and $CSP(plan)$ succeed, then $plan$ is returned.
- In step 4, when establishing a precondition $p(X)$ of an operator $\alpha$, a successor plan is generated with an existing or new operator $\beta$ containing an effect $p(Y)$ unifiable with $p(X)$. Such a successor plan contains new ordering $(\beta < \alpha)$ and binding $(X = Y)$ constraints, and is generated if and only if the intersection of $Domain(X)$ and $Domain(Y)$ is not empty. This intersection becomes the domain of both variables $X$ and $Y$. If $\beta$ is the dummy action $Start$, the domain of $X$, in the successor plan, is represented by the intersection of all literals satisfying $p(X)$ in the initial state and the original domain of $X$. Thus, domains are reduced thanks to constraint propagation and commitments are delayed as much as possible.

Another approach which uses the CSP paradigm in order to perform active postponement on planning decisions is that described in [9]. In this approach, as far as open condition achievement is concerned, the CSP variables are represented by the open conditions themselves whose domain contains possible actions that achieve the open condition. Since domains only contain actions already introduced in the plan, not all the domain values can be know from the beginning.

---

[1] The effects of $Start$ represent all the true facts in the initial state.

Thus, the planning problem is mapped onto a Dynamic Constraint Satisfaction Problem (DCSP) [15] handling dynamic changes of the constraint store such as addition and deletion of values and constraints.

Unfortunately, in general, constraint propagation algorithms are not powerful enough to reduce variable domains to a single consistent value. Thus, when no further propagation is possible, some commitments are performed in order to go on in the computation, possibly resulting in failures. However, the described methods allows the planning algorithm to explore a drastically limited number of branches (only one in the best case).

## 3    Coping with Incomplete Knowledge

The approach proposed by Yang and Chan considers the information about the world to be known and static (Close Word Assumption [19]). Traditional constraint satisfaction techniques are then applied that assume that all variable domains are completely known at the beginning of the plan construction.

In many application domains, however, not all the values that a variable can assume are known in advance. In some cases, it might be inconvenient to retrieve them all because data acquisition can be expensive and, on the other hand, only a small subset of data might be significant for the problem to be solved. Also, some domain objects can be created or destroyed during the system's life and it is not convenient to retrieve all of them and handle frequent changes. Consider, for instance, a planner in charge of computing configuration plans in a networked computer system [4,11]. The domain knowledge is composed by many different types of objects (e.g., machines, users, printers, services, files, processes), their attributes (e.g., sizes, availability, location) and relations among them (e.g., user $u$ is logged on machine $m$). In this case, we have an enormous amount of knowledge to store. In addition, this information can change during the system's life due to actions performed on the objects (e.g., removing or creating files, connecting or disconnecting machines, adding or deleting users, starting or killing processes). Thus, we need to keep this knowledge up-to-date.

As an example, consider data on users logged in the distributed system. In a standard constraint-based approach, the planner has to collect all the users having an account on all the machines, create a unique list of users and use the resulting list as the domain of variable $U$. Finally, it will prune from the domain of $U$ those users which are not currently logged on any machine. Suppose also that we are interested in a particular machine $m$, the planner will *a posteriori* prune the domain of $U$ by removing users not logged on $m$.

More efficiently, we can start the computation with an unknown domain for $U$ represented by a variable, $Domain(U) = D_U$ and as soon as the information on users logged on $m$ is needed, perform a data acquisition to retrieve only relevant information. As a consequence, the domain of $U$ is significantly smaller if compared with that containing all users having an account.

To this purpose, we propose an extension of a constraint-based planner able to start the plan construction process even when the domain knowledge is not complete. The planner makes the Open World Assumption [5,16], i.e., data not contained in the planner initial state is unknown and needs to be acquired on demand. In this setting, only significant data will be retrieved in order to avoid the acquisition and processing of useless information.

As FSNLP [21], our planner exploits CS techniques for the variable binding process. While in [21] the planning problem is mapped onto a CSP, we need to map the planning problem to a CSP extension, called Interactive CSP (ICSP), defined in [12] and recalled in section 4, for coping with incomplete knowledge. ICSP variables are those appearing in the action schemata introduced into the plan. Variable domains can be either (*i*) completely known, containing objects which can be assigned to the corresponding variable; (*ii*) partially known, containing some values already at disposal and a variable representing intensional future acquisitions; (*iii*) totally unknown, when no information has already been retrieved for the variable. Constraints are represented by the usual nocodesignation and codesignation constraints imposed among problem variables during the planning process. In order to retrieve domain data, also predicates defined as action preconditions and final goal conjuncts are treated as constraints. They are called *Interactive Constraints* (ICs) and are in charge of "completing" variable domains by retrieving consistent values from the underlying system. ICs behave as standard constraints when they link completely known domain variables. Clearly, since not all the predicates can be used to sense the possible values for arbitrary variables, only selected preconditions are associated with ICs and used for sensing.

Suppose the domain contains the following action schemata: *Logoff(user:U, machine:M)* with a precondition *logged(U,M)*. This condition represents an IC linking variables $U$ and $M$ which is satisfied by the couple $(u, m)$ if the user $u$ is logged on the machine $m$. $U$ domain represents all the users in the system, while $M$ domain contains all the machines available in the networked system. If $U$ and $M$ are associated with a completely known domain, the propagation of the IC *logged(U,M)* removes couples of inconsistent values. Otherwise, it retrieves values consistent with the constraint itself once the machine $m$ (or the set of machines) is known. The IC embeds a data retrieval procedure, e.g., the UNIX command `finger@m`, that returns only users logged on $m$. Note that the resulting domain for $U$ is the same as that obtained by a standard constraint based approach after constraint propagation.

The method we propose embeds knowledge acquisition activity into the constraint solving mechanism, thus simplifying the planning process in two points. First of all, while existing OWA planners perform knowledge acquisition by adding sensing action to the plan [1,7,11], we provide a sensing mechanism where no further declarative action is needed apart from the causal actions. Also, only information significant for the planner is retrieved. As a consequence variable domains are significantly smaller than in the standard case.

## 4   Interactive Constraints Satisfaction Techniques

An *Interactive CSP* framework has been defined [12] where interactive constraints link variables ranging on partially or completely unknown domains. As in traditional CSPs, a solution to the ICSP is found when all the variables are consistently instantiated. A formal definition of the ICSP framework follows.

**Definition 1.** *An ICSP is a triple $(V, D, C)$ where $V$ is a finite set of variables $X_1, \ldots, X_n$, $D$ is a set of interactive domains $D_1, \ldots, D_n$ and $C$ is a set of interactive constraints. An interactive domain $D_i$ of a variable $X_i$ is $D_i = [Known_i \cup UnKnown_i]$ where $Known_i$ represents the set of known values for variable $X_i$, and $UnKnown_i$ is a domain variable itself representing (intensional) information which is not yet available for $X_i$. Both $Known_i$ and $UnKnown_i$ can possibly be empty[2]. Also $\forall i\ Known_i \cap UnKnown_i = \emptyset$.*
*A (binary) interactive constraint $IC$ on variables $X_i$ and $X_j$, i.e., $IC(c(X_i, X_j))$, defines a possibly partially known subset of the cartesian product of variable interactive domains $D_i \times D_j$ representing sets of possible assignments of values to variables.*

Given an interactive constraint $IC(p(X, Y))$, its operational behaviour is the following:

1. **If** both variables are associated to a partially or completely unknown domain, the constraint is suspended;
2. **else**, if both variables range on a completely known domain, the constraint is propagated as in classical CSPs;
3. **else**, if variable $X$ (resp. $Y$) ranges on a fully known domain and $Y$ (resp. $X$) is associated to a fully unknown domain a knowledge acquisition step is performed; this returns either (i) a finite set of consistent values representing the domain of $Y$ (resp. $X$), or (ii) an empty set representing failure.
4. **else**, if variable $X$ (resp. $Y$) ranges on a fully known domain and $Y$ (resp. $X$) is associated to a partially known one, $Y$ (resp. $X$) domain is pruned from values non consistent with $X$ (resp. $Y$). If $Y$ domain becomes empty a new knowledge acquisition step is performed for $Y$ driven by $X$.

Concerning the data acquisition (step 3i), in general, we can retrieve all values consistent with the IC, while in other cases it could be more convenient to retrieve just one value and go on with the computation. In case of failure, the backtracking mechanism leads to the acquisition of further consistent values. In the application domain considered, we have used the technique of acquiring all values consistent with the IC. The investigation of the other approach is subject of current work.

The ICSP is a general framework which can be used in many applications. It is particularly suited for all the applications that process a large amount of constrained data provided by a lower level system (see for instance [2]).

## 5   An ICSP-Based Planner

We embed a Constraint Solver suitably extended for dealing with Interactive Constraint Satisfaction in a Partial Order Planner. Like all the POP algorithms,

---

[2] When both are empty an inconsistency arises.

our planner starts by generating the *NullPlan* with the dummy actions *Start* and *End*. *Start* has no preconditions and its postconditions consist of a set of logic predicates, called *InitialState* (hereafter referred to as *IS*), representing the current state of the world. The *End* action has no effect, and its preconditions are the conjuncts of the final goal of the planning problem. The final goal conjuncts are initially put into an *Agenda* representing the set of open conditions the planner has to satisfy. During the planning process, as soon as a new action is introduced into the plan, its preconditions are added to the *Agenda*.

POP algorithms, as well as their constraint-based version, FSNLP, consider *IS* as a complete and static set of information representing **all** the true facts in the real world. In our approach, we suppose that the initial state represents partial information, thus *IS* is a "cache" storing the information the planner is aware of. When, during the plan construction, the planner needs to test for which objects a particular property $p$ holds, it tests the initial state via the interactive constraint associated to $p$. Thus, as soon as an open condition $p(X, Y)^3$ is selected from *Agenda*, the constraint solver will test if there exists at least one value for $X$ and $Y$ that already satisfies $p$ in the initial state (i.e., there is no need to add any action to the plan in order to achieve $p$). Such test is performed by means of ICs propagation whose operational behaviour has been described in section 4. If the needed information is not available in *IS*, a knowledge acquisition from the real system is performed, guided by the constraint itself. The retrieved data are stored in *IS* and become available for future requests. Note that it is often not possible that only *consistent* information with constraints is retrieved and loaded in *IS*. In fact, the available sensors might not always be able to weed out objects which do not satisfy a given constraint $p(X)$. However, even if more objects are retrieved in *IS*, the variable $X$ domain will only contain objects that are consistent with $p$ thus simplifying the CSP resolution. On the other hand, by associating appropriate sensors to the ICs, it is still possible to reduce the retrieved information to a minimum. At the beginning of the computation, *IS* can either be empty or contain only static information, i.e., data which is not frequently changed or updated. It is worth noting that the acquired information is always referred to the initial state since we assume that no action changes the state of the system during the plan construction.

If the constraint fails (i.e., a variable's domain becomes empty), it means that the corresponding precondition is not satisfied in the initial state, thus, the planner has to select an existing action (different from *Start*), or a new operator, to achieve the goal.

The main steps of the planning algorithm are the same of FSNLP (see section 2) apart from the step 4 (i.e., *Action Selection*) that becomes as follows:

4.1 Call the Interactive Constraint associated with Q, (IC(Q)) in order to verify Q in the initial state:

    **if** IC(Q) fails, (i.e., *Start* does not satisfy Q)

---

[3] So far only binary predicates (i.e., constraints) are considered. The possibility of extending the ICSP framework to ternary and n-ary constraints is subject to current work.

**then** non-deterministically choose an action $A_j$ (already in $A$ or newly instantiated)
whose effects contain a conjunct $R$ that unifies with $Q$.
**else** $A_j = Start$.

4.2 Retrieved information is loaded in $IS$ in terms of binary predicates.

The reason why we keep a cache $IS$ of facts retrieved by the real system is to avoid *sensor abuse* [5], i.e., we want to reduce as much as possible the number of interactions with the real system. Thus, when possible, constraint propagation infers information from $IS$ rather than running low level functions. This mechanism can be improved by expliciting the information that $IS$ contains **all** the possible values for a variable. This reasoning methodology is based on the Local Closed Word ($LCW$) assumption described in [5,7,6].

We argue that the ICSP framework can be applied also to partially solve the problem of dynamic changes. When we cope with frequently updated data, a good approach can be delaying as much as possible data acquisition in order to increase the probability to have consistent information. Obviously, we still can retrieve data which can change soon after the retrieval, thus we do not completely solve the problem of dynamic changing knowledge. A further step in this direction could be to classify system resources in accordance with their changing frequency. In this way, ICs on highly volatile data can be delayed until execution time. Clearly, we run the risk of producing incorrect plans, but we can interleave planning and execution in a more *reactive* fashion. This topic is subject of a current investigation.

The planner has been implemented by using the finite domain library of the constraint logic programming language $ECL^iPS^e$ [3] properly extended to cope with the interactive model [2]. Unlike in FSNLP [21], we use a third-party constraint solver to propagate constraints rather than the planner itself. Thus, the constraint solver is in charge of removing inconsistent combinations of assignments and performing knowledge acquisition.

# 6     Examples in the Field of Networked Systems

In order to clarify the different behaviour of our planner and FSNLP, we consider a simple example of plan construction in the field of computer system management (see Fig. 1).

Suppose the goal is to print the job $j$ from a given machine $m$ by using a laser printer (i.e., goal: **laserprinted(j,m)**). Let us suppose that in the real world there are $n$ printers in the system; $m$ is connected to three printers $p1$, $p2$ and $p3$; there are two laser printer $p4$ and $p5$ (none of them is connected to $m$); $p1$, $p2$, $p3$, $p4$ and $p5$ are all on line. In the domain theory there are the following two action schemata:

**A1** *ConnectPrinter(printer : P1, machine : M)*
 **Preconditions:** *online(P1)*
 **Postconditions:** *connected(P1, M)*.

**A2** *LaserPrint(printer : P2, integer : JobID, machine : M)*
    **Preconditions:** *type(P2, laser), connected(P2, M).*
    **Postconditions:** *laserprinted(JobID, M).*

The former connects an on line printer to a given machine. The latter is devoted to print a given job from a given machine by means of any connected laser printer.

FSNLP would collect **all** the $n$ printers in the domain description in the printer variable domain. Once selected the action $A2$ in order to print $j$, the codesignation constraints $M = m$ and $JobID = j$ are stated and its two preconditions $(type(P2, laser), connected(P2, m))$ become the current goals. The printer domain becomes empty (i.e., no printer satisfies both conditions in the initial state). An $A1$ action is needed in order to connect a laser printer to $m$.

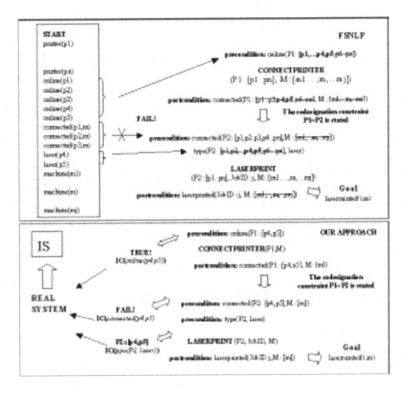

**Fig. 1.** An Example of plan construction

In our framework the printer's domain is initially fully unknown. When $A2$ is added to the plan, the interactive constraint $connected(P2, m))$ is propagated. Through the appropriate UNIX sensing command[4], it retrieves **only** consistent printers (i.e., only printers connected to the machine $m$: $p1$, $p2$ and $p3$ in

---

[4] `lpstat -p`

the example depicted in figure 1) along with all the related information. Those information are loaded in *IS*. Then, since none of those printers satisfies the constraint *type*(*P2*, *"laser"*) with *P2* :: [*p1*, *p2*, *p3*], *A1* is selected in order to connect one of the laser printers (i.e., *p4* or *p5*). Since both *p4* and *p5* are on line laser printers, the printer variable will be non deterministically labelled to either *p4* or *p5*.

## 7   Performance Comparison

In this section, we present some comparisons of our planner and FSNLP in real environments. In order to get the following experimental results we have solved a file system management problem by means of both FSNLP and our approach.

As far as our approach is concerned, the real world has been simulated by a file (hereafter referred to as *world*) containing n-ary facts according to the "type" of the object to which the information is referred. For instance, given the object type file, the fact file(*rights*, *name*, *path*, *disk*, *dim*, *type*, *date*) is contained in *world*. Our planner initial state is empty at the beginning of the plan generation process. On the other hand, in order to test the FSNLP approach, a complete initial state description (hereafter called *domain*) has been defined in terms of binary predicates. Computational times (obtained on a Sun Sparc 10) reported below, refer to the planning time (in seconds) necessary to produce one plan. *World* and *domain* are identified by the triple ⟨*n*, *m*, *t*⟩. *n* represents the number

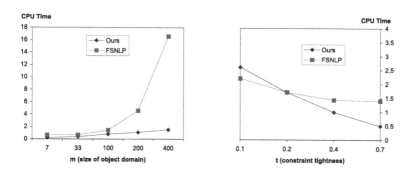

**Fig. 2.** Performance evaluation

of objects (i.e., files, disks, printers, machines and so on), that correspond to the CSP variables; *m* is the size of object domains (e.g., number of files in the system) corresponding to the number of values in the CSP variable domains; finally, *t* represents the constraint tightness, i.e., the inverse of the conditional probability that two values in a constraint are consistent. We have considered real environments with different combinations of those parameters. In the specific example we have varied *t* by changing the number of files satisfying some of the problem constraints. Different $m$ parameters have been obtained by changing the number of file in the file system. In figure 2 on the left, we consider a fixed $t$ (about 0.4) and a varying $m$. We can see that when $m$ grows our planner outperforms FSNLP. This is due to the fact that within FSNLP the involved CSP works on variable domains whose size is proportional to $m$, while in our approach only few values are retrieved from the real world (i.e., those consistent with constraints) and inserted in the domains.

By the same token, if we consider a fixed number of files ($m = 100$) and we vary the constraint tightness, as shown on the right in figure 2, we can see that the more $t$ increases the better our planner performs, since values consistent with the constraints (i.e. the retrieved values) are relatively few with respect to those managed by FSNLP. Instead, for low tightness, the interactive constraint based approach is not convenient since the number of values retrieved (and consequently the domain size) is comparable with that of FSNLP, and we introduce the overhead of managing incomplete information. In these cases, it is more convenient to retrieve only one value (and not all values) consistent with constraints. We are currently investigating this possibility.

## 8    Conclusion and Future Work

We have presented an extension to the work of Yang and Chan [21] for the application of CS techniques to the variable binding process in presence of incomplete knowledge. We have implemented a constraint solver based on the ICSP framework, defined in [12], and embedded it in a POP framework. The ICSP mechanism allows constraint satisfaction algorithms to cope with unknown variable domains by embedding a knowledge acquisition activity in the constraints propagation process.

The main advantages of the extension proposed are that: ($i$) our approach allows the planner to gather needed information without increasing the complexity of the planning problem itself with additional sensing actions; ($ii$) the data acquisition can be limited to useful information thanks to Interactive Constraints that allow the planner to gather only consistent data; ($iii$) ICSP variable domain size is consequently smaller that in standard CSP approaches thus improving efficiency.

We are currently working on how to embed the ICSP model in a reactive planner able to cope with dynamic knowledge and changes, as briefly suggested in section 5.

**Acknowledgments.** Authors' work has been partially supported by Hewlett Packard Laboratories of Bristol-UK (Internet Business Management Department) and CNR, (Project 40%).

# References

1. N. Ashish, C.A. Knoblock, and A. Levy. Information gathering plans with sensing actions. In *Proceedings of the 4th European Conference on Planning*, 1997.
2. R. Cucchiara, M.Gavanelli, E.Lamma, P.Mello, M.Milano, and M.Piccardi. Extending CLP(FD) with interactive data acquisition for 3d visual object recognition. *Proceedings of PACLP'99*, pages 137–155, 1999.
3. ECRC. *ECL$^i$PS$^e$ User Manual Release 3.3*, 1992.
4. O. Etzioni, H. Levy, R. Segal, and C. Thekkath. Os agents: Using ai techniques in the operating system environment. Technical report, University of Washington, 1993.
5. K. Golden. *Planning and Knowledge Representation for Softbots*. PhD thesis, University of Washington, 1997.
6. K. Golden, O. Etzioni, and D. Weld. Xii: Planning with universal quantification and incomplete information. In *Proceedings of the 4th International Conference on Principles of Knowledge Representation and Reasoning, KR'94*, 1994.
7. K. Golden and D. Weld. Representing sensing actions: The middle ground revisited. In *Proceedings of 5th Int. Conf. on Knowledge Representation and Reasoning*, 1996.
8. P. Van Hentenryck. *Constraint Satisfaction in Logic Programming*. MIT Press, 1989.
9. D. Joslin. *Passive and Active Decision Postponement in Plan Generation*. PhD thesis, University of Pittsburgh, 1996.
10. S. Kambhampati. Using disjunctive orderings instead of conflict resolution in partial order planning. Technical report, Department of Computer Science and Engineering Arizona State University, 1996.
11. C.T. Kwock and D.S. Weld. Planning to gather information. Technical report, Department of Computer Science and Engineering University of Washington, 1996.
12. E. Lamma, M. Milano, P. Mello, R. Cucchiara, M. Gavanelli, and M. Piccardi. Constraint propagation and value acquisition: why we should do it interactively. *Proceedings of the IJCAI*, 1999. to appear.
13. J. Lever and B. Richards. parcplan: a planning architecture with parallel actions, resources and constraints. In *Proceedings of 8th ISMIS*, 1994.
14. D. McAllester and D. Rosenblitt. Systematic nonlinear planning. In *Proceedings of the 9th National Conference on AI, Anaheim, CA*, 1991.
15. S. Mittal and B. Falkenhainer. Dynamic constraint satisfaction problems. In *Proceedings of AAAI-90*, 1990.
16. D. Olawsky and M. Gini. Deferred planning and sensor use. In *Proceedings DARPA Workshop on Innovative Approaches to Planning, Scheduling, and Control*, 1990.
17. E. Pryor and P. Collins. Planning for contingencies: A decision-based approach. *Journal of Artificial Intelligence Research*, 4:287–339, 1996.
18. A. Tate, B. Drabble, and J. Dalton. Reasoning with constraints within o-plan2. Technical report, AI Applications Institute Univeristy of Edinburgh, 1994.
19. D.S. Weld. An introduction to least commitment planning. *AI Magazine*, 15:27–61, 1994.
20. Q. Yang. A theory of conflict resolution in planning. *Artificial Intelligence*, 58:361–392, 1992.
21. Q. Yang and A.Y.M. Chan. Delaying variable binding commitments in planning. In *The 2nd International Conference on AI Planning Systems (AIPS)*, 1994.

# Scaling Up Planning by Teasing Out Resource Scheduling

Biplav Srivastava and Subbarao Kambhampati

Department of Computer Science and Engineering
Arizona State University, Tempe, AZ 85287-5406, USA
Email: {biplav,rao}@asu.edu
FAX: (+1) 602-965-2751

**Abstract.** Planning consists of an action selection phase where actions
are selected and ordered to reach the desired goals, and a resource allocation phase where enough resources are assigned to ensure the successful
execution of the chosen actions. In most real-world problems, these two
phases are *loosely coupled*. Most existing planners do not exploit this
loose-coupling, and perform both action selection and resource assignment employing the same algorithm. We shall show that this strategy
severely curtails the scale-up potential of existing planners, including
such recent ones as Graphplan and Blackbox. In response, we propose
a novel planning framework in which resource allocation is teased apart
from planning, and is handled in a separate "scheduling" phase. We ignore resource constraints during planning and produce an abstract plan
that can correctly achieve the goals but for the resource constraints.
Next, based on the actual resource availability, the abstract plan will
be allocated resources to produce an executable plan. Our approach not
only preserves both the correctness as well as the quality (measured in
length) of the plan but also improves efficiency. We describe a prototype
implementation of our approach on top of Graphplan and show impressive empirical results.

## 1  Introduction

Planning comprises of causal reasoning and resource reasoning. Given a domain,
a set of actions to change states in the domain, an initial state and the desired
goal state, the planning problem is to find a sequence of actions (also known
as a plan) such that when it is executed from the initial state, a goal state can
be reached. Causal reasoning ensures that for every action in the plan, its preconditions can be satisfied from the effect of another action preceding it within
the plan. Causal relationships force sufficient orderings among actions to achieve
the goals and furthermore, determine the extent of concurrency[1] possible in a

---

[1] Borrowing from operating system terminology, concurrency refers to the potential of
executing actions in parallel. The parallelism (or lack of it) exhibited in the final plan
is dependent on the actual number of resources available to exploit this concurrency.

S. Biundo and M. Fox (Eds.): ECP-99, LNAI 1809, pp. 172–186, 2000.

plan. Resource reasoning ensures that all the resources needed for the execution of an action are available for allocation without any resource conflicts. A resource conflict occurs when two actions cannot be assigned the same resource, either due to resource characteristics (non-sharable resources) or due to domain characteristics (actions interfere). If resources are scarce, the resource allocation may involve freeing and reallocating the limited resource which can add more ordering relationships among actions and effectively serialize the plan.

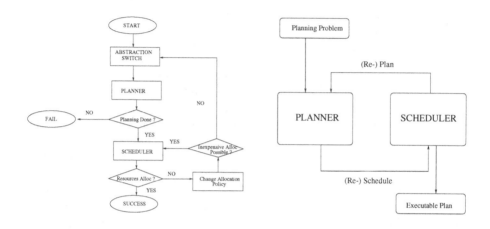

**Fig. 1.** *The left figure shows our unified planning-scheduling framework. In the right figure, we compare it with a generalized plan model for separate planning and scheduling (see discussion in Section 6).*

Most planners do not distinguish between these two forms of reasoning and handle them within the same planning algorithm. Discrete resources (sharable or non-sharable) like robots, trucks and planes have traditionally been straightforwardly handled by logic-based planners like UCPOP[15] and Graphplan in the same way as other objects in the domain. We will show (in Section 2) that this strategy severely curtails the scale-up potential of existing planners, including such recent ones as Graphplan [2] and Blackbox [5]. In particular, these planners exhibit the seemingly irrational behavior of worsening in performance with increased resources. For continous resources like time and fuel, planning systems have additionally employed time/resource map managers to ensure resource consistency (SIPE[17], IxTeT[10], IPP[14], LPSAT[16]). But such an integration explodes the search space for the planner beyond action sets that are minimal with respect to the logical goals. Actions may be added to achieve the resource goals but may not be necessary for logical goals. To handle this, IPP restricts expressivity by avoiding explicit temporal modeling while other planners take a performance drop with slower flaw resolution.

In this paper, we introduce a novel approach where causal-reasoning (planning) is used to generate an abstract plan ignoring all resource conflicts. The abstract plan is then post-processed to allocate the required resources without altering the causal structure of the plan. Separating planning and scheduling is quite the normal practice in project planning scenarios in the industry, where planning is done by the humans, and scheduling is done by a variety of software packages. We are proposing a similar flow – except that both planning and scheduling phases will be automated.

Figure 1 (left) provides a general overview of our approach. Our unified framework accepts a domain description along with optional annotations for resources, finds a plan modulo the choice of resource abstraction, and then allocates resources to produce the correct final plan (if necessary). In this paper, we focus on discrete reusable as well as non-sharable resources. But we argue in Section 6 that the approach can be extended for continous resources as well. Resources are either declared by the domain expert, or deduced through automated methods discussed later. After planning is complete, a scheduler can decide which resources to actually allocate. We have implemented this approach on top of Graphplan algorithm, and the resulting planner is not only more rational in its treatment of resources (i.e., does not worsen with increased resources), but also significantly outperforms Graphplan on several benchmark problems.

There are a number of technical challenges that arise in making our approach work. First, we have to identify resources in a given domain. Second, we have to decide about optimization criteria during scheduling. Third, we need to allocate resources to an abstract plan without transferring the full complexity of planning to the scheduling phase. The planning phase produces an abstract plan of shortest length in terms of number of steps (where each step may contain several concurrent actions)[2]. Our aim in scheduling phase is to use the least number of resources for producing a final plan of the same length as the abstract plan and with minimum number of additional actions and causal relationships, without changing the relative positions of actions. When resource allocation is not possible without increasing the plan length, we currently consider it as a hard resource allocation problem and revert back to traditional planning. Our ongoing work is however directed at relaxing this restriction. In particular, we are addressing the issue of how the scheduling (resource allocation) failure is intimated to the planner, and is then used to guide the planner's replanning effort (see Section 6).

Here is an outline of the rest of this paper. In Section 2 we provide an empirical motivation of the need for separation of planning and scheduling. Section 3 presents a new planning formalism to abstract resources away during planning and allocate them in a separate phase. Section 4 elaborates on the resource scheduling phase. Section 5 presents an empirical evaluation of the effectiveness of our approach. In Section 6, we make some observations on the overall nature of the problem and put our work in perspective. Section 7 discusses related work and Section 8 presents the conclusions.

---

[2] Such a plan may not be the optimal if actions have differing costs, but this is how Graphplan works.

## 2 An Empirical Motivation

To motivate the need for separating resource scheduling, we shall see the behavior of a state-of-the-art planner, Graphplan, in a modified blocks world domain that contains multiple robot hands. If we run Graphplan multiple times on the same problem, while increasing the number of robot hands available, we would expect that the performance would improve with increased resources. Figure 3 (left) shows the performance of Graphplan on the *"shuffle"*[3] problem, where a 6-block stack needs to be shuffled in a symmetric way to form a new stack. Notice that the total planning time, GP-TOT, *increases* quite steeply with the increase in the number of robots. In fact, by providing 8 robots instead of 1 robot, the planning time is slowed down by an order of magnitude! Lest the reader suspect that the increase is just due to the increased cost in constructing a planning-graph, the figure also plots the time for building the planning graph (GP-G) and the time for searching the planning graph (GP-S). We note that *both of them* increase with the number of robots.

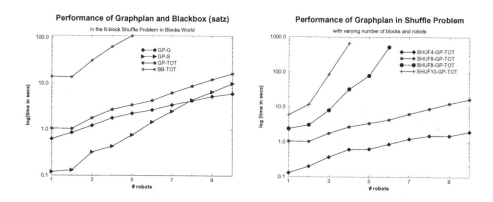

**Fig. 2.** *Performance of Graphplan and Blackbox(satz) on the 6-block* Shuffle *problem with varying number of robots in the left, and comparative performance of Graphplan in* shuffle *problem of 4, 6, 8 and 10 blocks in the right. Performance degrades with increase in size of the domain as well as resources.*

We wanted to further check if the results are consistent when the problem size is scaled independent of number of resources. In Figure 3 (right), we show the performance of Graphplan on *shuffle* problems of 4, 6, 8 and 10 blocks as the number of robots are varied from 1 to 10. We note that planning performance degrades with increase in size of the domain as well as resources.

---

[3] Shuffle problem is the multiple robots version of the 6-block *blocks_facts_shuffle* problem in the GRAPHPLAN system. Later we consider k-block *shuffle* versions also.

This rather counter-intuitive behavior of the planning algorithm can be deciphered once we realize that *every causal failure is being needlessly rediscovered multiple times* with different identities of the robot hands. Specifically, the asymptotic cost of planning in Graphplan like planners is $O(w^l)$, where $w$ is the width of the planning graph and $l$ is the length of the graph. As the resources (e.g. number of robots in blocks world) increase, $l$ tends to reduce while $w$ increases. However, $l$ does not reduce indefinitely, while $w$ does increase monotonically with resource increase. Thus, the net effect is that the performance degrades with increased resources.

To ensure that this behavior is not peculiar to Graphplan, we also experimented with Blackbox [5], which uses SAT techniques for searching the plan graph; and UCPOP [15], a traditional partial-order planner. We found similar behavior in both cases. The plot titled BB-TOT in the left graph in Figure 3 illustrates the behavior of Blackbox.

## 3   A New Planning Formalism

One can reduce both graph expansion and search overheads by abstracting the resources needed by actions during planning and ignoring all interactions between them, thereby obtaining a maximally concurrent plan. This plan will then be post-processed to allocate resources to actions. See Figure 1 (left) for a schematic overview of our approach.

We start with identification of resources. Resources are either declared by the domain expert using a resource specification language that we have developed, or derived through a set of automated methods (under development). One such method involves considering as resources all object types such that no object of that type is mentioned in the goals. Such a procedure can, for example, be used to detect that a robot hand is a resource in the blocks world domain. Although we are currently extending our automated resource identification methods, resource identification methodology is orthogonal to the main focus of this paper–viz., abstracting the identified resources from planning.

Once the resources are identified, and resource abstraction switch is set, planning proceeds in the normal fashion, but with two important differences:

- Dummy values are assigned to resource variables in the initial state such that equivalent[4] resources have the same dummy value.
- Interference relationships (mutexes in the case of Graphplan) between otherwise resource equivalent operators are ignored. Operators may still interfere due to other preconditions/ effects.

---

[4] In complex domains like logistics, all objects of a resource type are not equivalent (e.g. trucks in Boston are not substitutable for trucks needed in Phoenix) and this can be handled either by recognizing them as different equivalence classes within a resource type or as altogether different resource types. For now, we choose the latter and consider all objects in a resource type as equivalent.

If the problem is unsolvable at this stage, we know that resource scheduling is not going to make it solvable. Otherwise we give the resultant plan to the scheduler for resource allocation. An example of the plan generated for the *shuffle* problem, by disregarding inter-resource conflicts during planning, is shown in Figure 4. The plan consists of 10 time steps (levels) with the number of resources left allocated at each level shown in the right column (marked "#Robots").

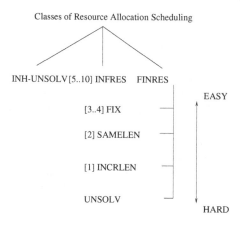

**Fig. 3.** *A Classification of resource allocation instances (with indication of resource quantities that make* shuffle *problem fall into each of the classes).*

The aim of resource scheduling is to assign actual resources to the dummy resource variables, without undoing any causal relations established during planning. A straightforward method for resource allocation is to assign a new or previously freed resource to any action that is involved in a resource conflict. Suppose that this method uses a maximum of $R$ resources. Now for all problems with resources $N \geq R$, the infinite resource assumption holds, and thus resource allocation is quite trivial. If this method fails, the allocation problem is solved through a set of progressively complex methods that modify the abstract plan by: (i) introducing actions to free unnecessary allocations and re-allocate the freed resources when needed again or (ii) moving actions from one step (level) to another (less-constrained) step. We restrict both of these modifications to keep the length of the plan (in terms of number of steps) unchanged, but allow adding of actions at different steps. When we have a hard resource allocation problem - currently interpreted as requiring increase in the plan length, our approach reverts back to traditional planning by resetting the abstraction switch and invoking itself recursively. Our motivation is to keep the cost of scheduling small enough that we are not revisiting the complexity of planning in the resource scheduling phase.

## 4    Scheduling Resources: The Details

We are now ready to delve deeper into the resource scheduling phase. Based on the level of resources available to the scheduler, and the way resource allocation phase interacts with the planning phase, the resource allocation problem can be classified into a variety of classes, as shown in Figure 3. We will start by describing the main classes briefly:

- Class INH-UNSOLV: If the problem is inherently unsolvable (for example, goals are *on_blockA_blockB* and *on_blockB_blockA* in blocks world), considering or ignoring resources during planning will not affect the solution but will help to create the planning graph faster. Hence, the problem can be handled by the planner more efficiently.
- Class INFRES: If indeed the resources are sufficient to overcome all resource conflicts, the scheduling view of the problem is the same as if there are infinite resources. For the *shuffle* problem, 5 robots are enough to overcome all resource conflicts in a plan and there is no reason why problems with 5 or more robots should take more time.
- Class FINRES: The remaining case is when the number of resources are small enough to cause resource conflicts but the problem is inherently solvable. This case can be decomposed, based on the difficulty of the resource scheduling problem, into a number of more specific sub-classes as shown in Figure 3. It turns out that we can handle many of these sub-classes through efficient (backtrackfree) methods.

| Level | Actions by level | # Robots |
|-------|------------------|----------|
| 1 | Unstack_R_blkF_blkE | 1 |
| 2 | Unstack_R_blkE_blkD | 2 |
| 3 | Unstack_R_blkD_blkC | 3 |
| 4 | Unstack_R_blkC_blkB | 4 |
| 5 | Putdown_R_blkC | |
| 5 | Unstack_R_blkB_blkA | 5 |
| 6 | Stack_R_blkF_blkC | |
| 6 | Pickup_R_blkA | 5 |
| 7 | Stack_R_blkB_blkF | 4 |
| 8 | Stack_R_blkE_blkB | 3 |
| 9 | Stack_R_blkA_blkE | 2 |
| 10 | Stack_R_blkD_blkA | 1 |

**Fig. 4.** *An resource-abstracted solution for* shuffle *problem. Curved lines show resource usage spans (see below). The number of resources needed at each level (which equals the number of spans crossing that level) is also shown.*

The complexity of resource scheduling instance, as well as the amount of modification needed to the original plan to allocate resources, increases from left to right and from top to bottom in Figure 3. Rather than use one general scheduling method for all classes, we cycle through scheduling methods tailored to each of the specific classes (from the easiest to the hardest). By using this approach, we are able to allocate resources with the least amount of modification to the given plan. This in turn ensures that the plans developed in our method are comparable in quality to those developed by the normal planner.

**Posing the resource allocation problem:** We setup the resource allocation problem as a constraint satisfaction problem (CSP) by treating actions needing resources as variables and posting codesignation and non-codesignation constraints according to whether the actions can have the same resource allocated or not (resource conflicts). For this, we only have to consider facts (effects and preconditions of actions) that refer to the resource type under consideration (e.g. $arm\text{-}empty\_R$ or $holding\_R\_blockA$ for ROBOT type in blocks world). This is because interactions involving facts with only non-resource objects (e.g. $clear\_blockA$) would have been handled during planning itself.

---

**Function:** $Allocate\_Resources$
**Parameters:** Problem, AbstractPlan, ResourceType[]
**Returns:** Plan


- LocalPlan=AbstractPlan
- For each $R_i$ =ResourceType,
    * LocalPool=NumberResources(ResourceType[$R_i$])
    * Span=GetResourceSpan(LocalPlan, $R_i$)
    * If Span is NIL, then continue with next ResourceType
    * Conflict=GetResourceConflict(LocalPlan, $R_i$)
    * For each Level $L_j$ in LocalPlan
        • Need,RelevantSpan=Number,Cutsets of Span at $L_j$
        • If $Need_j$ is positive, then for each $RelevantSpan_k$
            1. If $RelevantSpan_k$ has been allocated, then continue with next RelevantSpan.
            2. Solve for the scheduling classes at Level $L_j$ until an allocated plan (not NILPLAN) is obtained.
                (a) LocalPlan=$Solve\_INFRES$(LocalPlan, $RelevantSpan_k$, Conflict,$R_i$, LocalPool,$L_j$).
                (b) LocalPlan=$Solve\_FIX$(LocalPlan, $RelevantSpan_k$, Conflict,$R_i$, LocalPool,$L_j$).
                (c) LocalPlan=$Solve\_SAMELEN$(LocalPlan, $RelevantSpan_k$, Conflict, $R_i$, LocalPool,$L_j$).
                (d) Return OriginalPlanner(Problem).
            3. Increase LocalPool by 1 if $RelevantSpan_k.E$ = Level $L_j$.
    * Continue with the next ResourceType
- Return LocalPlan

---

**Fig. 5.** *Pseudo-code for allocating resources*

We setup the CSP problem as follows. We visit the plan level by level and determine the span (similar to a causal link in partial-order planners [12]) of allocation of resources. A span $S$ is a tuple $\langle A_i, A_j, F, B, E \rangle$ where the effect $F$ of an action $A_i$ is produced at level $B$ and consumed at level $E$ for the precondition of another action $A_j$. Actions $A_i$ and $A_j$ must have the same resource allocated[5]. The resource conflicts are that two actions $A_j$ and $A_k$ at the same level cannot have the same resources if resources are non-sharable. Moreover, any action $A_k$ that falls within a span $S$ (i.e. levels $S_B$ and $S_E$) and which either deletes or produces fact $S_F$ must be allocated differently from actions $S_{A_i}$ and $S_{A_j}$.

---

[5] Information about variable bindings could also have been supplied by the planner from causal dependencies. We re-visit this issue in Section 6.

Examples of spans in Figure 4 are S1: $\langle A_1, A_6^1, holding\_R\_block F, 1, 6\rangle$ and S2: $\langle A_2, A_8, holding\_R\_block E, 2, 8\rangle$.

**Solving the resource allocation problem: INFRES case:** The resource allocation problem posed as a CSP is now ready for solving. Pseudo-code of our resource allocation algorithm is shown in Figure 5. This method traverses the levels of the abstract plan and computes the spans that are relevant to a level $L_j$ by finding spans that pass through $L_j$. In the *shuffle* example, the spans $S1$ and $S2$ are relevant at $L_2$. For each unallocated relevant span, it checks to see if there is a way to assign a resource. The check is made from the easiest to the hardest allocation instance in terms of the change to the abstract plan. Class INFRES is compatible with the conventional CSP formulation because no new actions (variables) are introduced. Hence, any standard CSP solver can be used in place of *Solve_INFRES*. If the CSP problem was solved for all the levels in the abstract plan by *Solve_INFRES*, this means that the bet made during planning (causal reasoning) that sufficient resources are available to handle all resource related interactions paid off and we have a solution. Otherwise, we have to free and reallocate resources as necessary.

**Solving the resource allocation problem: FINRES case:** Based on the amount of resources, we can divide Class FINRES into a number of sub-classes as summarized in Figure 3. These sub-classes are currently detected during scheduling itself. The general idea is that we traverse the plan level-by-level and go on allocating the resources from the resource pool until we face a resource scarcity at some level $i$. A scarcity suggests a greater demand for a resource than its availability. It can be resolved, provided the number of resources are not too low, by re-arranging the resource usage pattern. Conflict resolution starts by going to a previous level (level $i - 1$) of the plan and introducing a freeing[6] action to de-allocate a resource assigned to an action whose effect is not immediately needed. In the process, *Solve_FIX* shrinks the resource demand by one for all the levels from level $i$ to level $j - 2$ where the effect is needed at level $j$ again (an action to reallocate the resource will be added at level $j - 1$). *Shuffle* problems with 3 and 4 robots can be handled by this method. In particular, with respect to Figure 4, the robot corresponding to span $\langle A_1, A_6^1, holding\_R\_block F, 1, 6\rangle$ can be freed at level 2 and re-allocated at level 5. The number of robots needed at levels 3 and 4 will then reduce by 1. Problem instances solvable by this method are in the class FIX.

If resource scarcity persists, an unallocated action is moved to a subsequent level where it can be potentially allocated. But the move may force the consumers of its effects and any other actions whose effects can be clobbered by the potential move, to be moved too. Since unrestricted movement of actions can be as complex as planning itself, *Solve_SAMELEN* is not allowed to increase the plan length

---

[6] In general, adding actions to a plan can change its causal structure but we assume that there are actions or known sub-plans in the domain (e.g.*pickup, putdown* in blocks world) that can free and re-allocate resources without doing so. While there are pathological cases where the assumption may not hold, it seems to hold in most normal domains.

and is required to maintain the relative action positions. *Shuffle* problem with 2 robots can be handled by this method. In the *shuffle* problem in Figure 4, the action $A_5^2$ can move down to level 6 to ease the scarcity. Since $A_6^2$ needs the $clear()$ effect of $A_5^2$, it then needs to move to level 8. Note that the length of the plan still remains the same. Problem instances solvable by this method are in the class $SAMELEN$ in Figure 3.

If the two above approaches fail, we are in Class INCRLEN where the length of the abstract plan must be increased during scheduling (i.e. plan serialization affects plan length). *Shuffle* problem with 1 robot belongs to this class. We give problems in this class back to the original planner for solving it without any resource reasoning in the normal way. Class UNSOLV occurs when the number of resources are too small for any resource allocation to be feasible at all. If there are no resources, we can identify this class at the start of scheduling; otherwise it cannot be determined until after Class INCRLEN.

We can handle Classes INFRES, FIX and SAMELEN without backtracking in time polynomial in the length of the abstract plan (since the plan is traversed only once). For Class INCRLEN, resource abstraction is a penalty because we go back to the original planner and solve the problem without abstraction. However, as empirically shown later, this penalty is small and easily offset by the savings in other classes. As mentioned earlier, the reason we use a series of methods in this order is to keep the number of additional actions as small as possible while maintaining the optimal plan length.

## 5   Empirical Evaluation

We have developed a prototype implementation of our approach on top of Graphplan. We now compare the performance of our prototype to standard Graphplan, as we vary the amount of resources. We consider the blocks world (where the number of robot hands is varied) and the logistics domain (where the number of trucks at different cities are varied).

Figure 6 (left) shows the results for the *shuffle* problems with 4, 6, 8 and 10 blocks as the number of robots are varied from 1 to 10. The plots clearly show that planning followed by scheduling (PS-TOT) is significantly better than original planning in the presence of resources (GP-TOT). Let us consider the 6-block *shuffle* problem in detail.

In our method, the planning time is constant and the scheduling time is dependent on the specific class (in Figure 3) that the problem falls into. In the *shuffle* case, problems with 5 to 10 robots are in class INFRES, problems with 3 and 4 robots are in class FIX, and problems with 2 robots are in class SAMELEN. The first class needs no modifications to the plan, the second class requires insertion of new actions, while the third also requires movement of actions across levels (steps). *Shuffle* problems with 1 robot are in class INCRLEN, and are sent back to the planner. This is reflected by the dip in the plot (SHUF6-PS-TOT) after 1 robot case.

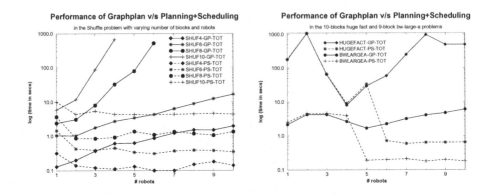

**Fig. 6.** *Comparative performance on shuffle problems of 4, 6, 8 and 10 blocks in the left and on huge-fact (9 blocks) and bw-large-a (10 blocks) problems in the right.*

Although not shown in the plots, the length of the plan with our approach is the same as with original Graphplan in terms of the number of levels and actions. As the number of resources (here robots) increase, our approach takes almost constant time whereas the performance of Graphplan is adversely impacted to a significant extent.

In Figure 6 (right), we see the performance of our method on the 10-block *huge-fact* problem and the 9-block *bw-large-a* problem. For *huge-fact*, problems with robots 1 to 5 are in Class INCRLEN and the remaining problems are in Class INFRES. Within Class INCRLEN, the search time of the original planner first increases with resources and then falls as the resource scarcity is eased. But across classes, the performance of the original planner degrades with resources. Our method relies on the original planner for Class INCRLEN and thus suffers a minor penalty in those instances, but it shows remarkable improvement later on. For *bw-large-a*, problems with robots 1 to 4 are in Class INCRLEN and the remaining problems are in Class INFRES. We obtain results similar to those in the *huge-fact* case. Notice that the amount of resources at which we transition from one class to another depends on the problem. This is why the algorithm in Figure 5 cycles through all the methods for each problem.

**Table 1.** *Runtime results from experiments in the logistics domain (in cpu sec)*

| # Trucks/city | Normal GP | GP+Sched |
|---|---|---|
| 1 | 1.0 | 2.7 |
| 2 | 2.4 | 1.0 |
| 3 | 4.6 | 1.6 |
| 4 | 10.0 | 1.5 |
| 10 | 500.0 | 1.0 |

**Multiple resources & the Logistics domain:** Note that the algorithm in Figure 5 can handle domains with multiple resources. Since a valid plan must be allocated resources with respect to all the resource types in the domain, the abstract plan is iteratively scheduled with respect to the resources of each type. The order in which the resources should be scheduled may be important for scheduling efficiency but not correctness or optimality of the final plan. To illustrate the multi-resource case, consider the results of our experiments in the logistics domain, shown in Table 1. The problem here involves 3 Packages at 3 cities which need to be delivered to cities other than the originating city using 3 airplanes. The number of trucks $(t)$ at each city is varied as shown. The resource declaration makes a truck at each city equivalent to other trucks at the same city. This ensures that trucks in different cities are not considered inter-changeable. The total number of trucks $(n)$ in the domain is $3t$ (thus the total number of trucks in the domain in the largest problem, $t = 10$, is 30). We do planning by abstracting all trucks first. The resource allocation will then have three phases, each corresponding to the allocation of trucks at a specific city. We see that separating planning from scheduling is again a very good idea in this domain too – leading to significant speedups as the number of resources (trucks/city) increases.

# 6   Discussion

We are exploring a planning model in which resource allocation is teased apart from planning, and is handled in a separate "scheduling" phase (See Figure 1 (right)). We observe that a necessary condition for a schedulable plan is that it should be causally correct irrespective of the nature of resources. We can produce an abstract plan $(P')$ which is correct sans the resource allocation and use it as a starting point for all planning problems that differ only in the number or amount of resources present. Next, based on the actual resource availability, the abstract plan will be allocated resources to produce an executable plan.

In most existing classical planning systems, sharable discrete resources are typically assumed to have infinite capacity (e.g. trucks can load any number of packages) and continous resources are assumed unlimited (e.g. fuel is available or not). Our causal plan $P'$ is also created under the most optimistic resource assumption (unlimited or infinite). While scheduling, the actual resources may be found insufficient to assign to $P'$ and this will force replanning to take place to honor the resource limits. The scheduler can aid the planner by informing it where re-planning is needed.

Although the communication between planner and schedule is limited to flagging failure to allocate resources, it can be improved further by having the scheduler "explain" the reason for its failure to allocate sufficient resource. This explanation, which is presumably in terms of resource limitations as well as the plan structure, is then regressed to just the restrictions on the plan structure. The regressed explanation can then be passed on to the planner, to be used in re-starting its search. This type of "multi-module dependency directed backtracking" approach is a variation on the hybrid planning methodology developed in [7], and is also akin to the approach used to link satisfiability and linear programming solvers in [16]. In our algorithm, the resource type and the level where transition from one scheduling class to another occurs, marks the failure information. The planner and scheduler share the common plangraph structure making regression straightforward.

In replanning, additional actions can be introduced for the limited resources to be prudently allocated, freed and later re-allocated for reusable resources or simply provided (e.g. re-fueling). The iteration between planning and scheduling will continue until the plan is executable as shown in Figure 1 (right). The current work has implemented the model for discrete sharable and non-sharable resources. In future, we will consider continous resources by modeling and handling linear constraints.

An inefficiency of our current implementation is that we do not propagate constraints among resource variables during planning and figure them aposteriori from causal dependencies and mutex constraints. This increases our scheduling setup time whereas the information was available during planning itself. We are addressing this problem as a *lifted planning* scenario where codesignation and non-codesignation constraints between variables are propagated during causal reasoning. Note that this is selective variablization because only resources have not been committed to.

As we mentioned earlier, the counter-intuitive behavior of performance worsening with more resources is not restricted to disjunctive planners such as Graphplan and Blackbox, and also afflicts more traditional planners including UCPOP [15]. Indeed, the work on O-Plan [4, pp. 73], has identified the inefficiency of combining resource scheduling with planning (although, to our knowledge, no specific steps were taken to address that inefficiency in the O-Plan work).

## 7    Related Work

Our work can be seen as abstraction of resources from planning phase. From this angle, our idea of keeping the structure of the causal plan intact during resource allocation phase is akin to the enforcement of ordered monotonicity property in ALPINE[9]. An important difference however is that our work is not dependent on the availability of strong abstractions, but is rather motivated by the desire to exploit the loose-coupling between planning and scheduling in most real world domains. If the abstract plan cannot be scheduled, we support interaction between the scheduler and planner to arrive at a schedulable plan.

Among planners that have considered resources, in SIPE[17], domain-specific operator ordering can be provided by defining what are resource objects in the domain. Work more closer to ours is by El-Kholy and Richards[6] and Cesta and Cristiano[3] who perform temporal and resource reasoning after a plan is obtained. They however do not consider the interactions between resource allocation and planning phases. The recent LPSAT planner by Wolfman and Weld[16] distinguishes between discrete and continous state variables, pushing the assignment of continous ones to an LP solver. Note that discrete/continous distinction is really orthogonal to resource/non-resource distinction. Abstraction of resources can be applied to both continous and discrete parts of LPSAT.

Fox and Long[11] have described a way of utilizing symmetry in domains to speedup planning. Symmetric domain objects are by definition functionally similar and cannot be usefully distinguished. The insight here is that any one of the symmetric objects is sufficient during solution verification to avoid equivalent failures. They keep track of symmetric objects during planning while we abstract out resources. There exist methods for improving the performance of Graphplan by removing irrelevant literals from the problem specification (c.f. [13]). Such methods however are not applicable for our problem as resources – however many of them there may be – are never irrelevant. Explanation-based learning (EBL) and dependency directed backtracking (DDB) techniques have been used by Kambhampati[8] to expedite Graphplan. Though these methods capture some of the regularities of the domain/problem, we found that they are still not competitive with our approach. Finally, the complexity of changing plans for scheduling and parallelization has also been studied by Backstrom [1]. While he focuses on parallelizing a complete and correct plan, we start with a maximally parallel resource-abstracted plan and add or shift actions across levels to handle resource constraints.

## 8    Conclusion

We have introduced a novel planning framework in which resource allocation is teased apart from planning, and is handled in a separate "scheduling" phase. The aim is to make planning efficient and scale it to large domains containing multiple resources. We described resources and discussed how they can be allocated to an abstract plan, which is obtained after causal reasoning, on the notion of plan length optimality. The runtime of our approach is much less sensitive to the resource quantity. It thus admits the paradigm of *plan once and schedule anytime*. If some allocated resource becomes unavailable during plan execution, we can handle the exception through resource re-allocation.

Our resource scheduling phase currently only aims to generate the shortest length plan – equating, in effect, the plan cost with plan length. While this is consistent with the current practice in systems like Graphplan and Blackbox, real world domains would need more general cost metrics that are a function of both the plan length and resource costs. We have handled discrete sharable and non-sharable resources until now. We plan to incorporate continous resources by modeling linear constraints. In planner-scheduler interaction, the planner can provide variable binding information to the scheduler while the latter can explore Class INCRLEN rather than falling back on planning without resource abstraction. Our ongoing work concentrates on these more general issues.

# References

1. Backstrom, C. 1998. Computational Aspects of Reordering Plans. *JAIR Vol.9* 99-137.
2. Blum, A., and Furst, M. 1995. Fast planning through planning graph analysis. *Proc IJCAI-95* 1636–1642.
3. Cesta, A. and Cristiano, S. 1996. A Time and Resource Problem in Planning Architectures. *Proc. ECP-96.*
4. Currie, K. and Tate, A. 1991. O-Plan: the open planning architecture. *AI, Vol 52, 49-86.*
5. Kautz, H., and Selman, B. 1998. BLACKBOX: A New Approach to the Application of Theorem Proving to Problem Solving. *Workshop Planning as Combinatorial Search, AIPS-98, Pittsburgh, PA, 1998.*
6. El-Kholy, A. and Richards, B. 1996. Temporal and Resource Reasoning in Planning: the *parc*Plan approach. *Proc. ECAI-96.*
7. S. Kambhampati, M.R. Cutkoksy, J.M. Tenenbaum and S. Lee. Integrating General Purpose Planners and Specialized Reasoners: Case Study of a Hybrid Planning Architecture. IEEE Trans. on Systems, Man and Cybernetics, Special issue on Planning, Scheduling and Control, Vol. 23, No. 6, November/December, 1993). (An earlier version appears in Proc. AAAI-91).
8. Kambhampati, S. 1999. EBL and DDB for Graphplan. *Proc. IJCAI-99.*
9. Knoblock, C. A. 1994. Automatically Generating Abstractions for Planning. *AI Journal, 68(2).*
10. Laborie, P., and Ghallab, M. 1995. Planning with sharable resource constraints. *Proc. IJCAI-95.*
11. Fox, M., and Long, D. 1999. The Detection and Exploitation of Symmetry in Planning Domains. *Proc. IJCAI-99.*
12. McAllester, D., and Rosenblitt, D. 1991. Systematic Nonlinear Planning. *Proc. 9th NCAI-91, 634-639.*
13. Nebel, B.; Dimopoulos, Y.; and Koehler, J. 1997. Ignoring irrelevant facts and operators in plan generation. *Proc. ECP-97.*
14. Koehler, J; Nebel, B.; Hoffmann, J.; and Dimopoulos, Y. 1997. Extending Planning Graphs to an ADL Subset. *Proc. ECP-97.*
15. Penberthy, J., and Weld, D. 1994. UCPOP: A sound, complete, partial order planner for ADL. *Proc. AAAI-94, 103-114.*
16. Wolfman, S., and Weld, D. 1994. The LPSAT Engine and its Application to Resource Planning. *Proc. IJCAI-99.*
17. Wilkins, D. E. 1988. Practical planning: Extending the classical AI planning paradigm. *Morgan Kaufmann Pub., San Mateo, CA.*

# Real-Time Scheduling
# for Multi-agent Call Center Automation

Yong Wang, Qiang Yang, and Zhong Zhang

Information Service Agents Lab
School of Computing Science
Simon Fraser University
Burnaby B.C. V5A 1S6 Canada
{ywangc,qyang,zzhang}@cs.sfu.ca

**Abstract.** In a call center, service agents with different capabilities are available for solving incoming customer problems at any time. To supply quick response and better problem solution to customers, it is necessary to schedule customer problems to appropriate service agents efficiently. We developed SANet, a service agent network for call center, which integrates multiple service agents including both software agents and human agents, and employs a broker to schedule customer problems to service agents for better solutions according to their changing capabilities and availability. This paper describes the real-time scheduling method in SANet as well as its architecture. There are two phases in our scheduling method. One is problem-type learning. The broker is trained to learn the problem types and hence can decide the type of incoming problems automatically. The other is the scheduling algorithm based on problem types, capabilities and availability of service agents. We highlight an application in which we apply SANet to a call center problem for a cable-TV company. Finally, we support our claims via experimental results and discuss related works.

## 1    Introduction

Our research is motivated by a realistic problem in call center environments. To solve customers' problems, many telecommunications companies, such as Cable-TV and telephone companies, maintain large call centers that are aimed at providing real-time solutions to their customers. In a Cable-TV call center environment, for example, a customer may phone in to ask about a solution to his fuzzy-picture problem. A human agent is selected for answering the question and interactively diagnoses the source of the problem. In this environment, the human agent's expertise is distributed and changing. They are not always available either due to time shifts, or changing interests and training. A solution is to create a number of software agents which can provide subsets of the expertise that human agents can provide, and cater to customers through a network of human and software agents. From the agent research point of view, a call center is a multi-agent system which includes customer service agents and an

S. Biundo and M. Fox (Eds.): ECP-99, LNAI 1809, pp. 187–199, 2000.

agent who schedules customer problems to service agents. Using agent development techniques, we have developed SANet — a service agent network for call center automation. SANet integrates both human and software service agents in providing customer service in real time, and employs a broker to schedule customer problems to service agents for better solutions according to the nature of the input problems as well as the changing capabilities and availability of service agents on the network. This paper mainly describes how SANet schedules the incoming customer problems to appropriate service agents, which includes a problem-type learning method and a problem scheduling algorithm based on the problem types, the capabilities and availability of service agents. As we will see, SANet is a flexible network on which service agents can be added or deleted at any time and their capabilities to solve problems can be tracked and utilized.

In this paper, we first describe the architecture of SANet. Then we present the problem-type learning method and the problem scheduling algorithm. Also we present an application in which we simulate a call center environment of a cable-TV company. Finally, we demonstrate through the experimental results that the proposed architecture and algorithms can provide high-quality service in real time.

## 2    The SANet Architecture

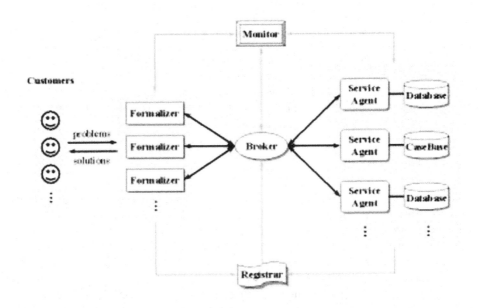

**Fig. 1.** The architecture of SANet

Figure 1 shows the architecture of SANet. There are two kinds of service agents on SANet. One is the software agent. A software agent is capable of solving one or more types of problems, and is associated with a solution database or case base. The other is the live agent. A live agent can be a human agent who is an expert in one or some specific areas. It may also be an integrated system of a human agent and a software agent. In this case, software agents act as a decision-support assistant to help human solve problems. Obviously, different service agents have different problem-solving capabilities. Their availability also change with time. Because problems in a call center environment may come in many manners, through telephone, fax, letter, e-mail etc. The formalizer agent transforms an unstructured incoming problem into a structure problem format using a set of attribute/value pairs. Any service agent or formalizer can be easily added to or deleted from the network by sending a registrar agent to the broker agent. The registrar is a mobile agent which carries information about the added/deleted agent, such as its most current capability, to the broker. The broker is an important agent on SANet, which is responsible for selecting the appropriate agents to solve problems according to the problem type, and capabilities and availability of the agents. The broker is trained to learn problem types and hence can decide the type of a given problem automatically. There are some subbrokers in the broker agent. Each subbroker corresponds to one type of problem and is created or deleted by the broker agent. Figure 2 shows the architecture of the broker agent.

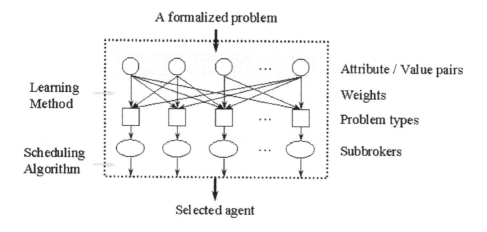

**Fig. 2.** The architecture of the broker agent

Each subbroker maintains a table of availability and capabilities of the agents which can solve the corresponding problem types. The broker maintains a table showing the correspondence between problem types and the subbrokers, a list of registered agents and a list of registered formalizers. When a formalized problem comes in, the broker has two steps to select an agent for this problem:

**Step 1:** decide the problem type using the result of problem type learning and then pass the problem to the corresponding subbroker.

**Step 2:** the subbroker selects an appropriate agent for the problem by using a scheduling algorithm.

In the next subsection, we discuss the learning method and the scheduling algorithm in details.

As soon as an agent receives a problem, it sends a "not available" message to the broker and then the broker can inform both the monitor and the subbroker. After an agent finishes working on a problem, it updates the quality of solving this type of problem by the following quality updating formula:

$$q'(t) = \frac{100q(t) + q(p)}{100 + 1}$$

where $q(t) \in [0, 1]$ is the average quality of solving the problems of type $t$ before solving the problem $p$, $q'(t) \in [0, 1]$ is the average quality after solving $p$ and $q(p) \in [0, 1]$ is the quality of solving $p$. One simple way to decide $q(p)$ is:

$$q(p) = \begin{cases} 1, & \text{if } p \text{ is solved}; \\ 0, & \text{if } p \text{ is not solved.} \end{cases}$$

Another way is to ask the user to provide a feedback ranking score on the quality of the solution provided by the agent.

An important feature of our SANet system is its ability to continuously update the capabilities and availability of different agents in real time. After a problem is solved, the solution and the quality at this time are added to the problem package. Then the agent sends back the problem package and the new quality to the subbroker, along with an "available" message to the broker. When a subbroker receives an availability changing message of an agent, it updates its availability table. When a subbroker receives a problem with a new quality returning from an agent, it updates its capability table. If the problem was not solved, then the subbroker delivers it again. Otherwise, the subbroker sends it back to the broker and the broker then relays it to the user through the corresponding formalizer. The monitor agent monitors the system performance and shows the changing capabilities and availability of service agents on real time.

## 3    Matching Problems with Software Agents through Learning

In order to pass a given problem to an appropriate subbroker, the broker has to decide what type the problem is of. We can relay this task to the outside end users, but this will impose an extra requirement that an end user have to know the problem type *a priori*. This will not only burden the users but also, more importantly, force a user to choose a problem type when in fact the user may not know exactly what the problem type is. Furthermore, the situation is made more complex by the dynamic nature of the capabilities and availability of the distributed agents. Obviously, we wish to alleviate such a burden, making automatic the decision-making process of problem types in the broker agent.

As discussed before, after a problem is input, the formalizer will convert it into a standard representation form — a set of attribute/value pairs. Essentially, the broker will decide the problem type based on these attribute/value pairs. Therefore, we can view the relationship between attribute/value pairs and problem types as that shown in Figure 2. ¿From the figure, it can be seen that each problem type is connected to a set of attribute/value pairs. We say that these attribute/value pairs decide this problem type. They distinguish a problem type from another. However, in real applications, not all attribute/value pairs have equal importance in this decision-making process. We thus consider assigning different weights to the connections between attribute/value pairs and problem types.

In [16], a learning model is applied to a network which is very similar to the one shown in Figure 2. The learning model aims to track its users' preferences dynamically, such that whenever the users' preferences change, the changes will be captured by the learning model and reflected in the following usage session. In our broker agent, we have the goal of learning problem types in the broker agent, given the attribute/value pairs. With the two-level architecture as shown above, the learning algorithm is a variation of the perception learning algorithm.

In particular, the training process of a problem type in the broker can be described as follows. After an input problem is formalized into a set of attribute/value pairs, these pairs will be fed into the network shown in Figure 2. The ranking scores for individual problem types will be computed. During the training, an expert critiques whether the highest-ranking problem type is the correct one. This will, in turn, be taken by the learning policy to update the relevant weights. In our broker agent, the training process is off-line assisted by domain experts. Our experiments show that after learning, the broker agent will choose, on behave of the user, the most appropriate problem type to be sent to the corresponding subbroker.

For the sake of simplicity, we do not discuss the learning model in detail here. Interested readers are referred to [16]. In Section 6, we will discuss the experiments we have conducted in order to demonstrate that the learning model fulfills our desired goal.

## 4    SANet's Agent Scheduling Algorithm

This section describes SANet's scheduling method for the subbroker to deliver problems to agents. The problems a subbroker may receive can be divided into three kinds:

**new-problem.** A new-problem is one delivered from the broker.

**problem-with-solution.** A problem-with-solution is one returned from an agent which solved the problem.

**problem-without-solution.** A problem-without-solution is one returned by an agent which did not solve the problem.

Besides the capability table, the subbroker also maintains a problem queue. Each problem which can not be sent to any agent at current time will be put into the queue for future delivery. To select an agent to solve the problem, we define a capability function for each problem type.

**Definition 1** *The capability function for a problem type $t$ is denoted by $c(a) = c_t(a, q(a), t(a))$, where $a$ is an agent which is capable to solve the problems of type $t$, $q(a) \in [0, 1]$ is the quality for this agent to solve the problems, and $t(a) \in [0, \infty]$ is the solving time. Let $\alpha$ be a user-defined factor. For agents $a$ and $a'$,*

1. *If $q(a) \geq q(a')$ and $t(a) \leq t(a')$, then $c(a) \geq c(a')$.*
2. *If $q(a) \geq q(a')$ and $t(a) > t(a')$, then*
   *if $(q(a) - q(a'))\alpha \geq (t(a) - t(a'))$ then $c(a) > c(a')$ else $c(a) < c(a')$.*
3. *If $q(a) < q(a')$ and $t(a) \geq t(a')$, then $c(a) < c(a')$.*
4. *If $q(a) < q(a')$ and $t(a) < t(a')$, then*
   *if $(q(a') - q(a))\alpha \geq (t(a') - t(a))$ then $c(a) < c(a')$ else $c(a) > c(a')$.*

$\alpha$ acts like a weight to make $|q(a) - q(a')|$ and $|t(a) - t(a')|$ comparable. For example, it can be defined as $\alpha = \max\{(t(a) - t(a')\}$.

We now describe the problem scheduling algorithm. We assume that we are given $N$ problems, each associated with a problem type $i = 1, \ldots, K$. There are $M$ agents $A_l$. We allow a problem to be of multiple types; in that case we assign a problem to a subbroker based on one of the types selected by the broker. Similarly, agents can have overlapping capabilities that are also changing with time. Also assume that there are $N_i$ agents initially available to problem of type $i$, where each agent $j$ has a capability defined by a value $c_{ij}$ which is a real number between zero and one. We wish is to assign problems $P_k$ to agents $A_l$ such that the total sum of the capability value is maximized.

We adopt a greedy algorithm, extended to handle concurrent agent-based problem solving. The greedy scheduling method is shown as follows:

---

**Algorithm** SANet Scheduling Algorithm

**Input:** $M$ agents $\{A_l, l = 1, \ldots, M\}$, $K$ problem types, Agent capabilities where agent $A_l$ has capability $\{c_{li}, i = 1, \ldots, K\}$, $N$ problems $P_{ij}$ where the problem is the $j^{th}$ problem of type $i$, and a number $N$ which is the maximum time a problem is allowed to be returned by agents without solution.

**Output:** An assignment of Agent $A_l$ to Problem $P_{ij}$.

1. Partition problems $P_{ij}$ into $K$ typed groups, one for each sub-broker $B_i$;
2. **concurrently for each $B_i$ $i := 1$ to $K$ do**
3.     Set a queue of problems $Q_i$ sorted by come-in time;
4.     **loop until $Q_i$ is empty do**

5.        **for** every problem $P_{ij}$ in the queue $Q_i$, **do**

6.          **if** all agents have looked at $P_{ij}$, **then**

            give it to a designated human agent.

          **else**

            Let $S_{ij}$ be the set of available agents which haven't worked on $P_{ij}$;

7.        Let $P_{im}$ be the first problem in $Q_i$ with non-empty $S_{im}$,

        and $n$ be the time $P_{im}$ is already returned by agents without solution.

8.        find an agent $A_l$ with maximal capability from suggested agents

        in $S_{im}$, if any; otherwise, find $A_l$ with maximal capability in $S_{im}$;

        then send $P_{im}$ to $A_l$;

9.        **if** $A_l$ is not available and $n < N$, **then**

        return $P_{im}$ to $B_i$ without working on it.

        **else**

          $A_l$ must work on $P_{im}$;

          update capability for $A_l$ for problem type $i$;

          update and return the problem package to $B_i$;

---

This algorithm is designed to have the following two properties. First, we can ensure that no problem will wait forever before it is solved by an available agent. This is ensured in **step 9** of the algorithm, which commits a problem to an agent. Second, we claim that this greedy algorithm returns solutions with good quality and solving time. We verify this point experimentally in the next section.

## 5 SANet in a Cable-TV Call Center Environment

We have applied SANet to a simulated call-center environment of a local cable-TV company. In this application, a software service agent is a case-based reasoning (CBR) system that can answer typical questions or frequently asked questions automatically by retrieving the most similar case in the case base. A live agent is a staff in the call center with an assistant CBR system. At this point, the system can solve 16 types of problems. Each problem type has one to ten cases in one or more case bases. To input a problem, the user has to answer some questions. The question list is:

**Q1** What's the problem type?
**Q2** Which channels have the problems?
**Q3** Is the account an active cable account?
**Q4** Is the problem affecting more than 1 outlet?
**Q5** What does the picture on the screen look like?
**Q6** Is the customer in an affected area?

These questions are optional to answer. If the user answered question 1, which means the user knows the problem type, then the broker just passes this problem to the corresponding subbroker. Otherwise, the broker has to decide the

problem type. In other words, the user has two choices — let the problem type
be decided by himself or the broker. For instance, the user may have answered
question 1, 2 and 3. Then the formalizer creates a problem package such as the
following:

| Problem ID | 111 |
| --- | --- |
| Description | the set of q/a pairs |
| Solution | Null |
| Suggested Agents | $\phi$ |
| Agent/Time List | $\phi$ |

The set of q/a pairs is:

| Q1 | Poor reception |
| --- | --- |
| Q2 | Channels 2 to 6 |
| Q3 | Yes |

If the problem is sent to a live agent, the solution part may be filled by a
staff who solved this problem. If it is routed to a CBR agent, a CBR system will
then solve this problem and the solution part is filled by the most similar case
in the case base. In this case, when the problem returns to the user, the solution
part looks like the following:

| Case Name | Poor reception low band |
| --- | --- |
| Q1 | Poor reception |
| Q2 | Channels 2 to 6 |
| Description | Channels 2 to 6 have poor reception |
| Solution | Sometimes caused by a loose connection. |
| | 1. Check channel outage note pad; |
| | 2. Check for a possible loose connection; |
| | 3. Try unplug power cord for television; |
| | 4. Try hook cable direct to Television; |

Also, the agent/time list will be filled. This list provides all agents which
have worked on this problem and the solving time. The quality of solution can
then be provided by the end customer based on his or her satisfaction with the
solution. This feedback value is used to update the capability matrix for each
agent. According to the solution quality, the end customer can also feedback a
evaluation value of the problem-type decision to the broker to adjust the learning
result of problem types.

This application system was developed by using Java JDK 1.1.6 and ASDK
V1.0.3 (IBM's aglets software development kit).

## 6    Experiments

This section describes the experiments we have conducted and discusses the experiment results. We have done two kinds of experiments. One is for the problem-
type learning in broker and the other is for the performance of SANet.

## 6.1   Experiment for the Problem-Type Learning

We introduce a learning model into our broker agent with the hope that it would automatically choose the most appropriate problem type on behalf the user. In the following we demonstrate our experiments with the learning model.

The learning process is off-line and mainly occurs in the broker agent. We create 20 problem types with each type being associated maximally with 20 attributes. In the experiment, on average each attribute can have 2 to 5 values. Accordingly, we also create a set of 100 user queries or problems represented by attribute/value pairs. The ranking of a problem type is between 0.0 and 1.0.

We feed these queries one by one into the learning model in the broker agent. The learning model then computes the rankings for the problem types and compares them with the desired ones specified by the queries. If there is a discrepancy between the computed ranking and desired ranking of a problem type, the learning process will be triggered and the corresponding weights will be updated. In the experiment, the learning process takes five rounds. Therefore, there is a total of 500 learning data points.

In Figure 3, we show the average error convergence chart for all the 100 queries after five learning rounds, where the X-axis represents the learning process which is composed of 500 learning data points while the Y-axis represents the average error among the 100 problem types in all the queries after each learning data point is learned. The error is defined as the absolute distance between the computed ranking and desired ranking of a problem type in individual queries. It can be seen that along the learning process, the average error for all the problem types tends to approximate zero and will be stabilized at learning rounds 4 and 5. We also see that because of the interactions among different queries, the learning error could not be zero no matter how long the learning process undertakes.

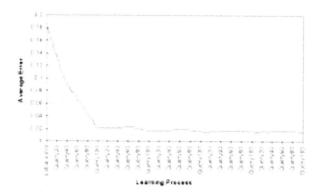

**Fig. 3.** Average Error Convergence Chart for 100 Queries

## 6.2   Experiment for the Performance of SANet

We have also conducted experiments to verify the performance of SANet. The application domain of the experiment is the call center of a cable TV company, as described above. In our experiment, we have 10 software agents which are CBR systems. Each software agent is associated with a case base in which there are 5 to 33 cases and 1 to 16 problem types. The solving time is set to $[n_c/300]$ seconds where $n_c$ is the number of cases in the case base. We set different qualities to the agents for the experiment. Setting 1 sets higher qualities to the agents with lower solving time. Setting 2 sets lower qualities to the agents with lower solving time. Setting 3 is a mixture of Settings 1 and 2. Table 1 shows the different settings of our experiment. For each problem type, there are four agents which are able to solve this type of problems.

**Table 1.** Settings of experiments

| Agent | Case Number | Type Number | Solving Time | $S_1$ | $S_2$ | $S_3$ |
|-------|-------------|-------------|--------------|-------|-------|-------|
| $A_1$ | 5 | 5 | 0.01 | 1.0 | 0.1 | 0.1 |
| $A_2$ | 6 | 2 | 0.02 | 0.9 | 0.2 | 1.0 |
| $A_3$ | 6 | 2 | 0.02 | 0.8 | 0.3 | 0.2 |
| $A_4$ | 6 | 6 | 0.02 | 0.7 | 0.4 | 0.9 |
| $A_5$ | 10 | 1 | 0.03 | 0.6 | 0.5 | 0.3 |
| $A_6$ | 11 | 7 | 0.03 | 0.5 | 0.6 | 0.8 |
| $A_7$ | 16 | 7 | 0.05 | 0.4 | 0.7 | 0.4 |
| $A_8$ | 18 | 10 | 0.06 | 0.3 | 0.8 | 0.7 |
| $A_9$ | 21 | 8 | 0.07 | 0.2 | 0.9 | 0.5 |
| $A_{10}$ | 33 | 16 | 0.11 | 0.1 | 1.0 | 0.6 |

$S_i$ refers to Setting $i(i = 1, 2, 3)$

Initially, 100 random problems are created for the experiment. Then we send some of them to the broker 100 times. Each time, the first $n$ problems are sent to the broker, where $n$ varies from 1 to 100. For each problem, we calculate the solving time which begins at the time it is sent to the broker and ends at the time it is returned with the solution. When a problem is solved, we get the quality of the agent which solved this problem. Then, for $n$ problems, we calculate the average solving time and the average quality. Figures 4 and 5 show the results.

These results highlight the following three main headlines.

- The system is stable for different settings.
  In our experiment, we used three different and typical settings as described above. We got very close results for these settings.
- The average solving time is nearly linear with the problem size.
  We observe that when the solving time of each agent for any problem is the same, the average solving time is linear to the problem number. Compared to this case, our result for the average solving time is reasonable.

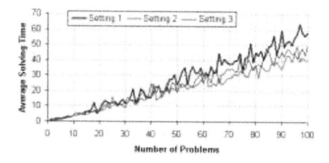

**Fig. 4.** Average Solving Time

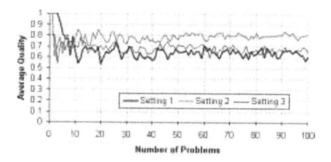

**Fig. 5.** Average Solving Quality

- The average solving quality is high and remains stable when the number of problems increases.

  These results show even when we have a large number of problems, the higher average solving quality can still be obtained.

## 7  Related Work

Our system is closely related to *information gathering systems* in which there is usually an information broker. In such systems, the information broker is to gather information and find solutions from heterogeneous resources on a network of resources. Currently, many of them have been developed or are under development. Among them are BIG [9], Disco [15], Garlic [13], HERMES [1], InfoSleuth [12], Infomaster [6], Information Manifold [11,10], SIMS [2,3], TSIM-MIS [4,7]. These systems take a user query and translate it into sub-queries that can be executed by various information agents attached to information sources. Most of them mainly focus on the integration of different knowledge representations. Therefore they do not take scheduling problem and the availability and the capabilities of resources or agents as main issues. But in our application domain, scheduling is an important problem because the primary concern in a a call center is to provide high quality service at a low service cost. The other major difference from information gathering systems is that, in our SANet system, a mechanism is designed so that after solving each problem, the quality and availability of the agent are updated and used by the broker to assign problems to agents in a better way.

Our work is also closely related to scheduling problems. In the past, scheduling has been widely studied by researchers [5,8,14,17]. The problem is to schedule jobs to processors under certain constraints. A novelty in our work is the assumption that processor capability and availability are a function of time.

## 8    Conclusions and Future Work

This paper describes SANet — a service agent network designed for call center automation. We have focused on its scheduling method for customer problems. Our contributions are mainly the following:

- Availability and capabilities of service agents are considered.
- The capability of each agent to solve a type of problem is maintained by the broker agent. In SANet, the broker can learn and use the newest capabilities to select appropriate agents for the customer problems.
- As a network, the agents can be easily added and deleted at any time. This makes the network very flexible and easy to be maintained. The broker scheduling algorithm maintains the overall competency of the network by intelligently choosing the right software agents for a problem.

In our future work, we are considering to add a manager agent to manage the agents on the network. For example, if all agents on the network are very busy and many problems are still coming in, then the management agent should add some appropriate agents automatically. Some other efficient task-scheduling methods need to be studied and compared. Also, decomposing a problem into some sub-problems and then integrating the solutions returned by agents are left as our future work.

## References

1. S. Adali and V. S. Subrahmanian. Amalgamating knowledge bases, ii - distributed mediators. *Journal of Intelligent and Cooperative Information Systems*, 3(4):349–383, 1994.
2. Y. Arens, C. Y. Chee, C.-N. Hsu, and C. A. Knoblock. Retrieving and integrating data from multiple information sources. *International Journal of Intelligent and Cooperative Information Systems*, 2(2):127–158, 1993.
3. Y. Arens, C. A. Knoblock, and C.-N. Hsu. Query processing in the sims information mediator. In A. Tate, editor, *Advanced Planning Technology*. AAAI Press, Menlo Park, CA, 1996.
4. S. Chawathe, H. Garcia-Molina, J. Hammer, K. Ireland, Y. Papakonstantinou, J. Ullman, and J. Widom. The tsimmis project: Integration of heterogeneous information sources. In *Proceedings of IPSJ Conference*, pages 7–18, 1994.

5. J. Dey, J. Kurose, and D. Towsley. On-line processor scheduling for a class of iris real-time tasks. *IEEE Trans. Computers*, 45(7), July 1996.
6. O. M. Duschka and M. R. Genesereth. Infomaster - an information integration tool. In *Proceedings of International Workshop: Intelligent Information Integration, during the 21st German Annual Conference on Artificial Intelligence*, 1997.
7. H. Garcia-Molina, Y. Papakonstantinou, D. Quass, A. Rajaraman, Y. Sagiv, J. Ullman, V. Vassalos, and J. Widom. The tsimmis approach to mediation: Data models and languages. *Journal of Intelligent Information Systems*, 1997.
8. J. Jonsson and J. Vasell. Evaluation and comparison of task allocation and scheduling methods for distributed real-time systems. Technical report, CTH, Dept. of Computer Engineering, Computer Architecture Laboratory (CAL), MicroMultiProcessor Group (MMP), 1996.
9. V. R. Lesser, B. Horling, F. Klassner, A. Raja, T. A. Wagner, and S. X. Zhang. Big: A resource-bounded information gathering agent. In *Proceedings of the Fifteenth National Conference on Artificial Intelligence (AAAI-98)*, January 1998.
10. A. Y. Levy, A. Rajaraman, and J. J. Ordille. Query-answering algorithms for information agents. In *Proceedings of the 13th National Conference on Artificial Intelligence, AAAI-96*, pages 40–47, 1996.
11. A. Y. Levy, D. Srivastava, and T. Kirk. Data model and query evaluation in global information systems. *Journal of Intelligent Information Systems, Special Issue on Networked Information Discovery and Retrieva*, 1995.
12. M. Nodine, B. Perry, and A. Unruh. Experience with the infosleuth agent architecture. In *Proceedings of AAAI-98 Workshop on Software Tools for Developing Agents*, 1998.
13. M. T. Roth and P. Schwarz. Don't scrap it, wrap it! a wrapper architecture for legacy data sources. In *Proceedings of the 23rd VLDB Conference*, Athens, Greece, 1997.
14. P. Sparaggis and D. Towsley. Optimal routing and scheduling of customers with deadlines. *Probability in the Engineering and Informational Sciences*, 8(1), January 1994.
15. A. Tomasic, L. Raschid, and P. Valduriez. Scaling heterogeneous databases and the design of disco. In *Proceedings of International Conference on Distributed Computing Systems*, 1996.
16. Z. Zhang and Q. Yang. Towards lifetime maintenance of case based indexes for continual case based reasoning. In F. Giunchiglia, editor, *Artificial Intelligence: Methodology, Systems, and Applications. 8th International Conference, AIMSA'98. Sozopol, Bulgaria, September 1998. Proceeedings*, volume 1480 of *Lecture Notes in Artificial Intelligence*, pages 489–500. Springer, 1998.
17. Y. C. Zhuang, C. K. Shieh, and T. Y. Liang. Centralized load balance on distributed shared memory systems. In *Proc. of the Fourth Workshop on Compiler Techniques for High-Performance Computing (CTHPC'98)*, pages 166–174, Mar. 1998.

# Task Decomposition Support to Reactive Scheduling *

Brian Drabble

Computational Intelligence Research Laboratory,
1269, University of Oregon,
Eugene, OR 97403, USA
drabble@cirl.uoregon.edu

**Abstract.** This paper describes the development of an intelligent tasking model which has been designed to enable complex systems, human agents and software agents, to be tasked and controlled within a reactive workflow management paradigm. The task models exploit recent advances within the AI community in reactive control, scheduling and continuous execution. The Dynamic Execution Order Scheduler (DEOS) extends the current workflow paradigm to allow tasking in dynamic and uncertain environments by viewing the planning and scheduling tasks as being integrated and evolving entities. DEOS is being applied to the domains of Air Campaign Planning (ACP) and Intelligence, Surveillance and Reconnaissance (ISR) management. These are highly reactive domains in which new tasks and priorities are identified continuously and plans and schedules are generated and updated within a temporal and resource constrained setting.

## 1 Introduction

Over the past few years there has been a great deal of interest in the use of advanced planning and scheduling technology to support workflow and process management systems. There have been a number of deployed systems which have used simple process templates and dispatch based scheduling techniques to develop small, linear process flows. These systems have been applied to fairly static environments in which the execution of activities follows a fairly predictable path, e.g. mortgage application processing. However, some groups e.g. DARPA are attempting to use workflow techniques on more reactive domains e.g. logistics, crisis management, mission planning, etc. These require technologies which are

---

* This research is supported by DARPA Contract: DABT63-98-C-0069 "Intelligent Workflow for Collection Management" and Contract: F30602-97-1-0294 "Understanding and Exploiting Hierarchy". The U.S. Government is authorized to reproduce and distribute reprints for Governmental purposes notwithstanding any copyright annotation hereon. The views and conclusions contained herein are those of the author and should not be interpreted as necessarily representing official policies or endorsements, either express or implied, of DARPA, Rome Laboratory or the U.S. Government.

capable of reacting quickly to changes occurring over time, e.g enacting new processes, editing and deleting existing ones and to re-balance the current resource assignments as new components are added and/or removed and new capabilities evolve. To tackle such domains it will require a quantum leap forward in the capabilities of process management systems.

Similar needs are also now starting to arise in the business community due to an increasingly competitive marketplace, widespread automation and the availability of online information. In addition there has been a great deal of interest in re-engineering and automating business processes and the field of workflow management (WFM) has emerged as an outgrowth of this interest in recent years. The workflow community advocates the use of explicit models and representations of processes, along with automated tools to support the activation and ongoing management of workflow processes. This has allowed tools to become more adaptive in the ways they work and interact with other systems. At the same time within the AI community, work on reactive control has led to the exploration of techniques for intelligent process management to meet the requirements of adaptivity for dynamic and unpredictable environments.

This paper describes a number of new techniques for representing and reasoning with task specifications which allow a process management system to gain a better understanding of the requirements of the task and the resources needed to satisfy the task. The task models described can be used at both the process level "what do I need to do" and the domain level "how do I carry out the chosen task". In addition the two levels can be seamlessly integrated thus allowing changes and requirements from the domain layer to be fed directly into the process layer without the need for expensive mappings. The system described is currently being developed for a number of different DARPA needs, e.g. ISR management, air mission planning and crisis management. The paper is structured as follows, Firstly, it describes the motivation behind the need to develop reactive models and gives a brief overview of the state of workflow systems. Secondly, it describes how rich models of activities and tasks are essential in building reactive workflow systems. Thirdly, it provides an overview of the ACP domain and fourthly, describes the component task models developed for reactive workflow. Fifthly, it describes the information model which has been developed to support the task models and sixthly, describes the DEOS scheduling engine developed to manipulate the task and information models. Finally, it provides a summary of current progress and describes two other domains in which the task models are being used.

# 2 Why Rich Task Models Are Important

In the ACP domain, as in many complex domains, the underlying process controlling the flow of information from user requirement to satisfaction defaults to a basic stove pipe. Requirements are pushed in one end, and after a fairly linear path, pushed out the other end. This encourages batch processing and creates an inflexible and unresponsive system. For a workflow manager to have a significant impact on the dynamic responsiveness of the system it must be able to manipulate and create context-dependent processes, thus, the requirement for rich task models.

Effective workflow management requires representations that makes the process logic explicit, thus allowing processes to be readily understood and adapted. [11]

Rich task models provide methods to enhance a Workflow Manager (WFM) capabilities to both select processes based on context and adapt processes to the dynamics of the environment. For example, hierarchical task models are desirable for modeling sophisticated processes because of their capacity to simplify complex tasks. They allow for the recording of attributes including, purpose, expected effects, applicability conditions, resource requirements, scheduling constraints, participants, and subprocesses that may be invoked. Task model representations can support the definition of metrics relating to the time, cost, or quality of performing the task, thus allowing comparisons and improvements to be made. Concepts such as authority and accountability are also useful and essential in many domains.

Another characteristic of a good task representation is the ability to support a rich set of control metaphors, including iteration, sequencing, concurrency, monitoring, testing, and suspension/resumption. Given the unpredictability of many domains, the ability to represent uncertainty is also critical. As discussed in the workflow literature [3,8], the task models must also be rich in low-level constructs allowing transactional capabilities, synchronization, and information exchange. The PAER and PRFER models described later have been developed with these characteristics in mind and thus allow WFM systems to tackle the dynamic and unpredictable domains of interest to DARPA and the business community.

## 2.1   Overview of the ACP Process

The ACP process defines the mechanism for translating high level objectives e.g. "gain and maintain air superiority by D+2", "destroy weapons of mass destruction by D-5", etc, set by the commander into actual targets and missions to be flown. This involves breaking down the objectives into sub-objectives and further refining these sub-objectives to tasks and missions. At each step a large number of agents (human and/or software) are involved to gather, refine and communicate information regarding the needs of the process. Within the current ACP process the main focus is the development of the Air Tasking Order (ATO) which describes which targets will be attacked, when and how. The USAF themselves have identified a number of problems and shortcomings in the ATO generation process and include:

- the need to move away from the "stove piped" way in which military plans are generated. For example, the generation of an ATO takes many hours with a new ATO issued every 12 hours.

- inability to rapidly respond to changes and requirements. The fixed nature of the ATO does not allow units to be re-tasked on the fly because there is little dependency and plan rationale recorded.
- lack of integration between different plans and forces into a fully integrated battle-space. Separate plans for ISR, tankering, ammunition, etc, mean that opportunities and problems can be missed leading to inefficient and potentially costly ATOs
- to make better use of the shrinking number of resources they have at their disposal. For example, by better allocating resources it means less tasks need to be redone and the quality of the final product increases.

Although the ATO planning process involves activities and their coordination, they are described in terms of the activities that take place in the planning process itself (such as plan, change, review, publish) rather than containing activities that relate to military effects (such as destroy, paralyze, delay, etc). Thus the ACP is a planning process whose outcome is a plan, i.e. the ATO. The ATO provides the low level detail which the pilots and mission planners need to plan at the domain level. At the domain level the planning involves allocating a number of resources, i.e. aircraft and weapons to specified targets at a time designated in the ATO. The problem is complicated by the ability of the aircraft to be reconfigured to suit the mission. For example, a flight of 4, A-10 Thunderbolts carrying AIM-65 Maverick anti-tank missiles would be a perfect match for missions against armored formations. However, the same flight of A-10s could be re-armed with MK-82 bombs and sent on a different mission. The problem is further complicated by the fact that weapons have a probabilities of both hitting a target and destroying it. The domain level planning process aims is to find an optimal assignment of aircraft to targets so as to avoid having to re-strike the target or causing unnecessary risk of aircraft loss or collateral damage. At the domain level there are a number of different ways in which the optimality can be measured, e.g. number of missions undertaken, number of targets destroyed, aircraft lost, etc.

The current aim is to turn both the process and the domain levels into workflow processes in which targets are assigned to aircraft on an as needed basis and the results and outcomes of the mission fed straight back into the process levels. If the target has been successfully dealt with then it is removed from the target list. If not then process level activities are initiated to deal with the outcome, e.g. re-strike immediately, integrate with a later mission, rerouting air to air refueling tankers and SEAD[1] aircraft, re-task an in-flight mission, etc. USAF refer to this as "just in time targeting" and has a lot of similarities with tasks found in manufacturing domains. This will be discussed in Section 4.

The original task models described in this paper were developed for the domain level (to improve coordination and feedback to the process level) but is was soon discovered that these models could be generalized to support planning and task management at the process levels as well. Details of both of these models are now provided.

---

[1] SEAD aircraft protect other aircraft from SAM attack

## 2.2    Task Models

Tasks models define a natural breakdown of a task into its constituent parts. Rather than have the task as a single component it is held as a series of sub-components which reflects the generic sub-activities necessary to support the task. This allows a tasking agent to create a better model of the processing the task needs and also allows a scheduling algorithm to better understand how to allocate resources, identify tradeoffs, asses changes and modify the associated workflow. To date two task models have been identified and applied to both the ACP domain and the ISR domain.

### Domain Task Model

For the ACP domain the generic domain task is divided into 5 sub-tasks or blocks and is referred to as the PRFER task model:

- **P**lan: Time taken for the pilot to plan the mission. Once a plan has been identified it is inserted in the slot for other workflow tasks to examine and check.
- **R**eady: Time taken to prepare the plane for the mission
- **F**ly: Time taken to get to the mission objective [2]
- **E**xecute: Time taken to execute the mission, e.g. drop weapons, unload food pallets, etc.
- **R**econstitute: Time taken to turn the aircraft round once it has returned to base.

Each task is associated with a task specification blocks (TSB) which is allowed to "breath" as changes in the domain are reflected as changes in one or more of the TSB's sub-blocks. For example, if the aircraft chosen for the mission develops a failure during its ready time then the "Ready" sub-task will expand and accommodate the extra time. To handle this change the the workflow engine may decide to substitute the aircraft for another if a spare aircraft exists or another can be re-weaponed in time. If no other aircraft is available then the workflow engine may try and reduce the time of the "Execute" block to recover the lost time. For example, if the aircraft is tasked with a food drop and the current method is to land and off load the supplies then the "Execute" block would be set to three hours. If it was changed to an air drop then the "Execute" block would drop to 30 minutes but the "Ready" time would increase due to the time taken to change the food pallets to an air drop configuration. By breaking the task into sub-components it allows the workflow engine to focus on ways to improve the schedule and recover from changes occurring in the domain and task. New TSBs can be added to the schedule as needed and removed just as easily should the decision be reversed. The DEOS scheduler described later in this paper keeps track of the criteria under which the task was spawned. On each cycle of

---

[2] This can be replaced by a "drive" or "sail" block for operations using land or sea transport

the scheduler it quickly reexamines the need and removes the task where appropriate. For example, if the food drop is being carried out using C-141 Starlifter aircraft then the food pallets must be loaded using a K-1 lifter However, if C-5 Galaxy aircraft are used then a K-1 is not needed. Once an aircraft is chosen then the workflow engine can examine the "Ready" block and examine needs for the aircraft type, in the case of a C-141 it requires a K-1 lifter. Other needs such as fuel can be attached to other sub-tasks e.g. "Fly".

By representing the schedule as a series of TSBs it becomes possible to create different perspectives and avoid the problem of resources being informed of changes which have no effect on them. Figure 1 shows a number of tasks being coordinated using the PRFER model.

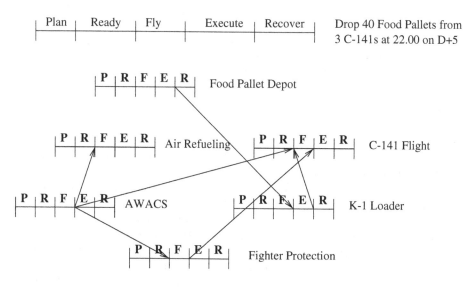

**Fig. 1.** PRFER Task Model Example

If the AWACS aircraft which is supplying air traffic control for the C-141s, the fighter protection and the air to air refueling tankers aborts on take off then the immediate effect is that its "Fly" block expands thus pushing its "Execute" block out further. This has the knock on effect of pushing the "Fly" blocks of the tankers, C-141 and fighters out as well. If the C-141s are on the ground then provided no later missions are impacted then they can remain there. If the C-141s are in the air they can loiter providing they have enough fuel and if not a new task should be spawned to provide them with enough fuel. The operator in the food depot is informed that he has longer to provide the food pallets and is hidden from why his "Execute" block has expanded. Should it be required then it is possible to follow the chain of reasoning back from the K-1 task to the AWACS problem. Normally, the K-1 operator can extract his own tasks without being concerned about the other tasks in the schedule.

The domain example presented here is defined for a number of strike and support aircraft. However, in the case of the K-1 lifter the model still applies accept that "Fly" has been replaced by "Drive" and the "Recover" step consists of refueling, cleaning and resetting the loader. Full details of the DEOS workflow engine which supports this and the PAER task models is provided in Section 3.

### Process Task Models

The generic process task model is broken down into 4 sub-tasks and is referred to as the PAER model:

- **P**lan: Time to plan the task. Once a plan has been identified it is inserted in the slot for other DEO tasks to examine and check.
- **A**cquire: Time to acquire the information e.g. process products [3] necessary to carry out the task. This also specifies the resources e.g. platforms, software packages, etc, needed to run the task.
- **E**xecute: Time to carry out the alloted task.
- **R**eport: Time to file or report the results of the task.

Again each task is associated with a TSB and can be handled with the same workflow engine as the PRFER model. During the "Plan" sub-task a number of requirements are identified and posted to the "Acquire" block. The workflow engine can generate new TSBs for these if there are no other tasks in the current schedule providing them. Alternatively, if another TSB is expected to generate the required document then its "Report" block can be modified by the workflow engine to provide an addition copy. In this way the execution of the PAER block is a partial order with some information gathering being carried out before all planning is complete. In addition, by identifying the information passing between process level TSBs and domain level TSBs it becomes possible to route the right information through the hierarchy. Figure 2 shows an example of how a PRFER task launched from an information product generated in a PAER task can be traced back should a problem occur with the PRFER task.

Information passing through the system is used to coordinate and trigger different TSBs. For example, once a target list moves from recommended to approved (through a vetting task) then other TSBs are triggered. However, the workflow engine could identify that a task can execute with the recommended target list to get a "jump start" and can finish its processing once a check has been made against the approved target list (in case changes have been made). This allows the workflow engine to trade off accuracy for time i.e. the recommended target list may change but if not the process can be considerably shortened. It is this type of adaptive workflow that the PAER and PRFER models are designed to support.

---

[3] Process products are the orders, documents, reports, letter, etc which are used to coordinate the workflow process

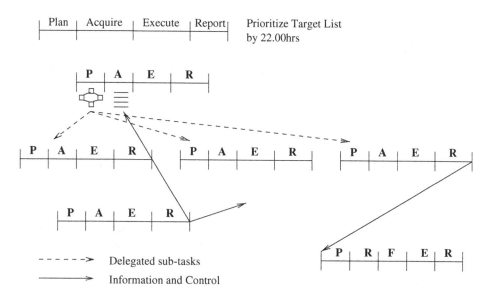

**Fig. 2.** PAER Task Model Example

## Examples of Task Model Use

The PRFER and PAER models can be used to handle a number of different situations occurring in the workflow process.

- **Change in a major process product e.g. commanders guidance**:
  This triggers an event in the workflow manager that a primary process product has changed status unexpectedly and that action should be taken. This means identifying the TSBs in the process which have the commander's guidance in their "Acquire" block. This would result in some tasks being suspended and others re-tasked to make use of the new information. If a secondary process product supporting the development of a new commanders guidance is delayed then the "Acquire" block of any task using it would be extended by the time required. This will have potential "knock on" effects with other tasks in the process. Should the support document be delayed indefinitely an alternative or inferior (the previous days) could be used instead.
- **Initial loss of a domain asset, e.g. U2 aircraft**:
  As with the previous example the assumption is that the change requires one or more current/planned tasks to be modified. This is achieved by identifying the tasks impacted by the lost asset. The workflow engine identifies the process products associated with the lost asset. These are then be updated and changed as appropriate (see above). The loss of the asset may also have an impact of the current processing of the system. For example, if an asset scheduler was carrying out a full schedule with data including the lost asset then the option might be to reduce the scheduler's task to a feasibility estimate to try and determine the consequence of the lost asset. This means

reducing the "execute" block to reflect a feasibility probe rather than the full schedule.

– **Loss of a process level asset, e.g. loss of asset scheduler:**
This involves adding news tasks and in some cases decomposing tasks to lower levels should there be no equivalent asset available. This can be handled simply by putting a new plan in the "Plan" slot and modifying its process product needs. For example, if some process products have been created or updated then the new task can ignore this need. If however, the task is yet to start then a simple substitution may be adequate.

– **Upgrade in the forces in the conflict and the need for feasibility estimates of the ISR needs:**
The above examples show workflow engine repairing events occurring in an already assigned series of tasks. There will be situations in which new tasks and requirements are added to the system. Such a situation would arise if a feasibility estimate is needed for future ISR requirements while trying to maintain the overall picture of the ISR process. This would mean identifying ways of scaling back current effort through changing "Plan" and "Execute" blocks of needed TSBs.

## 2.3    Information Model

Each of the tasks at either the process or domain level is specified using "task verbs" indicating the task to be performed. The tasks are described in the following form:

– **Verb**: the task to be carried out (e.g., analyze, develop, refine)
– **Noun Phrase(s)**: one or more noun phrases describing the object(s) or products on which the activity is being performed (e.g., prioritized target list)
– **Qualifier(s)**: zero, one or more qualifiers constraining how the activity is performed (e.g., time/resource limits)

The process products are modeled as resources that are created, modified, used, and authorized within the process. Examples of process products are the documents, reports, orders, letters, and communications (formal or informal). Authority relationships and other conditions are also modeled and can be used as an extension of the basic mechanism. The current task models and process products used by DEOS are encoded in the ACT representation [9], which can be directly executed by the Procedural Reasoning System (PRS) [10,6]. It is hierarchical and provides a rich scheme for both the representation of normative processes and the derivation of new processes based on AI reasoning and planning. The DEOS system has been incorporated in the SWIM workflow systems [1, 2] which uses a PRS system to decide which tasks the DEOS should work on and DEOS in return provides feedback on the types of processes that the PRS should select next.

Details of the development of the verb and process product models can be found in [1]. This framework is general enough to be applicable across various

**Table 1.** Part of the Verb/Noun(s)/Qualifier(s) Table

| Verb | Noun Phrase(s) | Qualifier Phrase |
|---|---|---|
| De-conflict | ACM Requests | |
| Finalize | Air Control Order<br>Special Instructions<br>Air Tasking Order | Quality Control |
| Produce | Air Tasking Order<br>Target Groupings<br>CAS Sortie Allocations<br>Potential Target List<br>Initial Target Nomination List<br>Weaponeering Assessment<br>Weaponeering Force Assessment<br>Mission Support Requirements | Broad |
| Release | Air Tasking Order | |
| Consider | Target and Route Threats | |

different military domains as well as those found in manufacturing, logistics and supply chain management. The *verb/noun(s)/qualifier(s)* (VNQ) model was originally developed for the ACP domain [5]. An overview of the verbs and process products are given in the following sections.

**Verb Descriptions**

A full list of the verbs and there definitions is given in [4] and a portion of the hierarchy is shown here which describes the decomposition of the major verbs **Analyze** and **Decide**.

- **Major Verb: Analyze**
  - **Notes**: **analyze** has notion of quantitative investigation
  - **Sub-Verbs**: predict, determine, monitor, diagnose, establish, measure.
- **Major Verb: Decide**
  - **Notes**: some of the meanings of **decide** imply finality (finalize, terminate), others seem to imply select by consideration (approve, calculate).
  - **Sub-Verb**: complete, finalize, approve, terminate, choose.

The verb model is also be use to describe the capabilities of the agents in the system. By having the capabilities and the tasks described in the same language it makes the resource assignment problem far easier.

**Process Objects and Products Model**

An example of part of the task to process product association is given in Table 2.3. Using the process product features identified from the CRUA matrices [4] it is possible to group features into classes and associate with each class a

---

[4] The CRUA matrix defines the process products, created, read, updated and authorized by the task.

descriptor type. For example, the features "available" and "not available" could be grouped together to form a single feature "availability" which can take one of these values. These features were used to form an ontology of primitive process product features which could in turn be used as the building blocks for more complex reasoning about the status of process products. For example, the status of the document could be (available, compound, published, draft). While this model has been developed from the ACP and ISR domains it is general enough to be applied to a wide range of workflow domains where systems (human and/or software) are required to work together.

Associated with each process product is a series of type attributes. A change in value of one of these attributes is used to trigger the start and finish of tasks in the schedule. Examples of product attributes are as follows

- **Availability: Boolean**: This describes the availability of the process product and simply defines whether the process product exists or not. Examples of this class are as follows: not_available or available.
- **Type-Information: Scalar**: The type-information describes the type of information contained within the process product. The types defined for each class of process product must be MIME compliant.
- **Contents_level: Scalar**: The contents_level defines the different levels at which the contents of a process product can be "measured".
- **Review-Status: Scalar**: This describes the review status of the process product as it is reviewed and passed through a process. Examples of review-status include: on-going, cut-off, final.
- **Approval/Recommendation-Status: Scalar**: This describes the status of a process product as it transitions from being recommended to approved. Examples of approval and recommendation status include: recommended, approved.

## 3    DEO Scheduler

The basic concept behind a DEO is to generate schedules quickly and to update them on the fly as new requirements and changes occur in the domain. A DEO schedule uses the PAER and PRFER models as it basic reasoning blocks. The basic technology behind the DEOS system is the Squeaky Wheel Optimization (SWO) system developed at CIRL. The basic SWO engine has been modified to accept the task models, handle uncertainty of task outcomes and the launching of new task requirements on the fly.

SWO is a scheduling technology developed for application to real-world scheduling applications [7]. The insight behind SWO is that in any real-world problem it is impossible to capture all associated constraints and that in most cases the context in which the constraints apply cannot be easily determined. SWO uses a priority queue to determine the order in which tasks should be released to a greedy scheduling algorithm. The priority queue is determined by how difficult the task is to deal with that is, i.e. higher the task is in the queue the harder it is to find a good resource assignment. The queue is not defined by external priorities identified by the user. On each iteration of the algorithm, SWO quickly creates a schedule and then examines it to identify the parts that were handled badly, for example, the task was completed too late or assigned to an unsuitable agent. Any task that "squeaks" is promoted up the priority queue, with the distance it is promoted determined by the extent of the problem. The new priority queue is then used to generate another schedule that is analyzed for problems. This process continues until no significant improvement in the schedule is noted over several iterations or a predefined limit is reached i.e. cycle count or elapsed time. SWO is extremely fast with each cycle of generate, analyze, and re-prioritize taking less than a second, even for large problems. In the ACP domain SWO is able to generate schedules every three or four seconds for schedules involving 2500 tasks. SWO has the advantage of allowing changes in the environment and task to be easily integrated into an ongoing solution. For example, a new task can simply be added to the priority queue and dealt with on the next cycle.

## 4  Summary and Further Work

This paper has presented an overview of a series of techniques for representing and reasoning with flexible and reactive task models. The task models are work in progress but already provide the foundation for workflow enabled reactive control. They allow for the explicit analysis of trade-offs in resource allocation, dynamic update of on going schedules, on the fly task spawning and for focussed impact analysis and repair. To date the task model has been applied to large scale ACP problems i.e. 2500 targets and 200 aircraft over a 10 day period and to process level tasks in the ISR domain. Further work with the Smart Workflow for ISR Manager (SWIM) will further validate the approach and techniques.

The PRFER and PAER models were originally developed for military applications. However, being generic models they are applicable across a wide number of domains including manufacturing, logistics and assembly. Work with Intel is intended to verify the models on a number of semi-conductor assembly problems

An example of the PAER model has also been studied in the area of mortgage application processing with a large Scottish Bank. Their existing problem was associating information arriving in support of an application to the original case (e.g., a credit report or the letter from the applicants employer). By explicitly adding information from the **P**lan, **A**cquire, and **R**eport phases to subsequent sub-tasks it becomes possible to create the necessary dependency chains. Further investigations in this domain are now being planned to examine resource utilization, information throughput and its implications on the Bank's development of call center operations.

# References

1. Berry, P.M. and Drabble, B.: SWIM: An AI-based System for Workflow Enabled Reactive Control, proceedings of the Workshop on Workflow and Process Management held as part of the International Joint Conference on Artificial Intelligence (IJCAI-99), (eds. B, Drabble and M. Ibrahim), Stockholm, Sweden, August, 1999.
2. Berry, P.M. and Drabble, B.: The AIM Process Modeling Methodology, AI Center, SRI International, Technical Report, SRI-AI-TR-120, Menlo Park, CA, 1999.
3. Cichocki, A., Helal, A.S., Rusinkiewicz, M. and Woelk, D.: Workflow and Process Automation: Concepts and Technology, Kluwer, 1998
4. Drabble, B. and Lydiard, T.J.: Model of the ACP Process Verbs and ACP Tool Capabilities, AIAI, Division of Informatics, University of Edinburgh, ISAT Technical Report, ISAT-AIAI/TR/1, 1996.
5. Drabble, B., Lydiard, T.J. and Tate, A.: Workflow Support in the Air Campaign Planning Process, proceedings of the Workshop on Interactive and Collaborative Planning held as part of Fourth International Conference on Artificial Intelligence Planning Systems (AIPS-98), (eds. K. Myers and G. Ferguson), CMU, Pittsburgh, PA, June, 1998.
6. Georgeff, M.P. and Ingrand, F.F: Decision-Making in an Embedded Reasoning System, proceedings of the Eleventh International Joint Conference on AI, Detroit, MI, July, 1989.
7. Joslin, D., and Clements, D.: Squeakywheel Optimization, proceedings of the Fifteenth National Conference on Artificial Intelligence, AAAI Press, Menlo Park, CA, August 1998.
8. Lawrance, P,: WfMC Workflow Handbook 1997, John Wiley and Sons, New York, ISBN: 0 471 969947 8, 1997.
9. Myers, K.L.: The ACT Editor User's Guide, AI Center, SRI International, Menlo Park, CA, 1993.
10. Myers, K.L.: A Procedural Knowledge Approach to Task-Level Control, proceedings of the Third International Conference on AI Planning Systems, (AIPS-96), AAAI Press, Menlo Park, CA, 1996
11. Myers, K.L. and Berry, P.M.: Workflow Management Systems: An AI Perspective, AI Center, SRI International, Technical Report, SRI-AI-TR-134, Menlo Park, CA, 1999.

# Greedy Algorithms for the Multi-capacitated Metric Scheduling Problem[*]

Amedeo Cesta[1], Angelo Oddi[1], and Stephen F. Smith[2]

[1] IP-CNR, National Research Council of Italy, Viale Marx 15, I-00137 Rome, Italy
{cesta,oddi}@ip.rm.cnr.it
[2] The Robotics Institute, Carnegie Mellon University, Pittsburgh, PA 15213, USA
sfs@cs.cmu.edu

**Abstract.** This paper investigates the performance of a set of greedy algorithms for solving the Multi-Capacitated Metric Scheduling Problem (MCM-SP). All algorithms considered are variants of ESTA (Earliest Start Time Algorithm), previously proposed in [3]. The paper starts with an analysis of ESTA's performance on different classes of MCM-SP problems. ESTA is shown to be effective on several of these classes, but is also seen to have difficulty solving problems with heavy resource contention. Several possibilities for improving the basic algorithm are investigated. A first crucial modification consists of substituting ESTA's pairwise analysis of resource conflicts with a more aggregate and thus more powerful Minimal Critical Set (MCS) computation. To cope with the combinatorial task of enumerating MCSs, several approximate sampling procedures are then defined. Some systematic sampling strategies, previously shown effective on a related but different class of scheduling problem, are found to be less effective on MCM-SP. On the contrary, a randomized MCS sampling technique is introduced, forming a variant of ESTA that is shown to be quite powerful on highly constrained problems.

## 1 Introduction

A number of efforts have recently focused on development of CSP scheduling algorithms for problems with cumulative resources (i.e., multi-capacity resources) [1,9] and generalized precedence relations (i.e., maximum and minimum metric separation constraints between activities) [6,10]. Dealing with such expressive modeling extensions provides a better match to real life situations and avoids abstracting away relevant aspects of practical scheduling domains.

In our recent research, we have addressed two distinct types of scheduling problems that simultaneously involve cumulative resources and generalized temporal relations:

[*] Amedeo Cesta and Angelo Oddi's work is supported by Italian Space Agency, by CNR Committee 12 on Information Technology (Project SCI*SIA), and CNR Committee 4 on Biology and Medicine. Stephen F. Smith's work has been sponsored in part by the National Aeronautics and Space Administration under contract NCC 2-976, by the US Department of Defense Advanced Research Projects Agency under contract F30602-97-20227, and by the CMU Robotics Institute.

- a machine scheduling problem called the *Multi-Capacitated Metric Scheduling Problem (MCM-SP)* [3]. Formally, MCM-SP is an extension of both the classical Job-Shop Scheduling Problem (JSSP) and the Multi-Capacitated JSSP introduced in [8]. The problem involves synchronization of a number of jobs. Each job consists of a linear sequence of activities with metric separation constraints on their execution. Resources have the capacity to simultaneously support multiple activities, and each activity requires a single unit of capacity of a single resource to execute. MCM-SP is characteristic, for example, of production scheduling problems in chemical plants.
- a project scheduling problem called *Resource Constrained Project Scheduling Problem with Time Windows (RCPSP/max)* [5]. RCPSP/max has a somewhat different problem structure, consisting of a partially ordered set of activities with minimum and maximum time lags between temporally related pairs. In this case, activities may require multiple resources to be executed and may require different amounts of capacity of each. RCPSP/max has several applications in make-to-order production environments.

In both investigations, attention has been given principally to CSP search control and the development of effective search heuristics. This can be contrasted with much of the other work in CSP approaches to cumulative scheduling problems, which has focused on complementary techniques for constraint propagation that exploit the structure of cumulative resource constraints and are hence capable of stronger inference.

A common aspect of the approach taken to both problems is the use of analysis of resource demand profiles to motivate heuristic decisions. The resolution algorithms of interest observe the resource consumption levels of a current (intermediate) solution and attempt to level contention peaks (i.e., resource conflicts) by posting additional ordering constraints between pairs of activities.

Our goal in this paper is to further analyze and enhance the performance characteristics of the best algorithm that has emerged from our prior research on multi-capacitated scheduling problems [3]. This algorithm, called Earliest Start Time Algorithm (ESTA), builds a solution to MCM-SP starting from a temporally consistent, infinite capacity schedule that is produced by ignoring resource constraints and scheduling activities as early as possible. We define new sets of problems (not considered in [3]) and show that, while ESTA performs well on several sets, it has difficulty with one problem set where resource availability is scarce and contention peaks are quite high. This set of problems provides a basis for investigating possibilities for overcoming this performance limitation of the ESTA approach. The variants of ESTA that we define import previous results from others [7] and our own prior work on RCPSP/max [5]. They also incorporate some new ideas motivated by the structure of the MCM-SP problem.

The previously defined ESTA procedure is distinguished by its use of a pairwise analysis of contention peaks, while all variants defined in this paper use an alternative conflict analysis based on Minimal Critical Sets (MCSs). A MCS is a set of activities of minimal size that violates a particular resource constraint. In all cases, a precedence constraint is posted between two activities contributing to the most critical conflict (a competing pair in ESTA; a MCS in each variant). Unfortunately, while computing all pairs of activities in a contention peak is inexpensive, computing all MCSs is not. Accordingly, an approximate MCS computation is defined for use in each ESTA variant. Each approximation (appropriately called "sampling strategy" here) gives rise to different performance of the greedy algorithm, and a newly defined random sampling technique is shown to be quite effective for MCM-SP.

Below, we first define the MCM-SP problem (Section 2) and then describe the experimental setting used in the paper (Section 3). In Section 4 we present a general algorithmic template for solving MCM-SP, the basic ESTA procedure, and its performance on new instances of problems. In Section 5 we introduce and evaluate a set of ESTA variants based on the MCS computation. Some conclusions close the paper.

## 2   The MCM-SP Problem

The Multiple-Capacitated Metric Scheduling Problem (MCM-SP) considered in this paper involves synchronizing the use of a set of resources $R = \{r_1 \ldots r_m\}$ to perform a set of jobs $J = \{j_1 \ldots j_n\}$ over time. The processing of a job $j_i$ requires the execution of a sequence of $m$ activities $\{a_{i1} \ldots a_{im}\}$. A resource $r_j$ can process at most $c_j$ activities at the same time (with $c_j \geq 1$) and execution of each activity $a_{ij}$ is subject to the following constraints:

- *resource requirement* – each $a_{ij}$ requires the use of a single unit of a single resource $r_{a_{ij}}$ for its entire duration.
- *processing time constraints* – each $a_{ij}$ has a fixed processing time $proc_{ij}$, such that $(e_{ij} - s_{ij}) = proc_{ij}^{max}$, where the variables $s_{ij}$ and $e_{ij}$ represent the start and end times respectively of $a_{ij}$. [1]
- *separation constraints* – for each pair of successive activities $a_{ij}$ and $a_{i(j+1)}$, $j = 1 \ldots (m-1)$, in job $j_i$, there is a minimum and maximum separation time, $sep_{ik}^{min}$ and $sep_{ik}^{max}$, such that $\{sep_{ik}^{min} \leq s_{i(k+1)} - e_{ik} \leq sep_{ik}^{max} : k = 1 \ldots (m-1)\}$.
- *temporal horizon* – all jobs must be executed within a temporal horizon of H time units. [2]

A solution to a MCM-SP is any temporally consistent assignment of start and end times which does not violate resource capacity constraints. Let $S$ be a current solution, $Act(S, k, t)$ be the set of activities which are in progress and using resource $r_k$ at time $t$, and $req_k(S, t) = |Act(S, k, t)|$ the requirement for $r_k$ at time $t$. A schedule is *resource-feasible* if all resources satisfy the constraint $req_k(S, t) \leq c_k$ for each $t$.

---

[1] Note that differently from [3,4] here we have fixed the durations of activities (as is usually assumed in the scheduling literature).

[2] In other words, release and due dates for jobs are not considered explicitly in the current experiments so they are fixed to 0 (the beginning of the temporal horizon) and $H$ (the end of the time-line) respectively for all the jobs.

MCM-SP was first introduced in [3], as an extension of the MCJSSP studied in [8,9] in which metric separation constraints between activities are additionally introduced. Such metric constraints are usually referred to as Generalized Precedence Relations (GPR) and are more constraining than the simple qualitative precedence used in classical Job-Shop formalization. [3] Scheduling problems with GPR are known to be very difficult (due in particular to the presence of maximal separation constraints); and in [2] it is shown that the feasibility problem alone is NP-Hard. In this paper, as in [3], we concentrate on the problem of generating feasible solutions to MCM-SP instances.

## 3   An Experimental Setting

In this section, we establish a common basis for experimental comparison of alternative solution procedures to the MCM-SP. We first describe a reproducible procedure for generating test problems. Second, we specify relevant evaluation criteria.

*Experimental Design.* To establish a reproducible experimental setting we utilize the "controlled" random number generator proposed in [11]. In this way we obtain a uniform distribution function $U[a, b]$ which generates a random number $n$, where $n$, $a$ and $b$ are positive numbers such that $a \leq n \leq b$ (if both $a$ and $b$ are integers then $n$ is an integer, if one of them is real $n$ is real). To generate problem instances we use the first 50 seeds reported in Figure 1 of [11].

The dimensions along which problem instances will be varied follows the usual format for formulating job shop scheduling problems. Using the terminology $N_{jobs} \times N_{res}$ we define problems with $N_{jobs}$ jobs and $N_{res}$ resources. Each job is composed of a sequence of $N_{res}$ activities, and visits all resources in a randomly generated order (i.e., each activity requires capacity of a different resource). For our purposes in this paper, we create problem sets of 50 instances at each of the following sizes: $15 \times 5$, $20 \times 5$ and $25 \times 5$.[4]

The problem sets generated at each size are further specified as follows:

- every resource $r_j$ has a capacity $c_j$, generated randomly as either $U[2, 3]$ or $U[2, 5]$, and is available over the entire horizon;
- the processing time of activities is drawn from a uniform distribution $U[10, 50]$;
- the separation constraints $[a, b]$ between every two consecutive activities in a job are generated with $a = U[0, 10]$ and $b = U[40, 50]$;
- the horizon H is computed as $H = m_v H_0$ where $H_0$ is adapted from [6]: $H_0 = (N_{jobs} - 1)p_{bk} + \sum_{i=1}^{N_{res}} p_i$, where $p_{bk}$ is the average processing time of the activities on the bottleneck resource, and $p_i$ is the average minimum

---

[3] An alternative name for MCM-SP could be MCJSSP with GPR but we prefer to keep the first shorter name.

[4] in [3] we also solved $5 \times 5$ and $10 \times 5$ problems but these sets do not add relevant information to the current investigation.

processing time of the activities on resource $r_i$. The bottleneck resource is the resource with the maximum total required processing time. The parameter $m_v$ is used to reduce the horizon and increase the problem constrainedness.

In [3] the scenario with $c_j \in [2,3]$ and $m_v = 1$ was studied and ESTA emerged as a reliable heuristic for MCM-SPs. We designate this scenario as Experiment 1 and specify three further experiments by varying both the range of resource capacity levels and the duration of the temporal horizon. Specifically, we define Experiment 2 to be the scenario $c_j \in [2,5]$ and $m_v = 1$, Experiment 3 to be the scenario $c_j \in [2,5]$ and $m_v = 0.7$, and Experiment 4 to be the scenario $c_j \in [2,3]$ and $m_v = 0.7$.

Reasoning qualitatively about these four scenarios, we expect that for a given temporal horizon, the problem becomes more difficult as resource capacity levels are lowered (since the required capacity is constant for a given number of jobs). Likewise, as the horizon is decreased, we also obtain more difficult problems, because there is "less temporal space" in which to schedule activities. Overall, we expect Experiment 4 to present the most constrained conditions.

*Evaluation Criteria.* Since our interest is in generating feasible solutions, we will evaluate the performance of a given solution procedure along two dimensions:

- the *number of problems solved* $(N_{ps})$ from a fixed set;
- the *average CPU time*, in seconds, required to solve each instance.

## 4   Constructing a Feasible Solution with ESTA

ESTA and all variants described in this paper (with one exception) are single-pass procedures. They share several common features:

- All strategies are constructive and greedy. All proceed by iteratively modifying a flawed (intermediate) solution, and base decisions about which flaw to resolve next on an analysis of current solution state.
- All start from an infinite capacity, earliest start time solution, i.e., a solution that satisfies all temporal constraints and starts every activity as soon as possible (ignoring resource capacity constraints).
- All construct and utilize "profiles" of resource usage over time to detect resource capacity conflicts (flaws) and identify competing activities; hence the name "profile-based" methods.
- All insert (or "post") new ordering constraints between activities as a means of leveling resource profiles and achieving a conflict-free solution. Ordering constraints are chosen so as to retain as much temporal flexibility as possible.

Given the similarities, we first introduce a general algorithmic template, and then describe each algorithm as an instantiation of this general schema.

**PBA**(Problem)
1. ESTsol ← Problem$_{EST}$
2  SetHorizon(ESTsol, Horizon)
3. **loop**
4.         Conflicts ← ComputeConflicts(ESTsol)
5.         **if** ResourceFeasible(Conflicts)
6.         **then return**(ESTsol)
7.         **else**
8.                 **if** ExistUnsolvableConflict(Conflicts)
9.                 **then return**(EmptySolution)
10.                **else**
11.                        Conflict ← SelectConflict(Conflicts)
12.                        PrecC ← SelectPrecedence(Conflict)
13.                        Post&Update(PrecC,ESTsol)
14. **end-loop**
15. **end**

**Fig. 1.** The One-Pass, Profile-Based Algorithm

### 4.1   A Basic Algorithmic Template

Figure 1 specifies the basic "conflict removal" procedure used to generate a feasible solution, referred to as a *Profile-Based Algorithm* (PBA). The algorithm accepts a problem instance (**Problem**). It first computes an earliest start time solution, assuming that all resources have "infinite capacity" (**Problem**$_{EST}$ at line 1). It then attempts to incrementally transform this initial time-feasible solution into a resource-feasible solution. On each iteration of the loop, a resource conflict still present in the current solution is selected and resolved, by posting a precedence constraint between two competing activities.

   Given a time-feasible, earliest start time schedule (**ESTsol**), it is straightforward to identify time instants $t$ where the resource capacity constraint of a given resource is violated. We say that there is a *contention peak* on resource $r_k$ at time $t$ if condition $req_k(\texttt{ESTsol}, t) > c_k$ holds. Intuitively, a contention peak on resource $r_k$ characterizes a conflict, identifying a set of activities that simultaneously requires $r_k$ but have a combined capacity requirement $> c_k$.

   Obviously feasible solutions do not contain peaks, and one possible way to transform **ESTsol** into a feasible solution is to detect and eliminate all peaks. This is the basic idea behind *profile-based* scheduling procedures [3]. A peak can be removed by "leveling" it, that is by posting one or more precedence constraints between pairs of activities contributing to the peak (such constraints are posted between the end-time of one activity and the start-time of the other to avoid overlap). In leveling a conflict, the choice of which pair(s) of activities to order can be quite crucial to the effectiveness of the algorithm, as it is possible to add precedence relations that do not have a strong peak leveling effect.

   The posting of a new precedence constraint triggers propagation in an underlying Temporal Constraint Network representation of **ESTsol** (step Post&Update

at Line 13). In this network, the nodes are time points that represent the start and end times of activities (as well as the begin and end of the temporal horizon), and the edges are (temporal) distance constraints that represent both the temporal constraints of the problem and the ordering constraints added by the algorithm. The propagation performed in response to posting a new ordering constraint is an incremental all-pairs shortest path computation, which updates the current distances between any pair of time points. This information provides a basis for confirming the continued temporal consistency of ESTsol as well as a basis for assessing current temporal flexibility (see below). In the following, we let $d(t_i, t_j)$ denote the length of the shortest path from $t_i$ to $t_j$.

## 4.2   ESTA's Pairwise Conflict Analysis

ESTA is obtained from PBA by specifying its conflict analysis and conflict resolution strategies. These strategies derive directly from those utilized in previous studies of disjunctive scheduling problems [6,10]; in ESTA [3], they are extended to cumulative scheduling problems.

More specifically, ESTA relies on the use of a pairwise conflict analysis to direct the search process. A conflict in ESTA is defined to be any pair $\langle a_h, a_l \rangle$ of activities in a given contention peak. Relative to any such *pairwise conflict set*, it is possible to define a set of *dominance conditions*, which identify unconditional decisions and promote early pruning of ordering alternatives. Four possible cases of conflict are defined (introduced in [6]):

**Condition 1:** $d(e_{a_h}, s_{a_l}) < 0 \land d(e_{a_l}, s_{a_h}) < 0$. In this case we have a *pairwise unresolvable conflict*. There is no way to order $a_h$ and $a_l$ without introducing a temporal inconsistency.

**Condition 2:** $d(e_{a_h}, s_{a_l}) < 0 \land d(e_{a_l}, s_{a_h}) \geq 0 \land d(s_{a_h}, e_{a_l}) > 0$. This is a case of a *pairwise uniquely resolvable conflict*. There is only one feasible ordering of $a_h$ and $a_l$ (namely $a_l\{before\}a_h$) and the decision of which constraint to post is thus unconditional.

**Condition 3:** $d(e_{a_l}, s_{a_h}) < 0 \land d(e_{a_h}, s_{a_l}) \geq 0 \land d(s_{a_l}, e_{a_h}) > 0$. Again a case of *pairwise uniquely resolvable conflict*. $a_h\{before\}a_l$ is the only feasible ordering.

**Condition 4:** $d(e_{a_h}, s_{a_l}) \geq 0 \land d(e_{a_l}, s_{a_h}) \geq 0$. In this case, we have a *pairwise resolvable conflict*. Both orderings of $a_h$ and $a_l$ remain feasible, and it is necessary to make a choice.

We can now specify ESTA by describing the three basic steps of the general PBA schema. First, lets consider the temporal feasibility check. Satisfaction of the predicate ExistsUnsolvableConflict (line 8 of Figure 1) corresponds to a contention peak where, for each pairwise conflict set $\langle a_h, a_l \rangle$, with $a_h \neq a_l$, Condition 1 holds.

The function SelectConflict (line 11 of Figure 1) is concerned with selection of a pairwise conflict $\langle a_h, a_l \rangle$ within a resolvable peak. The general least-commitment principle behind this decision is to retain the maximum amount of temporal flexibility possible in the solution at each step of the search. In this regard, the distances $d(e_{a_h}, s_{a_l})$ (or $d(e_{a_l}, s_{a_h})$) characterize the number of mutually consistent time assignments remaining for a given pair of activities $\langle a_h, a_l \rangle$. Two cases are distinguished. When all pairwise conflicts in the current solution satisfy Condition 4, then the conflict set $\langle a_h, a_l \rangle$ with the minimum value $\omega_{res}(a_h, a_l)$ is selected [5], where:

$$\omega_{res}(a_h, a_l) = min\{\frac{d(e_{a_h}, s_{a_l})}{\sqrt{S}}, \frac{d(e_{a_l}, s_{a_h})}{\sqrt{S}}\}$$

with $S = \frac{min\{d(e_{a_h}, s_{a_l}), d(e_{a_l}, s_{a_h})\}}{max\{d(e_{a_h}, s_{a_l}), d(e_{a_l}, s_{a_h})\}}$. Alternatively, when one or more pairwise conflicts satisfies Condition 2 or 3, the heuristic selects the conflict with the minimum (and negative) value $\omega_{res}(a_h, a_l) = min\{d(e_{a_h}, s_{a_l}), d(e_{a_l}, s_{a_h})\}$. This corresponds to the pair of activities that is closest to having its ordering decision forced.

Finally, SelectPrecedence (line 12 of Figure 1) simply returns the ordering constraint which leaves the most temporal flexibility. If $d(e_{a_h}, s_{a_l}) > d(e_{a_l}, s_{a_h})$ the leveling constraint chosen is $a_h \leq a_l$, otherwise $a_l \leq a_h$.

### 4.3   Experiments with ESTA

Figure 2 shows the results of the extended performance analysis of ESTA. Results on Experiment 1 differ from those obtained in [3] for two reasons: (1) activity durations are now assumed fixed (even though an equivalent problem generation scenario was used) and (2) a more efficient software implementation has been developed. Some interesting observations can be made: (a) ESTA performs well not only on Experiment 1 and 2 (which vary only in resource capacity levels) but also on Experiment 3, where a a 30% decrease in the temporal horizon significantly increases the level of contention; (b) as expected, the problem sets in Experiment 4 are more difficult, particularly as the number of jobs are increased. These problem sets provide a basis for investigating possible improvements of ESTA. The rest of the paper is dedicated to this goal.

## 5   Introducing Analysis of Minimal Critical Sets

An alternative approach to conflict analysis has been proposed in [7] where, for each resource, an activity "intersection graph" is constructed and systematically searched for cliques which represent *Minimal Critical Sets* (MCSs). A MCS speci-fies *a set of activities that simultaneously require a resource $r_k$ with a combined*

---

[5] As suggested in [6] a balancing factor $\sqrt{S}$ is used. It is possible to see that $S \in [0, 1]$: $S = 1$ when $d(s_{a_h}, s_{a_l}) = d(e_{a_l}, s_{a_h})$ and it is close to 0 when $d(s_{a_h}, s_{a_l}) >> d(e_{a_l}, s_{a_h})$ or $d(e_{a_l}, s_{a_h}) >> d(s_{a_h}, s_{a_l})$. The aim of this balancing factor is to first select *conflicts* in which both choices are highly constrained and close to the failure state (dominance condition 1).

| Experiment 1 | | | Experiment 2 | | |
| --- | --- | --- | --- | --- | --- |
| $m_v = 1,\ c_j \in [2,3]$ | | | $m_v = 1,\ c_j \in [2,5]$ | | |
| Problem | $N_{ps}$ | CPU | Problem | $N_{ps}$ | CPU |
| $15 \times 5$ | 50 | 8.1 | $15 \times 5$ | 50 | 6.8 |
| $20 \times 5$ | 50 | 21.2 | $20 \times 5$ | 50 | 16.7 |
| $25 \times 5$ | 50 | 44.6 | $25 \times 5$ | 50 | 35.0 |
| Experiment 3 | | | Experiment 4 | | |
| $m_v = 0.7,\ c_j \in [2,5]$ | | | $m_v = 0.7,\ c_j \in [2,3]$ | | |
| Problem | $N_{ps}$ | CPU | Problem | $N_{ps}$ | CPU |
| $15 \times 5$ | 48 | 6.7 | $15 \times 5$ | 50 | 8.5 |
| $20 \times 5$ | 49 | 17.2 | $20 \times 5$ | 43 | 21.5 |
| $25 \times 5$ | 48 | 36.2 | $25 \times 5$ | 40 | 62.0 |

**Fig. 2.** Evaluating ESTA in different problem classes

*capacity requirement* $> c_k$, *such that the combined requirement of any subset is* $\leq c_k$. The important advantage of isolating a MCS is that a single precedence relation between any pair of activities in the MCS eliminates the resource conflict. In this section, we define several ESTA variants that exploit this more aggregate approach to conflict analysis.

## 5.1 Computing the Most Critical MCS

Similar to the approach taken in ESTA, we can utilize information about the temporal flexibility (or lack thereof) of various MCSs as a means of determining the most critical conflict to resolve at each step of the search and, once determined, to decide which ordering constraint to post. In [7], a heuristic estimator $K$ is proposed which provides a direct basis for generalizing ESTA's conflict selection and conflict resolution heuristics to apply to candidate MCSs. Given a candidate MCS and a set $\{pc_1 \ldots pc_k\}$ of precedence constraints that could be posted between pairs of activities in the MCS, [6] $K(\text{MCS})$ is defined as follows:

$$\frac{1}{K(\text{MCS})} = \sum_{i=1}^{k} \frac{1}{1 + commit(pc_i) - commit(pc_{min})}$$

where $commit(pc_i)$ ranges from 0 to 1 and estimates the loss in temporal flexibility as a result of posting constraint $pc_i$, and $pc_{min}$ is the precedence constraint with the minimum value of $commit(pc)$. [7]

Note that $K(\text{MCS})$ takes on its highest value of 1 in those cases where only one specific precedence constraint can be feasibly posted to resolve the conflict.

---

[6] Notice that $pc_k$ represents a possible constraint $a_h \leq a_k$. To be posted such a constraint requires that $d(e_{a_h}, s_{a_k}) \geq 0$ holds.

[7] $commit(pc)$ is computed as in [7]. It is a ratio between two temporal distances.

In general, the closer an MCS is to being unresolvable, the higher the value of $K(\text{MCS})$. The conflict selection heuristic (`SelectConflict`) chooses the MCS with the highest $K$ value, and the conflict resolution heuristic (`SelectPrecedence`) simply chooses $pc_{min}$.

## 5.2  Deterministic Selection of MCSs

The principal obstacle to use of MCS analysis is the computational expense associated with enumerating MCSs in the current `ESTsol`. In fact, the exponential nature of a complete search of the intersection graph (as utilized in [7]) prohibits its use on scheduling problems of any interesting size. In [4], it is shown that much of the advantage of this approach to conflict analysis can be retained by using an approximate procedure for computing MCSs on the "intersection graphs". However, the cost of recomputing MCSs across "intersection graphs" of all resources at each iteration of the CSP resolution procedure nonetheless remains high and significantly limits scalability.

In PBA, we can achieve an even better computational tradeoff by instead integrating the use of MCS analysis into the profile-based scheduling framework. On each iteration of the search, we first compute contention peaks (which is quadratic in the number of activities) to isolate those areas of the solution where conflicts (i.e., MCSs) should be computed. Next we generate a set of MCSs for each peak. (These two steps are embedded in the `ComputeConflicts` function of line 4 in the algorithm description of Figure 1.) The number of MCSs contained in a given peak can still be quite large (in the worst case $\binom{|V|}{c_k+1}$). Accordingly, we still require an approximate (heuristic) scheme for computing (or "sampling") MCSs.

Given the uniform resource requirement of all activities in MCM-SP instances, we specify two simple polynomial algorithms for extracting a subset of MCSs from a given contention peak

**Linear sampling.** A queue Q is used to select an MCS from a contention peak P. Activities $j \in P$ are considered sequentially and inserted in Q until the sum of the resource requirements exceed the resource capacity. At this point, the set Q (the current MCS) is saved in a list of MCSs and the first element from $Q$ is removed. The previous steps are iterated and MCSs are collected until there are no uninserted activities in $P$.

**Quadratic sampling.** This scheme can be seen as an extension of linear sampling in which the second step is expanded as follows. Once the current MCS has been collected, instead of immediately removing the first element from Q, a forward search through the remaining uninserted activities in $P$ is first performed to also collect all MCSs that can obtained by dropping the last item placed in Q and substituting with single subsequent activities in $P$ until an MCS is still composed.

## 5.3   Randomized Selection of MCSs

As we will see in the experimental section, MCS variants of ESTA using either of above deterministic sampling strategies outperform the basic "pairwise analysis" approach of ESTA on Experiment 4. However, neither variant is able to find a solution to a subset of the most difficult problems in these sets. The reason seems to stem from the fact that MCSs are sampled rather arbitrarily. In fact, both sampling strategies were originally designed for domains like RCPSP/max where advantage can be taken of the fact that activities require variable amounts of capacity. In such cases, even a linear sampling procedure can be an effective heuristic sampling strategy if the list of competing activities is first ordered by decreasing size of the capacity requirements [5]. In the case of MCM-SP, however, all activities have the same requirement (one unit of capacity), and there is no information to productively bias the sampling. Both strategies become simply a way to sample a subset of MCSs at a low polynomial cost.

Given that all activities uniformly require one unit of capacity, the number of MCSs in a given peak is actually quite high. For example, if a peak contains 20 activities and the resource capacity is 5, the number of MCSs is $\binom{20}{6} = 38760$!. However, we can expect that the associated values of K are distributed in the interval [0,1]. Hence it is reasonable to assume that there are several MCSs with maximal values of K, as well as a number of MCSs with values of K quite close to the maximal. If one of these MCSs could be efficiently computed it would clearly constitute a critical MCS and likely be a good heuristic choice.

To attempt to compute such an MCS (and in the process overcome the limitations of our deterministic sampling strategies), we define a simple, non-deterministic procedure: for each contention peak we randomly sample a sequence of MCSs, each time computing the corresponding K value. We continue to randomly sample MCSs until either the value of K "stabilizes" (i.e., some pre-specified number of random samples are made without finding an MCS with a better K value) or a pre-specified maximum number of samples have been generated. With regard to previously defined MCS variants of ESTA, this procedure replaces the conflict selection heuristic defined in Section 5.1 (Line 11 of the PBA algorithm description in Figure 1). Note that since the procedure is randomized, each time it is called with a given contention peak it will return a different critical MCS.

**An Iterative ESTA Algorithm.** The presence of a randomized step in the PBA resolution procedure, as defined above, makes it possible to restart the search in the event of failure, to obtain different but "heuristically equivalent" search paths, and ultimately to increase the probability of finding a solution. This possibility leads to definition of *restartPBA* shown in Figure 3. This restarting procedure, coupled with the randomized conflict selection step, constitutes our final variant of ESTA.

It is worth noting that the use of randomization here is different from its use in [10,5], where it is used to select among similar choices (e.g. similar pre-

restartPBA(Problem, MaxRestart)
1.       Sol ← EmptySolution
2.       FoundSoution ← NIL
3.       Counter ← 1
4.       **While** (Counter ≤ MaxRestart **and not**(FoundSolution)) **do**
5.              Sol ← **PBA**(Problem)
6.              if IsSolution(Sol)
7.                      **then** FoundSolution ← T
8.                      **else** Counter ← Counter + 1
9.       **return**(Sol)
10.     **end**

**Fig. 3.** The Restarting PBA Procedure

computed $K$s in [5]). Here it is used to repeatedly sample MCSs until a local optima is encountered and no further improvement in $K$ value is possible.

### 5.4   Comparing Algorithms on Difficult Instances

Figure 4 compares the performance of the different solution strategies on the most difficult Experiment 4 problem sets. Here ESTA refers to basic ESTA, ESTA-M-L to the variant that uses MCS analysis with linear sampling, ESTA-M-Q to the variant with quadratic sampling of MCSs, and ESTA-RND to the algorithm with randomization and multiple restarts (with "stabilization" checks after every 5 samples and a maximum of 10 checks). We observe the following:

- all MCS-based variants outperform basic ESTA on the $5 \times 20$ and $5 \times 25$ problem sets;
- neither deterministic sampling strategy is able to solve all instances;
- ESTA-RND solves all problem instances in an acceptable amount of time, and, as problem size increases, becomes more efficient than ESTA-M-Q.

| | ESTA | | ESTA-M-L | | ESTA-M-Q | | ESTA-RND | |
|---|---|---|---|---|---|---|---|---|
| | $N_{ps}$ | Cpu | $N_{ps}$ | Cpu | $N_{ps}$ | Cpu | $N_{ps}$ | Cpu |
| $5 \times 15$ | 50 | 8.5 | 50 | 7.7 | 49 | 7.9 | 50 | 18.2 |
| $5 \times 20$ | 43 | 21.5 | 47 | 19.7 | 46 | 21.7 | 50 | 72.3 |
| $5 \times 25$ | 40 | 62.0 | 44 | 134.0 | 47 | 214.6 | 50 | 192.9 |

**Fig. 4.** Comparison on Experiment 4: $m_v = 0.7$, $c_j \in [2,3]$

## 6   Conclusions

This paper presents a range of greedy algorithms for solving the MCM-SP problem. All algorithms considered have a structure similar to the ESTA procedure introduced in [3], but substitute an analysis of minimal critical sets (MCSs) for ESTA's pairwise conflict analysis. This enhancement is tested on difficult benchmark problems.

One interesting result that emerges is confirmation that the structure of scheduling problems involving cumulative (multi-capacitated) resources is well captured by MCS conflict analysis; in this sense this paper complements the results given in [5] for RCPSP/max with a comparable analysis for MCM-SP.

The main contribution of the paper is development of a randomized procedure for MCS conflict selection. This variant of ESTA is particularly well-suited for MCM-SP, where the resource requirements of activities are uniform, and it is shown to efficiently solve all "difficult" problem instances. It is interesting to notice that randomization is used in a different manner than in [10,5] to achieve the same goal of exploring different but "heuristically equivalent" paths in the search space.

# References

1. P. Baptiste, C. Le Pape, and W.P.M. Nuijten. Satifiability Tests and Time-Bound Adjustments for Cumulative Scheduling Problems. Technical report, Univerity of Compiégnie, 1997. to appear in *Annals of Operations Research*.

2. M. Bartusch, R. H. Mohring, and F. J. Radermacher. Scheduling Project Networks with Resource Constraints and Time Windows. *Annals of Operations Research*, 16:201–240, 1988.

3. A. Cesta, A. Oddi, and S.F. Smith. Profile Based Algorithms to Solve Multiple Capacitated Metric Scheduling Problems. In *Proceedings of the Fourth Int. Conf. on Artificial Intelligence Planning Systems (AIPS-98)*, 1998.

4. A. Cesta, A. Oddi, and S.F. Smith. Scheduling Multi-Capacitated Resources under Complex Temporal Constraints. Technical Report CMU-RI-TR-98-17, Robotics Institute, Carnegie Mellon University, 1998.

5. A. Cesta, A. Oddi, and S.F. Smith. An Iterative Sampling Procedure for Resource Constrained Project Scheduling with Time Windows. In *Proceedings of the 16th Int. Joint Conference on Artificial Intelligence (IJCAI-99)*, 1999.

6. C. Cheng and S.F. Smith. Generating Feasible Schedules under Complex Metric Constraints. In *Proceedings 12th National Conference on AI (AAAI-94)*, 1994.

7. P. Laborie and M. Ghallab. Planning with Sharable Resource Constraints. In *Proceedings of the International Joint Conference on Artificial Intelligence (IJCAI-95)*, 1995.

8. W.P.M. Nuijten. *Time and Resource Constrained Scheduling - A Constraint Satisfaction Approach*. PhD thesis, Eindhoven University of Technology, The Netherlands., 1994.

9. W.P.M. Nuijten and E.H.L. Aarts. A Computational Study of Constraint Satisfaction for Multiple Capacitated Job Shop Scheduling. *European Journal of Operational Research*, 90(2):269–284, 1996.

10. A. Oddi and S.F. Smith. Stochastic Procedures for Generating Feasible Schedules. In *Proceedings 14th National Conference on AI (AAAI-97)*, 1997.

11. E. Taillard. Benchmarks for Basic Scheduling Problems. *European Journal of Operational Research*, 64:278–285, 1993.

# Automata-Theoretic Approach to Planning for Temporally Extended Goals

Giuseppe De Giacomo[1] and Moshe Y. Vardi[2]

[1] Dip. Informatica e Sistemistica, Univ. Roma "La Sapienza",
Via Salaria 113, I-00198 Roma, Italy
degiacomo@dis.uniroma1.it
[2] Department of Computer Science, Rice University,
P.O. Box 1892, Houston, TX 77251-1892, U.S.A.
vardi@cs.rice.edu

**Abstract.** We study an automata-theoretic approach to planning for temporally extended goals. Specifically, we devise techniques based on nonemptiness of Büchi automata on infinite words, to synthesize sequential and conditional plans in a generalized setting in which we have that: goals are general temporal properties of desired execution; dynamic systems are represented by finite transition systems; incomplete information on the initial situation is allowed; and states are only partially observable. We prove that the techniques proposed are optimal wrt the worst case complexity of the problem. Thanks to the scalability of the nonemptiness algorithms, the techniques presented here promise to be applicable to fairly large systems, notwithstanding the intrinsic complexity of the problem.

## 1 Introduction

Artificial Intelligence has always been interested in the analysis and synthesis of dynamic systems behavior. In particular, the research area of reasoning about actions has been concerned with representing and reasoning on such systems in order to *analyze* interesting properties of their behavior; the area of planning has instead been concerned with the *synthesis* of devices (*plans*) in order to control the system behavior so as to achieve desired conditions (*goals*). The two areas have developed their research in quite different directions. In reasoning about actions a lot of work has been done in finding ways to represent and reason on dynamic systems and dynamic properties of increasing generality [39, 42,41]. In the area of planning the focus has been in achieving effectiveness of the planning process (and lately notable results have been obtained [29,5]), while being contented with limited capabilities both in modeling dynamic system and in the kind of goals considered.

In this paper, we study synthesis of *sequential* and *conditional* plans[1] in a setting which is close to that considered in the area of reasoning about actions.

---

[1] We call *sequential plans* plans that are sequences of primitive actions, and *conditional plans* plans that include choice points to be resolved by ascertaining given conditions at runtime.

S. Biundo and M. Fox (Eds.): ECP-99, LNAI 1809, pp. 226–238, 2000.

In particular following, in spirit, Reiter's formalization of dynamic systems in the Situation Calculus [39], we assume (i) *incomplete information* on the initial situation (several initial states are compatible with the information available on the initial situation); (ii) *deterministic actions*, that is, performing an action in a state brings about a univocally determined next state; (iii) actions with *conditional effects*, that is, the effects of an action depend on the state in which the action is performed. Also, we allow for *partially observable states*. In other words, the agent can observe only part of the state and hence its choices on the action to perform next may depend only on that part. The kind of goals we consider are *temporally extended goals*, i.e., goals that specify acceptable *sequences of states* as in [2,27]. This kind of goal subsumes the usual goals expressing *reachability* of desired conditions, as well as generalized goals as *don't-disturb* and *restore* requirements [47]. More generally, complex temporal properties typically used in the specification of processes can be expressed [16,31,46]. Observe that as we deal with goals expressing general temporal properties, even sequential plans may in fact involve loops, since goals of this form may require *infinite executions*. Consider, for example, a plan to satisfy the requirement that whenever certain triggering conditions are met within a finite (but undetermined) number of steps, a specified state of affairs must be brought about in which the triggering conditions are met again. The major compromise we accept, in order to get effective planning techniques in the outlined setting, is to restrict our attention to systems that have a *finite number of states*. Although this is a radical simplification wrt [39], it is a compromise that is widely accepted in planning.

Formally, we model dynamic systems as finite *transition systems*. A transition system can be thought as a graph, where *nodes* represent states and are labeled by the part of the state that is observable, while *edges* represent state transitions and are labeled by actions that cause the state transitions. Several representation formalisms in AI are based on transition systems. For example, both STRIPS-like formalisms [18,3] and Action Languages [22] are formalisms to compactly represent transition systems (of different generality). Furthermore, formalizations based on logics such as the Situation Calculus [34,39], or Dynamic Logics [40,13], are also tightly related to transition systems.

As goal specification we adopt automata on infinite words: the desirable traces correspond to the language accepted by the automaton. This way of specifying goals is very close to adopting linear time temporal logic (LTL) [16] as goal specification language, since every LTL-based specification can be translated into an automaton-based one [46].

In such a framework we establish techniques and characterize the worst-case computational complexity of synthesizing sequential and conditional plans by adopting an approach based on the theory of automata on infinite objects (infinite words in our case), an approach that is widely used in verification of hardware and control-intensive (as opposed to data-intensive) software [45,44].

The rest of the paper is structured as follows. We first introduce transition systems and Büchi automata on infinite words. Then, we study sequential planning when the initial state is unique (we have complete information on the initial situation) and states are fully observable. In Section 4 and Section 5, we study sequential planning and conditional planning when the initial state is not unique and states are only partially observable. Then, we briefly discuss algorithmic techniques, and related works. Finally, we draw some conclusions.

# 2   Preliminaries

## 2.1   Transition Systems

A (finite) *transition system* $\mathcal{T}$ is defined as $\mathcal{T} = (W, w_0, Act, R, Obs, \pi)$ where:

- $W$ is the finite set of possible states.
- $W_0 \subseteq W$ is the finite set of possible initial states
- $Act$ is the set of possible actions.
- $R : W \times Act \to W$ is the transition function (actions are deterministic), i.e., a function that given a state and an action return the next state.[2]
- $Obs$ is the finite set of possible observations, which model the observable part of states.
- $\pi : W \to Obs$ is the *observability* function, which returns the current observation, i.e., the observable part of the current state.

An *execution* on the transition system is an infinitive sequence of states $w_0, w_1, w_2, \ldots$ s.t. $w_0 \in W_0$ and $w_{i+1} = R(w_i, a)$ for some $a \in Act$. A *trace* is what we can observe of an execution. For example, $\pi(w_0), \pi(w_1), \pi(w_2), \ldots$ is the trace corresponding to the execution $w_0, w_1, w_2, \ldots$. The *observable behavior* of the dynamic system is the set of all possible traces of the transition system.[3]

## 2.2   Automata on Infinite Words

Given a finite nonempty alphabet $\Sigma$, an *infinite word* is an element of $\Sigma^\omega$, i.e., an infinite sequence $a_0, a_1, a_n, \ldots$ of symbols from $\Sigma$.

A *Büchi automaton* is a tuple $\mathcal{A} = (\Sigma, S, S_0, \rho, F)$ where:

- $\Sigma$ is the alphabet of the automaton.
- $S$ is the finite set of possible states.
- $S_0 \subseteq S$ is the set of possible initial states.
- $\rho : S \times \Sigma \to 2^S$ is the transition function of the automaton (the automaton need not to be deterministic).
- $F \subseteq S$ is the set of accepting states.

---

[2] For simplicity and wlog, we assume that $R$ is a total function. We can model the case in which a transition does not exists by making $R$ return a special dummy state.

[3] This way of describing the observable behavior of a dynamic system corresponds to the so called *linear time view of dynamic system*, and is to be contrasted to the so called *branching time view* –see [16] for a discussion.

The *input words* of $\mathcal{A}$ are infinite words $a_0, a_1, a_2, \ldots \in \Sigma^\omega$. A *run* of $\mathcal{A}$ on a infinite word $a_0, a_1, a_2, \ldots$ is an infinite sequences of states $s_0, s_1, s_2, \ldots \in S^\omega$ s.t. $s_0 \in \mathcal{S}_0$ and $s_{i+1} \in \rho(s_i, a_i)$. A run $r$ is *accepting* iff $lim(r) \cap F \neq \emptyset$, where $lim(r) = \{s \mid s \text{ occurs in } r \text{ infinitely often}\}$. In other word a run is accepting if it gets into $F$ infinitely many times, which in turn means, being $F$ finite, that there is at least one state $s_f \in F$ that is visited infinitely often. The *language* accepted by $\mathcal{A}$, denoted by $L(\mathcal{A})$, is the set of words for which there is an accepting run.

The *nonemptiness* problem for an automaton is to decide given an automaton $\mathcal{A}$ whether $L(A) \neq \emptyset$, i.e., if the automaton accepts at least one word.

**Proposition 1.** [46] *The nonemptiness problem for Büchi automata is NLOGSPACE-complete.*

Algorithms for nonemptiness are based on *fair reachability* on graphs. The idea behind the algorithms is best explained by the following three line Prolog implementation:

```
nonempty :- ini(X),cn(X,Y),acc(Y),cn(Y,Y).
cn(X,Y)  :- rho(X,A,Y).
cn(X,Y)  :- rho(X,A,Z),cn(Z,Y).
```

where `ini` denotes the elements in $S_0$, `acc` denotes the elements in $F$, `rho` denotes the relation corresponding to the transition function, and `cn` denotes that two states are connected by a `rho`-chain (`cn` is the transitive closure of `rho`).[4]

In other words an automaton is nonempty if starting from some initial state we can reach an accepting state from where there is a cycle back to itself.

A *nondeterministic algorithm for nonemptiness* can then work as follows: it nondeterministically chooses an initial state $x$ and an accepting state $y$ and then checks that $x$ is connected to $y$ and $y$ is connected to itself. To run the algorithm we only need to store the state $y$, as well as the current and next states, plus a constant number of control bits. To encode states as bit vectors we need only $O(\log(|S|))$ bits. This gives the NLOGSPACE bound.

A *linear time deterministic algorithm for nonemptiness* is the following (i) decompose the transition graph of the automaton into *maximally strongly connected components (mscc)* (linear cost [11]); (ii) verify that one of the mscc's intersects with $F$ (linear cost).

Büchi automata are widely used in verification to specify properties of dynamic systems [31,46]. Given transition system representing a dynamic system,

---

[4]   Observe the strong similarity with the following naive algorithm to check plan existence in more traditional approaches (which is in fact *reachability* on graphs):

```
planexis :- ini(X),cn(X,Y),goal(Y).
cn(X,Y)  :- result(X,A,Y).
cn(X,Y)  :- result(X,A,Z),cn(Z,Y).
```

where `ini` denotes the initial states (typically one), `goal` the states where the goal is satisfied, `result` corresponds to the result function that return a state resulting from executing an action (operator) in the current state, `cn` its transitive closure.

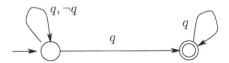

**Fig. 1.** Automaton for "eventually always $q$"

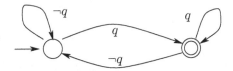

**Fig. 2.** Automaton for "always eventually $q$"

the words accepted by the automaton can be put in correspondence with traces of the transitions system that have specified properties. Two examples of such specifications are shown in Figure 1 and Figure 2. The automaton in Figure 1 accepts traces where from a certain point on the property $q$ will hold forever. The automaton in Figure 2, instead, accepts traces where at every point of the trace it is guaranteed that sooner or later a certain property $q$ will hold. More generally any property expressible in propositional linear time temporal logic (LTL) can be expressed as a Büchi automaton, but not vice-versa.[5]

In the following, we will also make use of generalized Büchi automata. A *generalized Büchi automaton* $\mathcal{A}_g = (\Sigma, S, S_0, \rho, \{F_0, \ldots, F_{k-1}\})$ is a variant of Büchi automata which has $k$ sets of accepting states $F_0, \ldots, F_{k-1}$ instead of one, and whose acceptance condition for a run $r$ is $lim(r) \cap F_i \neq \emptyset$ for $i = 0, \ldots, k-1$. Given a generalized Büchi automaton $\mathcal{A}_g = (\Sigma, S, S_0, \rho, \{F_0, \ldots, F_{k-1}\})$, it can be transformed into an equivalent[6] Büchi automaton $\mathcal{A}_b = (\Sigma, S', S'_0, \rho', F')$ where:

- $S' = S \times \{0, \ldots, k-1\}$
- $S'_0 = S_0 \times \{0\}$
- $\rho'((s, i), a) = \rho(s, a) \times \{i\}$ if $s \notin F_i$, and $\rho'((s, i), a) = \rho(s, a) \times \{(i+1 \bmod k)\}$ if $s \in F_i$
- $F' = F_0 \times \{0\}$

## 3   Planning with Complete Information

We start our investigation by considering a simplified case. We assume, that we have *complete information* on initial situation and that we have *full observability* on the state. The only kind of plans of interest in this case are sequential ones (sequences of actions), since a conditional plan exists iff a sequential plan does.

---

[5] There are standard techniques to transform LTL formulas into Büchi automata. The size of the resulting automaton is worst-case exponential wrt the formula [46].

[6] In the sense that $L(\mathcal{A}_g) = L(A_b)$.

We model the dynamic system of interest as a transition system $\mathcal{T} = (W, W_0, Act, R, Obs, \pi)$ where:

- $W_0 \subseteq W$ is a singleton set containing the initial state (which is unique since we are assuming complete information on the initial situation).
- $Obs = W$, and $\pi : W \to Obs$ is simply the identity function (since we are assuming full observability).

Let $\mathcal{A}$ be a Büchi automaton specifying the behavior of the desired executions of the system. Formally, $\mathcal{A} = (Obs, S, S_0, \rho, F)$ where:

- $Obs$ plays the role of the alphabet of the automaton.
- $S$ is the finite set of possible states of the automaton.
- $S_0 \subseteq S$ is the set of possible initial states.
- $\rho : S \times Obs \to 2^S$ is the transition function of the automaton (the automaton need not to be deterministic).
- $F \subseteq S$ is the set of accepting states.

A *plan* $p$ for $\mathcal{T}$ is an infinite sequence of actions $a_0, a_1, a_2, \ldots \in Act^\omega$. The *execution* of $p$ (starting from the initial state $w_0$) is the infinite sequence of states $w_0, w_1, w_2, \ldots \in W^\omega$ s.t. $w_0 \in W_0$ and $w_{i+1} = R(w_i, a_i)$. The *trace*, $tr(p, w_0)$, of $p$ (starting from the initial state $w_0$) is the infinite sequence $\pi(w_0), \pi(w_1), \pi(w_2), \ldots$. A plan $p$ *realizes* a specification $\mathcal{A}$ iff $tr(p, w_0) \in L(\mathcal{A})$.

How can we synthesize such a plan? We check for nonemptiness the following Büchi automaton $\mathcal{A}_\mathcal{T} = (Act, S_\mathcal{T}, S_{\mathcal{T}0}, \rho_\mathcal{T}, F_\mathcal{T})$ where:

- $Act$ is the alphabet of the automaton
- $S_\mathcal{T} = S \times W$
- $S_{\mathcal{T}0} = S_0 \times \{w_0\}$
- $(s_j, w_j) \in \rho_\mathcal{T}((s_i, w_i), a)$ iff $w_j = R(w_i, a)$ and $s_j \in \rho(s_i, \pi(w_i))$
- $F_\mathcal{T} = F \times W$

For $\mathcal{A}_\mathcal{T}$ we get the following result:

**Theorem 1.** *A plan $p$ for $\mathcal{T}$ realizing the specification $\mathcal{A}$ exists iff $L(\mathcal{A}_\mathcal{T}) \neq \emptyset$.*

Notably the nonemptiness algorithm can be easily modified to return a plan if a plan exists. The plan returned always consists of two parts: a sequence arriving to a certain state, and a second sequence that forms a cycle back into that state. Thus, such plans have finite representations.

As an immediate consequence of the construction we get:

**Theorem 2.** *Planning in the setting above is decidable in NLOGSPACE.*

*Proof.* The automaton $\mathcal{A}_\mathcal{T}$ can be built on the fly, thus for checking nonemptiness using a nondeterministic algorithm we only need $O(\log(|W|) + \log(|S|))$ bits.

Observe that if we adopt a compact (i.e., logarithmic) representation of the transition system, for example by using propositions to denote states and computing the transitions directly on such propositions,[7],then planning in the above setting becomes PSPACE. This is the complexity of planning in STRIPS [8], which can be seen as a special case of the setting considered here – reachability of a desired state of affairs is the only kind of goal considered in STRIPS; moreover, only certain transition systems are (compactly) representable.

Moreover considering that STRIPS is PSPACE-hard [8], we can conclude:

**Theorem 3.** *Planning in the setting above is NLOGSPACE-complete (PSPACE-complete wrt a compact representation of $\mathcal{T}$).*

## 4    Sequential Planning with Incomplete Information

Next we consider the more general case. We assume to have only partial information on the initial situation, and we assume that only part of the state is observable. In this section we consider generating sequential plans, in the next section we turn to conditional plans.

We model the dynamic system of interest as a general transition system $\mathcal{T} = (W, W_0, R, Act, Obs, \pi)$ defined as in Section 2.1. Such a transition system has several initial states $W_0 = \{w_{00}, \ldots, w_{0k-1}\}$, for $k > 1$, reflecting our uncertainty about the initial situation.

As in the previous section we specify the behavior of the desired executions of the system by a Büchi automaton $\mathcal{A}$.

A *plan* $p$ for $\mathcal{T}$ is an infinite sequence of actions $a_0, a_1, a_2, \ldots \in Act^\omega$. The *execution* of $p$ starting from $w_{oh}$ is the infinite sequence of states $w_{0h}, w_{1h}, w_{2h}, \ldots \in W^\omega$ s.t. $w_{0h} \in W_0$ and $w_{i+1h} = R(w_{ih}, a_i)$. The *trace*, $tr(p, w_{0h})$, of the plan $p$ in $\mathcal{T}$ is the infinite sequence $\pi(w_{0h}), \pi(w_{1h}), \pi(w_{2h}), \ldots$. A plan $p$ *realizes* a specification $\mathcal{A}$ iff $tr(p, w_{0h}) \in L(\mathcal{A})$ for $h = 0, \ldots k-1$.

How can we synthesize such a plan? Again we check for nonemptiness a Büchi automaton. This time, however, the construction is slightly more involved.

We first build the generalized Büchi automaton $\mathcal{A}_\mathcal{T} = (Act, S_T, S_{T0}, \rho_T, F_T)$ where:

- $S_T = S^k \times W^k$
- $S_{T0} = S_0^k \times \{(w_{00}, \ldots, w_{0k-1})\}$
- $(\boldsymbol{s_j}, \boldsymbol{w_j}) \in \rho_T((\boldsymbol{s_i}, \boldsymbol{w_i}), a)$ iff $w_{jh} = R(w_{ih}, a)$ and $s_{jh} \in \rho(s_{ih}, \pi(w_{ih}))$ for $h = 0, \ldots, k-1$.
- $F_T = \{F \times S^{k-1} \times W^k, \ldots, S^{k-1} \times F \times W^k\}$

---

[7] We want to stress that assuming that there are formalisms able to represent *every transition system* compactly is not realistic. Indeed, the number of possible transition functions is $|W|^{|W|}$, while the number of transition functions distinguishable with $O(\log(|W|))$ bits is $2^{O(\log(|W|))} = |W|^{O(1)}$. In many cases, however, compact representations of transitions systems do exist, e.g., digital circuits are often described compactly by means of hardware description languages.

From such an automaton we get an equivalent Büchi automaton $\mathcal{A}_T^b = (Act, S_T^b, S_{T0}^b, \rho_T^b, F_T^b)$ where:

- $S_T^b = S^k \times W^k \times \{0, \ldots, k-1\}$
- $S_{T0}^b = S_0^k \times \{(w_{00}, \ldots, w_{0k-1})\} \times \{0\}$
- $(s_j, w_j, \ell_j) \in \rho_T^b((s_i, w_i, \ell_i), a)$ iff $w_{jh} = R(w_{ih}, a)$ and $s_{jh} \in \rho(s_{ih}, \pi(w_{ih}))$ for $h = 0, \ldots, k-1$, and $\ell_j = (\ell_i + 1) \bmod k$ if $s_{ip_i} \in F$ and $\ell_j = \ell_i$ otherwise.
- $F_T^b = F \times S^{k-1} \times W^k \times \{0\}$

**Theorem 4.** *A plan p for $T$ realizing the specification $\mathcal{A}$ exists iff $L(\mathcal{A}_T) = L(\mathcal{A}_T^b) \neq \emptyset$.*

Again the nonemptiness algorithm can be easily modified to return a plan if a plan exists. The plan again consist of two parts: a sequence arriving to a certain state, and a second sequence that forms a cycle back into that state. Note that the possibility of expressing the plan as a finite sequence and a cycle is guaranteed in spite of the uncertainty about the initial state.

Building the automaton $\mathcal{A}_T$ on the fly, we can check nonemptiness with a nondeterministic algorithm needing $O(k \cdot \log(|W|) + \log(|S|))$ bits, where $k$ is bounded by the size of $|W|$. Considering that NPSPACE=PSPACE, we get:

**Theorem 5.** *Planning in the setting above is decidable in PSPACE.*

If we adopt a compact representation of the transition system, then planning in the above setting becomes EXPSPACE. What about lower bounds? The following theorem says that our upper bounds are tight.

**Theorem 6.** *Planning in the setting above is PSPACE-complete (EXPSPACE-complete wrt a compact representation of $T$).*

*Proof.* We only need to prove the hardness. Consider that the problem of finding a string that is accepted by the intersection of $k$ deterministic finite state automata over the same alphabet is PSPACE-complete [19]. It is easy to reduce such a problem to planning in the above setting. In particular, the reduction works even with the following two restrictions: (i) $Obs = S$ and $\pi$ is the identity function; (ii) the specification automaton denotes an achievement goal. When the transition system is represented compactly, techniques from [25] can be used to lift the PSPACE lower bound to EXPSPACE lower bound.

Note that plan existence in STRIPS with incomplete information on the initial situation is PSPACE-complete [3] — polynomial reduction to the case where the initial situation is completely known. This means that we do pay a price this time in generalizing the setting wrt more traditional approaches. Observe that the reduction used in the proof of Theorem 6 tells us that the increase in the complexity is essentially due to coping with the general form of transition systems once we allow for several possible initial states, and not to the partial observability of states or the more general form of goals considered here.

# 5   Conditional Planning with Incomplete Information

Now we turn to synthesis of conditional plans in the general setting introduced in the previous section.

Let $\mathcal{T}$ be the transition system and $\mathcal{A}$ be the specification automata both defined as in the previous section.

A *vector plan* $\boldsymbol{p}$ is an infinite sequence of vectors of actions $\boldsymbol{a_0}, \boldsymbol{a_1}, \boldsymbol{a_2}, \ldots \in (Act^k)^{\omega}$. The *execution*, $ex_h(\boldsymbol{p}, w_{0h})$ of its $h$-component (starting from the initial state $w_{0h}$) is the infinite sequence of states $w_{0h}, w_{1h}, w_{2h}, \ldots \in W^{\omega}$ s.t. $w_{0h} \in W_0$ and $w_{i+1h} = R(w_{ih}, a_{ih})$. The *trace*, $tr_h(\boldsymbol{p}, w_{0h})$, of its $h$-component is the infinite sequence $\pi(w_{0h}), \pi(w_{1h}), \pi(w_{2h}), \ldots$. A vector plan $\boldsymbol{p}$ *realizes* a specification $\mathcal{A}$ iff $tr_h(\boldsymbol{p}, w_{0h}) \in L(\mathcal{A})$ for $h = 0, \ldots k-1$.

A vector plan is not a conditional plan yet, it is simply the parallel compositions of $k$ sequential plans, one for each initial state. *Conditional plans* are vector plans whose actions agree on executions with the same observations.

To formally define conditional plans, we introduce the following notion of equivalence on finite traces. Let $w_{0l}, \ldots, w_{n1}$ and $w_{0m}, \ldots, w_{nm}$ be two finite traces, then

$$\langle w_{0l}, \ldots, w_{n1} \rangle \sim \langle w_{0m}, \ldots, w_{nm} \rangle \text{ iff}$$
$$\langle \pi(w_{0l}), \ldots, \pi(w_{n1}) \rangle = \langle \pi(w_{0m}), \ldots, \pi(w_{nm}) \rangle.$$

A *conditional plan* $\boldsymbol{p}$ is a vector plan such that given the executions $w_{0l}, w_{1l}, w_{2l}, \ldots$ and $w_{0m}, w_{1m}, w_{2m}, \ldots$ of a pair of components $l$ and $m$, we have that $a_{nl} = a_{nm}$ whenever $\langle w_{0l}, \ldots, w_{n1} \rangle \sim \langle w_{0m}, \ldots, w_{nm} \rangle$.

Intuitively a conditional plan can be though of as composed by an (infinite) sequence of *case* instructions that at each step on the base of the observations select how to proceed.

How can we synthesize a conditional plan? We follow the line of the construction in the previous section, checking nonemptiness of a Büchi automaton which this time has $Act^k$ as alphabet. Specifically, we build the generalized Büchi automaton $\mathcal{A}_{\mathcal{T}} = (Act^k, S_T, S_{T0}, \rho_{\mathcal{T}}, F_{\mathcal{T}})$ where:

- $S_T = S^k \times W^k \times \mathcal{E}_k$, where $\mathcal{E}_k$ is the set of equivalence relations on the set $\{0, \ldots, k-1\}$,
- $S_{T0} = S_0^k \times \{(w_{00}, \ldots, w_{0k-1})\} \times \equiv_0$, where $i \equiv_0 j$ iff $w_{0i} = w_{0j}$,
- $(\boldsymbol{s_j}, \boldsymbol{w_j}, \equiv') \in \rho_{\mathcal{T}}((\boldsymbol{s_i}, \boldsymbol{w_i}), \boldsymbol{a}, \equiv)$ iff
  - $w_{jh} = R(w_{ih}, a_h)$ and $s_{jh} \in \rho(s_{ih}, \pi(w_{ih}))$
  - if $l \equiv m$ then $a_l = a_m$
  - $l \equiv' m$ iff $l \equiv m$ and $\pi(w_{jl}) = \pi(w_{jm})$
- $F_{\mathcal{T}} = \{F \times S^{k-1} \times W^k \times \mathcal{E}_k, \ldots, S^{k-1} \times F \times W^k \times \mathcal{E}_k\}$

Such automaton can be transformed into a Büchi automaton $\mathcal{A}_{\mathcal{T}}^b$ as before.

**Theorem 7.** *A conditional plan $\boldsymbol{p}$ for $\mathcal{T}$ realizing the specification $\mathcal{A}$ exists iff $L(\mathcal{A}_{\mathcal{T}}) = L(\mathcal{A}_{\mathcal{T}}^b) \neq \emptyset$.*

The nonemptiness algorithm can again be immediately modified to return a plan if a plan exists. The plan returned again consists of two parts: a sequence arriving to a certain state and a second sequence that forms loop over that state (however, this time the element of the sequences are $k$-tuples of actions). Observe that even if formally we still deal with vectors of sequential plans, the conditional plan returned can be put in a more convenient form using *case* instructions and loops.

Finally, It is easy to verify that the same complexity bounds of the previous case still hold.[8]

**Theorem 8.** *Finding a conditional plan in the setting above is PSPACE-complete (EXPSPACE-complete wrt a compact representation of $\mathcal{T}$).*

# 6   Practical Algorithms

The results above show that planning can be reduced to nonemptiness of Büchi automata. Algorithms for checking nonemptiness of Büchi automata have proved to be well suited for scaling up to very large systems [7]. A breakthrough technology has been the use of *symbolic methods* [35], based on the idea being of encoding states as bit vectors, representing sets of states and transitions symbolically as Boolean functions on the encoding, and using ordered binary decision diagrams (OBDDs) to efficiently manipulate Boolean functions [6]. Industrial strength system used in hardware and protocol verification have been developed and used commercially with success [24,26].

This indicates that notwithstanding the worst-case complexity, it should be actually possible to implement planners even for the most general setting considered here. The experimental results in [9,10] on adopting symbolic techniques for planning are quite promising (see also [12]). The focus there is on attaining propositional goals when actions can be nondeterministic, but the symbolic techniques can be adapted to our framework.

# 7   Related Work

The need of dealing with incomplete information has often put forward in the area of planning, e.g., [33,36,17,23,32], as has the need of going beyond goals that specify the reachability of desired state of affair, e.g., [15,47,27,2]. In particular, in [27] a planning setting close to the one considered here is studied, where: dynamic systems are represented by transition systems with a single initial state, nondeterministic actions (which allow for modeling incomplete information), and fully observable states; goals are temporally extended goals, expressed in a variant of LTL that includes a metric over time; plans generated are reactive (conditional) plans. In [2,4] an analogous planning setting is studied,

---

[8] Observe that we only need $k$ bits to represent an equivalence relation on $\{0, \dots, k-1\}$.

under the additional assumption of deterministic actions: [2] focuses on generating finite sequential plans only, while [4] considers plans consisting of possibly infinite sequences of actions. The approach adopted in [27,2,4] for obtaining planning algorithms is somewhat ad-hoc (it is based on formula decomposition). The approach proposed here is based on the fundamental relationship between LTL and Büchi automata. The automata-theoretic approach separates the logical and the algorithmic aspects of the planning problem, resulting in clean and optimal algorithms. As we demonstrated, our approach is quite flexible and can be easily adapted to various planning scenarios. Also, neither of [27,2,4] studies the intrinsic complexity of the specific planning problem they tackle. In particular, no complexity lower bounds are established.

It is also worth mentioning that there are some similarities between the automata theoretic approach adopted here and approaches to planning based on techniques from operations research, such as MDPs and POMDPs, which are considered quite promising in dealing with incomplete information and generalized goals in stochastic domains [28,1,21,20]. Precise relationships, however, are yet to be established. In particular, to the best of our knowledge, encoding general temporally extended goals, as those expressible with Büchi automata or LTL, as a MDPs/POMDPs rewarding function still remains an open problem.

Finally, automata on infinite objects have already been studied for synthesis of hardware and control-software [37,43,30]. Also, automata theoretic techniques have been used in synthesizing discrete controllers [38,14]. The setting studied here (incomplete information on the initial situation plus deterministic actions), however, which naturally arise in planning and reasoning about actions, is simpler than the general synthesis framework, enabling us to obtain algorithms that are both simper and of lower computational complexity.

## 8    Conclusions

In this paper we have studied planning for temporally extended goals when incomplete information on the initial situation is available, states are only partially observable, and the number of possible states is finite. We have devised techniques based on nonemptiness of Büchi automata on infinite words, to synthesize sequential and conditional plans, and have characterized the worst case computational complexity. The techniques introduced here in an abstract framework can be easily specialized to a wide range of formalisms for reasoning about actions that are based on transition systems. Moreover, in spite of the high worst-case complexity, the scalability of the practical algorithms involved promises to make the automata-theoretic approach to planning actually feasible even in the most general setting considered here.

**Acknowledgments.** We would like to thank Yves Lesperance, who participated to the discussion that set the stage for this work at ESSLLI'98, and Amedeo Cesta, who gave us precious comments on how to relate our material to traditional planning. This work was partially supported by the NSF grants CCR-9628400 and CCR-9700061, by ASI, by MURST, and by ESPRIT LTR Project No. 22469 DWQ.

# References

[1]  F. Bacchus, C. Boutilier, and A. Grove. Structured solution methods for non-markovian decision processes. In *Proc. of AAAI'97*, 112–117, 1997.

[2]  F. Bacchus and F. Kabanza. Planning for temporally extended goals. *Ann. of Math. and AI*, 22:5–27, 1998.

[3]  C. Backstrom. Equivalence and tractability results for SAS+ planning. In *Proc. of KR'92*, 1992.

[4]  M. Barbeau, F. Kabanza, and R. St-Denis. Synthesizing plant controllers using real-time goals. In *Proc. of IJCAI'95*, 791–798, 1995.

[5]  A. Blum and M. Furst. Fast planning through planning graph analysis. *Artif. Intell.*, 90(1-2):281–300, 1997.

[6]  R. Bryant. Symbolic boolean manipulation with ordered binary-decision diagrams. *ACM Computing Surveys*, 24(3):293–318, 1992.

[7]  J. Burch, E. Clarke, K. McMillan, D. Dill, and L. Hwang. Symbolic model checking: $10^{20}$ states and beyond. *Information and Computation*, 98(2):142–170, 1992.

[8]  T. Bylander. Tractability and artificial intelligence. *J. of Experimental and Theoretical Computer Science*, 3:171–178, 1991.

[9]  A. Cimatti, E. Giunchiglia, F. Giunchiglia, and P. Traverso. Planning via model checking. In *Proc. of the ECP'97*, 1997.

[10]  A. Cimatti, M. Roveri, and P. Traverso. Automatic OBDD-based generation of universal plans in non-deterministic domains. In *Proc. of AAAI'98*, 875–881, 1998.

[11]  T. H. Cormen, C. E. Leiserson, and R. L. Rivest. *Introduction to Algorithms*. The MIT Press, 1990.

[12]  M. Daniele and P. T. M. Y. Vardi. Strong cyclic planning revisited. submitted, 1999.

[13]  G. De Giacomo, L. Iocchi, D. Nardi, and R. Rosati. Moving a robot: the KR&R approach at work. In *Proc. of KR'96*, 1996.

[14]  A. Deshpande and P. Varaiya. Sementic tableau for control of PLTL formulae. In *Proc. of the 35th Conf. on Decision and Control*, 2243–2248. IEEE, 1996.

[15]  M. Drummond. Situated control rules. In *Proc. of KR'89*, 103–113, 1989.

[16]  E. Emerson. Temporal and modal logic. *Handbook of Theoretical Computer Science*, 997–1072, 1990.

[17]  O. Etzioni, S. Hanks, D. Weld, D. Draper, N. Lesh, and M. Williamson. An approach to planning with incomplete information. *Proc. of KR'92*, 1992.

[18]  R. Fikes and N. J. Nilsson. A new approach to the application of theorem proving to problem solving. *Artif. Intell.*, 2(3/4), 1971.

[19]  M. R. Garey and D. S. Johnson. *Computers and Intractability—A guide to NP-completeness*. W. H. Freeman and Company, San Francisco, 1979.

[20]  H. Geffner. Classical, probabilistic and contingent planning: Three models, one algorithm. In *Proc. AIPS'98 Work. on Planning as Combinatorial Search*, 1998.

[21]  H. Geffner and B. Bonet. High-level planning and control with incomplete information using POMDPs. In *Proc. AIPS'98 Work. on Integrating Planning, Scheduling and Execution in Dynamic and Uncertain Environments*, 1998.

[22]  M. Gelfond and V. Lifschitz. Action languages. *Linköping Electronic Articles in Computer and Information Science*, 3(16), 1998.

[23]  K. Golden and D. S. Weld. Representing sensing actions: The middle ground revisited. *Proc. of KR'96*, 174–185, 1996.

[24]  R. Hardin, Z. Har'el, and R. Kurshan. COSPAN. In *Computer Aided Verification, Proc. 8th Int'l Conf*, LNCS 1102, 423–427. Springer-Verlag, 1996.

[25] D. Harel, O. Kupferman, and M. Y. Vardi. On the complexity of verifying concurrent transition systems. In *Proc. 8th Int'l Conf. on Concurrency Theory*, LNCS 1243, 258–272, Warsaw, July 1997. Springer-Verlag.

[26] G. Holtzmann. Tutorial: proving correctness of concurrent systems with spin. In *Proc. 6th Int'l Conf. on Concurrency Theory*, 453–455. Springer-Verlag, 1995.

[27] F. Kabanza, M. Barbeau, , and R. St-Denis. Planning control rules for reactive agents. *Artif. Intell.*, 95(1):67–113, 1997.

[28] L. Kaelbling, M. L. Littman, and A. R. Cassandra. Planning and acting in partially observable stochastic domains. *Artif. Intell.*, 101:99–134, 1998.

[29] H. Kautz, D. McAllester, and B. Selman. Encoding plans in propositional logic. In *Proc. of KR'96*, 1996.

[30] O. Kupferman and M. Y. Vardi. Synthesis with incomplete informatio. In *2nd Int'l Conf. on Temporal Logic*, 91–106, Manchester, July 1997. Kluwer Academic Publishers.

[31] R. Kurshan. *Computer Aided Verification of Coordinating Processes*. Princeton Univ. Press, 1994.

[32] H. J. Levesque. What is planning in presence of sensing? In *Proc. of AAAI'96*, 1139–1149. AAAI Press/The MIT Press, 1996.

[33] Z. Manna and R. J. Waldinger. How to clear a block: A theory of plans. *J. of Automated Reasoning*, 3(4), 1987.

[34] J. McCarthy and P. Hayes. Some philosophical problems from the standpoint of artificial intelligence. *Machine Intelligence*, 4:463–502, 1969.

[35] K. McMillan. *Symbolic Model Checking*. Kluwer Academic Publishers, 1993.

[36] E. Pednault. ADL: exploring the middle ground between STRIPS and the situation calculus. In *Proc. of KR'89*, 324–332, 1989.

[37] A. Pnueli and R. Rosner. On the synthesis of a reactive module. In *Proc. 16th ACM Symp. on Principles of Programming Languages*, Austin, January 1989.

[38] P. Ramadge and W. Wonham. The control of discrete event systems. *Proc. of IEEE*, 77(1):81–98, Jan. 1989.

[39] R. Reiter. *Knowledge in Action: Logical Foundation for Describing and Implementing Dynamical Systems*. 1998. In preparation.

[40] S. J. Rosenschein. Plan synthesis: A logical perspective. In *Proc. of IJCAI'81*, 331–337, 1981.

[41] E. Sandewall. *Features and Fluents. The Representation of Knowledge about Dynamical Systems. Volume I*. Oxford University Press, 1994.

[42] M. Shanahan. *Solving the Frame Problem: A Mathematical Investigation of the Common Sense Law of Inertia*. The MIT Press, 1997.

[43] M. Y. Vardi. An automata-theoretic approach to fair realizability and synthesis. In P. Wolper, editor, *Computer Aided Verification, Proc. 7th Int'l Conf.*, LNCS 939, 267–292. Springer-Verlag, Berlin, 1995.

[44] M. Y. Vardi. An automata-theoretic approach to linear temporal logic. In F. Moller and G. Birtwistle, editors, *Logics for Concurrency: Structure versus Automata*, LNCS 1043, 238–266. Springer-Verlag, Berlin, 1996.

[45] M. Y. Vardi and P. Wolper. An automata-theoretic approach to automatic program verification. In *Proc. 1st Symp. on Logic in Computer Science*, 322–331, Cambridge, June 1986.

[46] M. Y. Vardi and P. Wolper. Reasoning about infinite computations. *Information and Computation*, 115(1):1–37, November 1994.

[47] D. S. Weld and O. Etzioni. The first law of robotics (a call to arms). In *Proc. of AAAI'94*, 1042–1047, 1994.

# Integer Programs and Valid Inequalities for Planning Problems

Alexander Bockmayr[1] and Yannis Dimopoulos[2]

[1] Université Henri Poincaré, LORIA
B.P. 239, F-54506 Vandœuvre-lès-Nancy, France
bockmayr@loria.fr
[2] Dep. of Computer Science, University of Cyprus
P.O. Box 20537, CY-1678, Nicosia, Cyprus
yannis@cs.ucy.ac.cy

**Abstract.** Part of the recent work in AI planning is concerned with the development of algorithms that regard planning as a combinatorial search problem. The underlying representation language is basically propositional logic. While this is adequate for many domains, it is not clear if it remains so for problems that involve numerical constraints, or optimization of complex objective functions. Moreover, the propositional representation imposes restrictions on the domain knowledge that can be utilized by these approaches. In order to address these issues, we propose moving to the more expressive language of Integer Programming (IP). We show how capacity constraints can be easily encoded into linear 0-1 inequalities and how rich forms of domain knowledge can be compactly represented and computationally exploited by IP solvers. Then we introduce a novel heuristic search method based on the linear programming relaxation. Finally, we present the results of our experiments with a classical relaxation-based IP solver and a logic-based 0-1 optimizer.

## 1 Introduction

Planning is a hard problem. Until recently, general-purpose planning algorithms had great difficulties in solving even small instances. Over the last few years a number of combinatorial search methods have been applied with remarkable success in solving AI planning problems that are purely Boolean, i.e., which can be adequately represented in propositional logic [11], [13], [3], [16]. However, there are domains where the constraints imposed by the environment, e.g. the capacity of the available resources, are numerical in nature. Usually, these constraints do not restrict the applicability of each action independently, but rather the sets of actions that can be applied at each step. Therefore, we refer to these constraints as *global constraints*. In many cases, they can be represented compactly by a set of linear inequalities in 0-1 variables. Each of these inequalities may translate into an exponential number of propositional clauses. It seems that the current general-purpose planning algorithms, which are based on propositional representations, cannot exploit such global constraints in order to effectively prune

S. Biundo and M. Fox (Eds.): ECP-99, LNAI 1809, pp. 239–251, 2000.

the search space. Moreover, current propositional planners lack the expressive power needed in order to accommodate complex objective functions in the domain representation. These planners usually can optimize only the length of the generated plans or the number of actions in these plans. Finally, while there is an increased interest in utilizing domain knowledge in the problem representation [12,10,19], propositional logic again imposes restrictions on the forms of this knowledge.

The above considerations naturally lead to the idea of extending the representation language from propositional logic to linear inequalities in 0-1 variables. The first results in this direction, which apply techniques from integer linear programming to planning problems, were presented by the authors in [5]. Since that, the possibility of employing integer programming in planning has received more attention. Vossen et al. [20] presented a computationally appealing domain independent translation of planning into integer programming. Kautz and Walser [14] described an integer programming framework that can handle plans with resources, action costs and complex objective functions. Already in 1997, Bylander [7] proposed a linear relaxation based heuristic for non-linear planning.

In this paper, we further explore the relation between planning and Integer Programming (IP). Our basic methodology is to build *state-based* models of the planning problems we consider, i.e., binary variables for the fluents and a set of inequalities that describe how the world can change from one state to the other. For some problems that involve capacity constraints, action variables need to be explicitly introduced. However, we do not provide an automatic translation of planning to IP. This is for two reasons. First, because selecting a suitable representation is a fundamental issue in problem-solving with IP techniques. Second, because one of the strengths of IP is that it can accommodate richer forms of declarative domain-specific information than a propositional representation. This information can be used in two different ways.

In the constraints part, domain knowledge is used to derive "tight" models for the planning problems. This is mainly accomplished by the addition of *valid inequalities* to the problem representation, i.e., constraints that do not take away any of the integer solutions of the problem. Their effect is that they increase the accuracy of the information that the IP solver obtains from the linear programming relaxation at each node of the search tree. Interestingly, these inequalities also increase the performance of logic-based optimizers as they achieve higher consistency in the problem description.

In the objective function, domain knowledge can be used to transform planning into an *optimization* problem. The solver searches for states that are as close as possible to the final state, using the "distance" measure provided by the objective function. Modeling planning as optimization forms the base for a heuristic search algorithm using the linear relaxation, which we present in this paper. The values of the heuristic function are the values of the objective function obtained if some of the binary variables are relaxed to take real values. Therefore, IP modeling can be also seen as a way of bringing together combinatorial and heuristic search methods [6,17].

The paper is organized as follows. First we discuss the basics of IP problem solving and briefly review the branch-and-bound method and logic-based 0-1 optimization. The next section introduces the basic translation method of planning into IP, and shows how global constraints and powerful domain knowledge can be easily incorporated into the problem specification. Then we introduce the linear programming heuristic and report on a number of computational experiments with the IP models of planning problems. Finally, we discuss related work and conclude.

## 2   Integer Programming

A *mixed integer program* (MIP) is of the form

$$z_{MIP} = \min\{c^T x \mid Ax \le b, \ x_1, \dots, x_p \in \mathbb{Z}, \ x_{p+1}, \dots, x_n \in \mathbb{R}\}, \qquad \text{(MIP)}$$

where $A \in \mathbb{R}^{m \times n}$ is a real matrix, $b \in \mathbb{R}^m, c \in \mathbb{R}^n$ are real vectors, and $\cdot^T$ denotes transposition. If $p = n$, we have a *(pure) integer program*, if $p = 0$, a classical *linear program*. If we require $x_1, \dots, x_p \in \{0, 1\}$, we get a (mixed) *0-1 program*. A vector $x^* \in \mathbb{R}^n$ with $Ax^* \le b$ and $x_1^*, \dots, x_p^* \in \mathbb{Z}$ is called a *feasible solution* of (MIP). If moreover $c^T x^* = \min\{c^T x \mid Ax \le b, x_1, \dots, x_p \in \mathbb{Z}\}$, then $x^*$ is called an *optimal solution* and $c^T x^*$ the *optimal value*. The function $c^T x$ is called the *objective function*. Note that maximizing $c^T x$ is the same as minimizing $-c^T x$.

Mixed integer programming is a more general representation language than propositional logic. Note that any propositional clause

$$x_1 \vee \cdots \vee x_k \vee \bar{x}_{k+1} \vee \cdots \vee \bar{x}_{k+l} \qquad (1)$$

is equivalent to one linear inequality

$$x_1 + \cdots + x_k - x_{k+1} - \cdots - x_{k+l} \ge 1 - l \qquad (2)$$

in 0-1 variables. However, to represent a general linear inequality $a^T x \ge \beta$, with $a \in \mathbb{R}^n, \beta \in \mathbb{R}$ and $x \in \{0, 1\}^n$, exponentially many clauses (in the number $n$ of variables) may be needed. 0-1 integer programming combines propositional logic with arithmetic and therefore allows for more compact formulations. For example, we may use the linear 0-1 inequality

$$x_1 + \cdots + x_n \ge m, \qquad (3)$$

which is called an *extended clause*, to state that at least $m$ out of $n$ variables $x_1, \dots, x_n$ have to be true. An equivalent representation, requiring $\binom{n}{n-m+1}$ classical clauses, would be

$$\bigwedge_{I \subseteq \{1, \dots, n\} : |I| = n-m+1} \ \bigvee_{i \in I} x_i. \qquad (4)$$

Mixed integer programming has been studied in mathematics, computer science, and operations research for more than 40 years. A key notion is the *linear relaxation* of the mixed integer program (MIP), which is given by

$$z_{LP} = \min\{c^T x \mid Ax \leq b,\ x_1, \ldots, x_n \in \mathbb{R}\}, \tag{LP}$$

Note that we have relaxed the integrality condition on the variables $x_1, \ldots, x_p$. Solving (LP), which can be done very efficiently in practice, may help us in solving (MIP). An optimal solution of the relaxation may guide the search for an optimal solution of (MIP). In particular, since $z_{LP} \leq z_{MIP}$, it provides us with a lower bound on the optimal value.

In *branch-and-bound* algorithms based on linear relaxations, we recursively split the set of feasible solutions $S$ into subsets $S^i$, $i = 1, \ldots, k$. Then we compute a lower bound on the objective function on $S^i$ by solving the linear relaxation. If this lower bound is larger than the value of some feasible solution that we have already obtained (e.g., by solving another subproblem), then we may discard $S^i$ from further consideration. In order to obtain good lower bounds, we need *tight* linear relaxations. Given two systems $Ax \leq b$ and $A'x \leq b'$ defining the same set of feasible solutions, we say that $Ax \leq b$ is *tighter* than $A'x \leq b'$ if $\{x \in \mathbb{R}^n \mid Ax \leq b\} \subseteq \{x \in \mathbb{R}^n \mid A'x \leq b'\}$. Tight descriptions can be obtained statically by adding valid inequalities to the problem formulation or dynamically by computing *cutting planes* that cut off some part of the linear relaxation without eliminating feasible solutions. Combining branch-and-bound with cutting plane generation, which is called *branch-and-cut*, is one of the most successful techniques in (MIP) [18,4].

There exist various powerful software packages for solving mixed integer programs. In our computational experiments, we used the state-of-the-art commercial software CPLEX [8], which provides a very efficient implementation of branch-and-bound resp. branch-and-cut based on linear relaxations. We also experimented with the logic-based 0-1 optimization software OPBDP developed at MPI Saarbrücken [1]. OPBDP (Optimizing Pseudo-Boolean Davis-Putnam) is an extension of the classical Davis-Putnam procedure able to handle general linear 0-1 inequalities and objective functions. The relaxation used by OPBDP is weaker than the linear relaxation. But, it can be solved more quickly and therefore many more nodes of the search tree can be enumerated in the same time. As one may expect, the overall performance of the two approaches is problem dependent.

## 3   Planning and Integer Programming

This section explains how we model planning problems in IP. The basic translation method resembles that used in Satplan and Graphplan [11,3], where *parallel actions* are allowed. We first choose the maximum length of the plan by fixing the number of time steps allowed. If no plan is found that satisfies all the goals, the plan length is incremented by one. For every fluent in the domain and each time instant $t$ we associate a binary, i.e. 0-1, variable $X_t$. For instance, in the blocks world domain, each ground instance of the *on* fluent is represented by a binary variable $X_{blt}$ meaning that block $b$ is on location $l$ at time $t$. Here, a location can be either another block $b'$ or the table $T$. To model a planning domain in a *state-based* representation, we write a set of constraints that allow only valid transitions from one state to another. In domains where capacity constraints are involved, we need to introduce explicitly action variables. In the blocks world domain for instance, a variable $Y_{blt}$ is used to represent the *move* action of block $b$ to location $l$ at time $t$. Finally, an objective function that takes its optimum value if all goals are satisfied turns planning into an optimization problem.

## 3.1   IP Models for Planning

The set of constraints that correctly model a planning domain is not unique. Consider for example the blocks world domain, where an object cannot move unless it is clear. To enforce this constraint, we can use the inequality

$$X_{blt} + X_{bl't+1} + X_{b'bt} \leq 2 \tag{5}$$

for every tuple $b, b', l, l', t$, where $b, b'$ are blocks with $b \neq b'$, $l, l'$ are locations with $l \neq l'$, and $t$ is a time instant. However, in the branch-and-bound algorithm, when the relaxation is solved, these variables will assume fractional values so this constraint will be easily satisfied. Contrast this with the constraint

$$X_{blt} + \sum_{l' \neq l} X_{bl't+1} + \sum_{b' \neq b} X_{b'bt} \leq 2 \tag{6}$$

for every triple $b, l, t$. This inequality is strictly stronger than the set of inequalities (5), in the sense that there are real solutions that satisfy (5) but not (6), and not vice-versa. In order to see this, first note that all variables are non-negative. Therefore, the left-hand side of (6) is greater than or equal to the left-hand side of (5), i.e., (6) implies (5). For the reverse direction, assume that there are two blocks $b', b''$ different from $b$ and two locations $l', l''$ different from $l$. If we assign all variables the value $1/2$, then (5) is satisfied, but not (6).

Assume now that our application domain is a blocks world where the number of places on the table is limited to 7, and let the variable $Y_{blt}$ denote moving block $b$ to location $l$ at time $t$. The limited table places constraint can be easily captured by the inequality

$$\sum_b X_{bTt} + \sum_b Y_{bTt} \leq 7 \tag{7}$$

for each time instant $t$, where $T$ denotes the table. This capacity constraint also limits the number of parallel actions. Indeed, it is not hard to see that at most 3 blocks can be moved at each time instant. The inequalities

$$\sum_b \sum_l Y_{blt} \leq 3, \tag{8}$$

for each time instant $t$, suffice to represent our knowledge of the domain. While this set of constraints will be satisfied by any integer solution, it is likely that it will be violated when the linear relaxation is solved. By making this constraint explicit, we further enhance the tightness of our IP model. Moreover, the move variables that are introduced in the model are real variables over the interval $[0, 1]$, and consequently do not increase the search space of the branch-and-bound algorithm. The integrality requirement of the action variables will be enforced by the integrality of the fluent (state) variables and the constraints. The resulting model is a mixed-integer rather than a pure integer program. In fact, we can go even further and also relax some of the state variables by defining that the state variables that correspond to even time instants are real variables, while the odd time instant state variables are integer. The constraints will enforce the real variables to take integer values. Space limitations prevent us from further discussing this issue.

The use of domain knowledge in the form of valid inequalities turns out to be more important in domains that are highly combinatorial. A good example is the rocket domain. This domain involves a number of locations (airports), packages and planes and the task is to find plans that transport the packages from their original location to their destination. As the size of the problem increases, finding a plan becomes a hard combinatorial problem, in the sense that local constraint propagation does not sufficiently restrict the search space. However, this domain exhibits a rather rich structure that can be effectively encoded in the form of valid inequalities. Let $X_{plt}$ be the variable that takes the value 1 if package $p$ is at location $l$ at time $t$. Moreover, assume that the variable $Y_{ll't}$ takes the value 1 if some package is at location $l$ at time $t$, while at time $t + 4$ the same package is at some other location $l'$. The meaning of this variable can be easily captured by a set of inequalities

$$X_{plt} + X_{pl't+4} - Y_{ll't} \leq 1, \tag{9}$$

for each package, pair of locations, and time instant. It is not hard to see that, if $k$ is the number of available planes the following inequality holds

$$\sum_l \sum_{l'} Y_{ll't} \leq k. \tag{10}$$

Moreover, if $Z_{ll't}$ has a meaning similar to variable $Y_{ll't}$, with the difference that it refers to time $t + 5$ (instead of $t + 4$), the following inequality is also valid

$$\sum_l \sum_{l'} Z_{ll't} \leq 3 * k. \tag{11}$$

The available resources also imply constraints on the "intermediate" goals that must be achieved. Assume that we search for a plan of length $t_{max}$. Let $L_l$ be a binary variable that takes the value 1 if *all* packages that have location $l$ as their final destination are already there at time $t_{max} - 2$. Then, if $L$ is the number of locations in our problem and $k$ the number of planes, the following inequality is also valid

$$\sum_l L_l \geq L - k. \tag{12}$$

Similarly, let $L'_l$ be a binary variable that takes the value 1 if all packages that have location $l$ as their final destination are already there at time $t_{max} - 4$. Then the following inequality holds

$$\sum_l L'_l \geq L - 2 * k. \tag{13}$$

The auxiliary variables $Y$ and $Z$ are real, and therefore they do not introduce new branching points in the branch-and-bound algorithm. They do, however, increase the size of the model and (together with the associated constraints) the cost of solving the relaxation at each node of the search tree. Nevertheless, their overall effect is clearly rewarding as the information obtained from the relaxation is substantially more accurate. The valid inequalities also improve the performance of the logic-based optimizer, despite the fact that in this case, the auxiliary variables increase the number of choice points.

## 3.2  Planning as Optimization

In the IP model of a planning domain, apart from the constraint part, domain knowledge can also be encoded in the objective function. In the blocks world for instance, the final state $on(a,b)$, $on(b,c)$ and $on(c,d)$ can be represented by the objective function

$$4 * X_{cdt} + 2 * X_{bct} + X_{abt}, \tag{14}$$

which has to be maximized. Here, domain-specific knowledge is given to the solver about the relative importance of the goals as well as information about the ordering of the goals. This information can be used to guide the search for a plan. Notice, however, that the usefulness of such simple objective functions is rather limited, as there are final states, e.g. $on(a,b)$, $on(c,d)$, $on(e,f)$, where no such goal ordering information can be derived. Even in these cases we can still improve the representation by using more complicated objective functions, but this issue will not be discussed further.

Modeling planning as optimization also allows one to search for "approximate" solutions. Approximate in the sense that the plans that are computed may not reach the final state but one that is not "far away". The IP solver obtains successive solutions that gradually improve the value of the objective function, i.e., the distance from the final state. Hence, even in cases where a plan that achieves all the goals is hard to find within certain resource limitations, we may still obtain plans that reach states that are "close" to the final state. The quality of the solutions we obtain in this way depends on the hardness of the problem, but also the quality of the objective function as a distance measure.

### 3.3 Relaxation-Based Heuristic Search

The IP models perform quite well on moderately sized problems in the domains we studied. However, for hard problems that have long plans with many actions, trying to solve them to optimality, i.e., finding a shortest plan, is often not feasible. In order to address scalability, we need ways of decomposing planning problems and searching for non-optimal but "good" solutions.

A first approach is to exploit the formulation of planning as optimization. Let $x^i$ denote the vector of state variables associated with time $i$. Then choose a number of time steps $n$, and solve the problem

$$z = \max\{c^T x \mid Ax \le b, \ x^0, \ldots, x^n \in \mathbb{Z}^s\}, \tag{15}$$

where $x$ is the concatenation of the vectors $x^0, \ldots, x^n$, and $s$ is the length of these vectors. The vector of the state variables $x^n$ in the optimal solution becomes the new initial state. Then the procedure is iterated until we find the optimal value of the objective function. This algorithm can work reasonably well in domains where we never need to undo goals that have been already achieved, in order to achieve other goals. The problem with this approach is that it computes *locally* optimal solutions. The effect is that plans may become unnecessarily long. A way around this problem is to use a better estimate of the distance of each state to the final state. This can be accomplished by replacing (15) with the following optimization problem

$$z = \max\{c'^T x' \mid A'x' \le b', \ x^0, \ldots, x^n \in \mathbb{Z}^s, x^{n+1}, \ldots, x^{n+m} \in \mathbb{R}^s\}, \tag{16}$$

where $x'$ is the concatenation of the vectors $x^0, \ldots, x^{n+m}$. Here, we consider $m$ additional time steps. The state variables that correspond to the first $n$ steps of the plan are integer, while the state variables associated with the next $m$ time instants are relaxed to assume real values. The objective function remains unchanged. Solving this optimization problem amounts to finding a set of actions that can be performed at the first $n$ time steps and minimize the distance to the final state. The objective function subject to the relaxed constraints on the $m$ steps provides a computationally cheap estimate of the distance to the goal of each state we can reach within $n$ steps. After exploring a fixed number of nodes of the search tree, the state that gives the best value to the objective function becomes the new initial state, and the overall procedure continues until the objective function value reaches a threshold. At this point, the current state is used as the new initial state and a search for an integer solution starts. The particular values for the above parameters $n$ and $m$ depend on the hardness of the problem (i.e. how difficult it is to solve the linear relaxation) and the quality of the solution that is needed.

The method we discuss here is similar to that presented in [6]. The differences are that we use the objective function and the relaxed variables as the heuristic method, and that the search is performed by the branch-and-bound algorithm. The advantage is that the relaxation-based heuristic function can take into account numerical constraints, domain knowledge and action cost information.

## 3.4   Algebraic Modeling

To generate our models, we used the algebraic modeling system PLAM (Pro-
Log and Algebraic Modeling) [2]. PLAM allows us to specify the constraints in
a high-level declarative way using an algebraic language that is based on logic
programming. For example, we may write

```
forall((package(P),time_1(T),location(L), T1 is T+1),
       x(P,T,L) + sum((location(L1),L\==L1),x(P,T1,L1)) =< 1).
```

to say that it is not possible for a package $P$ to be at time $T$ at location $L$, and
at time $T + 1$ at a location $L1$ different from $L$. From such algebraic statements,
PLAM produces automatically the input for CPLEX or OPBDP.

# 4   Computational Experiments

We ran a number of experiments with IP models for the blocks world and the
rocket domains. The first is a classical AI planning problem, while the second
resembles the transportation problems considered in the Operations Research
and Constraint Programming literature. We assume that the reader is familiar
with these domains. We compare the logic based 0-1 optimizer OPBDP [1], the
branch-and-bound CPLEX algorithm [8] and BLACKBOX [13], a powerful classical
planning algorithm that combines techniques from graph-based planning and
propositional satisfiability. We present only a few experiments that highlight the
differences in their performance.

The experimental setup is the following. All experiments were run on a SUN
ULTRA Enterprise 10000, with 12 GB of RAM and a 333 MHz processor. We
used CPLEX 6.0.1 with the following setting. The value of the dual gradient used
in pricing was set to steepest-edge, while the variable selection heuristic was
set to pseudo reduced cost. For OPBDP we used the "fixingcheck" preprocessing
option while the search heuristic was set to 13. For BLACKBOX (version 3.3) we
used the most effective solver in each domain, that is, RELSAT in blocks world
and SATZ in the rocket domain. All rocket domain problems were preprocessed
with the "compact" simplification procedure. In the blocks world domain this is
not always possible, as it often leads to excessive memory requirements. In both
tables, running times are given in seconds, while a dash (-) denotes computation
time that exceeds 5 hours of CPU.

Table 1 shows the results for some of our experiments with blocks world
problems where the table has a limited number of places. To encode the table
capacity in the STRIPS representation, we explicitly introduce the different table
places in the problem description. This way of representing the limited table
capacity seems to have serious effects on BLACKBOX propositional solvers. The
IP model can very compactly accommodate this constraint in the form of linear
inequalities. Both OPBDP and CPLEX were run for solutions that achieve all the
goals of the planning problem. For each of these two solvers, the column entries
were obtained by adding the running times resp. the numbers of nodes needed

**Table 1.** Results for blocks world problem with limited table places.

| Problem | blocks | places | length | BLACKBOX Time | OPBDP Time | Nodes | CPLEX Time | Nodes |
|---|---|---|---|---|---|---|---|---|
| bw1.6 | 12 | 6 | 7 | 9577 | 0.5 | 150 | 2700 | 2248 |
| bw1.5 | 12 | 5 | 8 | - | 12 | 1372 | 212 | 85 |
| bw1.4 | 12 | 4 | 10 | - | 5972 | 2426010 | 6070 | 1146 |
| bw2.7 | 13 | 7 | 7 | 14490 | 0.2 | 57 | 72 | 70 |
| bw2.6 | 13 | 6 | 8 | - | 6 | 3371 | 1950 | 336 |
| bw2.5 | 13 | 5 | 10 | - | 335 | 117137 | - | - |

for finding optimal length plans and for proving that no shorter plan exists. It is clear that CPLEX explores fewer nodes than OPBDP. This however does not always translate into better running times, since the cost per node for the relaxation-based solver is substantially higher than for the logic-based algorithm.

Our experience is that while CPLEX can sometimes do better than OPBDP in finding a plan if one exists, it is in general less effective than OPBDP in proving infeasibility when no plan exists, i.e., the allowed plan length is not sufficient. In problem bw1.4 for instance, CPLEX finds a 10 time steps plan in 370 secs (after exploring 86 nodes) while OPBDP needs 5800 secs and explores 2347031 nodes. However, for proving that no plan of length 9 exists, CPLEX needs 5700 secs (for exploring 1060 nodes) and OPBDP 172 secs (for 78979 nodes). As it has been explained before, adding these numbers yields the corresponding entries in Table 1. It should be noted that a search heuristic different from the one we used for OPBDP improves its performance in the blocks world domain. For example, by setting the search heuristic to 11, problem bw1.4 can be solved in 540 secs.

Experimentally, the relaxation-based heuristic turns out to be an interesting alternative to optimal planning. Consider for example the hard problem bw2.5. We applied the heuristic with the number of real and integer time steps both set to 3. After two iterations a new initial state is identified and then CPLEX searches for an integer solution which happens to be 5 time instants long. The whole computation takes less than 5 minutes of CPU time while the plan length is 11, i.e., one time instant longer than the optimal. A detailed discussion of the relaxation-based heuristic as well as other ways of exploiting the linear relaxation in planning problems, will be included in an extended version of this paper.

Table 2 presents computational results for the rocket domain. The IP models encode domain knowledge through valid inequalities as those discussed in Section 3.1. Moreover, the rocket domain model differs form the block worlds encoding in that we introduce extra variables for the 'fly' actions and solve the problem of minimizing the number of these actions. This objective function significantly improves CPLEX performance. A similar observation has been made in [20]. The problems were generated in such a way that no pair of packages has the same origin and destination. The computation time for each problem is again the time required to find the first integer solution and to prove that no shorter plan exists. As in the blocks world domain, CPLEX enumerates fewer nodes than OPBDP, but, in terms of computation time, OPBDP can be significantly faster than CPLEX. For large problems solving the linear relaxation may become extremely expensive. CPLEX for example finds a plan for problem r8.18 after exploring 1227 nodes, but this computation requires about 7 hours of CPU time.

**Table 2.** Results for the rocket domain. The *pln* column shows the number of planes, *loc* the number of locations and *pack* the number of packages.

| Problem | pln | loc | pack | length | BLACKBOX Time | OPBDP Time | Nodes | CPLEX Time | Nodes |
|---|---|---|---|---|---|---|---|---|---|
| r1.13 | 2 | 5 | 13 | 9 | 535 | 2 | 227 | 303 | 55 |
| r2.16 | 2 | 5 | 16 | 9 | 760 | 3 | 400 | 183 | 18 |
| r3.16 | 2 | 5 | 16 | 9 | - | 4 | 454 | 301 | 35 |
| r4.18 | 2 | 5 | 18 | 10 | - | 4 | 438 | 2780 | 65 |
| r5.15 | 3 | 6 | 15 | 7 | 913 | 18 | 3438 | 449 | 64 |
| r6.16 | 3 | 6 | 16 | 7 | - | 20 | 3559 | 274 | 41 |
| r7.16 | 3 | 6 | 16 | 8 | - | 1282 | 211917 | 2705 | 314 |
| r8.18 | 3 | 6 | 18 | 8 | - | 1463 | 215285 | - | - |
| r9.20 | 3 | 6 | 20 | 9 | - | 18253 | 2407576 | - | - |

## 5   Related Work

The success of the planning as satisfiability framework motivated recently the application of integer and linear programming techniques to planning problems with numerical constraints and complex objective functions. The LPSAT planner [21] supports a language capable of expressing numerical constraints on real variables, and interleaves propositional satisfiability with calls to a simplex algorithm. Our approach instead, translates planning into linear inequalities on both integer and real variables. This enables us to use knowledge that is not expressible either by propositional clauses or by linear inequalities over real variables.

Closer to our approach is the work of Vossen et al. [20] and the ILP-PLAN framework of Kautz and Walser [14]. ILP-PLAN focuses on translating planning problems that are described in the resource-extended IPP language [15] into a set of linear inequalities on integer and real variables. In addition, the objective function is used to encode optimization criteria like minimization of resource usage or minimization of the sequential length over plans of minimal parallel length. Our work focuses on a more conservative extension of the classical AI planning paradigm. We consider problems that can be adequately represented by a set of inequalities in 0-1 variables, i.e., problems that can be seen as purely "logical". The usefulness of integer programming techniques is less obvious in this case. This is because, in general, such problems can be modeled in the STRIPS language, and therefore powerful graph and satisfiability based planning algorithms are directly applicable. Despite this, we showed that the extended language and its supporting algorithms can accommodate compact representations of planning problems as well as powerful domain knowledge. We should note here that while OPBDP can model domains with action costs and objective functions, it cannot handle problems where general (i.e. non 0-1) variables need to be introduced.

The work of Vossen et al. also remains within the classical STRIPS planning paradigm. The authors present a translation of classical planning problems into mixed integer programming and show that this is much stronger than a direct translation of the SATPLAN encodings [11] into linear inequalities.

We should note that none of these works is exclusive of the others. They are rather complementary. For example, one could start with the domain independent translation of Vossen et al., then add valid inequalities similar to the ones we propose, and finally extend the problem model with resource constraints as described in ILP-PLAN.

## 6   Conclusions

Despite the recent success in deriving new effective general-purpose planning algorithms, solving hard planning problems is unlikely without the use of domain-specific knowledge or algorithms. In the middle ground between using a general-purpose planning algorithm and a specialized one lies the possibility of expressing in a declarative way domain knowledge which can be exploited by a general-purpose solver. In this work, we discussed how integer programming models can incorporate domain knowledge and the different possibilities that are open for developing algorithms based on these methods. The main motivation for considering IP methods is (a) global constraints that involve numbers can be easily expressed (b) the representation language facilitates the use of rich forms of domain knowledge (c) non-optimal solutions can be obtained when optimal planning is not computationally feasible. Our experiments in the blocks world and the rocket domain gave encouraging results.

Many interesting problems remain open for future research. Firstly, it is possible that our models, especially for the rocket domain, can be further improved. In fact, transportation problems are a standard application domain for operations research techniques. Another interesting direction for further investigation is the enhancement of the IP models developed independently in [20] with valid inequalities like those used in our models. Moreover, we need to study more planning domains in order to verify that IP modeling has a wide applicability. More specifically, we plan to develop models for the logistics [11] and the trains domain [9], and extend our experiments to domains that require general real-valued resource variables, like those described in [14]. The study of the simpler but instructive rocket domain is a first step in this direction. Finally, we plan to investigate ways for automatically deriving valid inequalities from problem specifications, for example by exploiting the conflict graph representing certain logical relations between 0-1 variables.

**Acknowledgements.** This work has been partially supported by the Max-Planck-Institut für Informatik, Saarbrücken, Germany.

# References

1. P. Barth. *Logic-based 0-1 constraint programming.* Operations Research/ Computer Science Interfaces Series. Kluwer, 1996.
   http://www.mpi-sb.mpg.de/~barth/opbdp/opbdp.html
2. P. Barth and A. Bockmayr. Modelling discrete optimisation problems in constraint logic programming. *Annals of Operations Research*, 81:467–496, 1998.
3. Blum A. and Furst M. Fast planning through planning graph analysis, *Artificial Intelligence*, 90, 281 – 300, 1997.
4. A. Bockmayr. Solving pseudo-Boolean constraints. In *Constraint Programming: Basics and Trends*, 22 – 38. Springer, LNCS 910, 1995.
5. A. Bockmayr and Y. Dimopoulos. Mixed integer programming models for planning problems. In *CP'98 Workshop on Constraint Problem Reformulation*, 1998.
6. B. Bonet, G. Loerincs, H. Geffner. A fast and robust action selection mechanism for planning. *AAAI'97*.
7. T. Bylander. A Linear Programming Heuristic for Optimal Planning. *AAAI'97*.
8. CPLEX Optimization, Inc. *Using the CPLEX callable library*, 1995.
   http://www.cplex.com
9. A. Gerevini and L. Schubert. Accelerating Partial-Order Planners: Some Techniques for Effective Search Control and Pruning. *Journal of Artificial Intelligence Research*, Vol. 5, 1996.
10. Y-C. Huang, B. Selman and H. Kautz. Control Knowledge in Planning: Benefits and Tradeoffs. *IJCAI'99*.
11. H. Kautz and B. Selman. Pushing the envelope: planning, propositional logic and stochastic search. *AAAI'96*.
12. H. Kautz and B. Selman. The Role of Domain-Specific Knowledge in the Planning as Satisfiability Framework. *AIPS'98*.
13. H. Kautz and B. Selman. Unifying SAT-based and Graph-based Planning. *IJCAI'99*.
14. H. Kautz and J. Walser. State-Space Planning by Integer Optimization. *AAAI'99*.
15. J. Koehler. Planning under Resource Constraints. *ECAI'98*.
16. J. Koehler, B. Nebel, J. Hoffmann and Y. Dimopoulos. Extending Planning Graphs to an ADL Subset. *4th European Conference on Planning, ECP'97*.
17. D. McDermott. A heuristic estimator for means-ends analysis in planning. *AIPS'96*.
18. G. L. Nemhauser and L. A. Wolsey. *Integer and Combinatorial Optimization*. Wiley, 1988.
19. P. van Beek and X. Chen. CPlan: A Constraint Programming Approach to Planning. *AAAI'99*.
20. T. Vossen, M. Ball, A. Lotem and D. Nau. On the Use of Integer Programming Models in AI Planning. *IJCAI'99*.
21. S. Wolfman and D. Weld. The LPSAT Engine and its Application to Resource Planning, *IJCAI'99*.

# Deductive Synthesis of Recursive Plans in Linear Logic

Stephen Cresswell, Alan Smaill, and Julian Richardson

Division of Informatics, University of Edinburgh,
80 South Bridge, Edinburgh, EH1 1HN, U.K.
Fax: +44 131 650 6516
{s.cresswell,a.smaill,julian.richardson}@ed.ac.uk

**Abstract.** Linear logic has previously been shown to be suitable for describing and deductively solving planning problems involving conjunction and disjunction. We introduce a recursively defined datatype and a corresponding induction rule, thereby allowing recursive plans to be synthesised. In order to make explicit the relationship between proofs and plans, we enhance the linear logic deduction rules to handle plans as a form of proof term.

## 1 Introduction

We are interested in the automated synthesis of recursive plans. In many planning domains, it is possible to construct plans at a more general level than is normally done, e.g. by building plans that can solve a whole class of problems, rather than solving specific ground instances of problems. Typically, a recursive plan is short, but can solve problem instances of arbitrary size — e.g. consider a plan to invert a tower of 1000 blocks.

In the deductive planning approach, plans are formed by constructing proofs in some appropriate logic. These formal approaches to planning, such as situation calculus and modal logic, bring with them a problem of representing frame conditions. In contrast, practical planning techniques are usually based on the STRIPS representation [5], which is hard to extend in a declarative way.

One approach to deductive planning is to make use of Girard's linear logic [7]. This is appealing for planning because it is inherently resource-sensitive. We can use a logical planning framework while retaining a STRIPS-like approach to dealing with the frame problem.

The use of linear logic in planning was first explored by Masseron et al. [15] and [14] and has also been considered by Jacopin [12] and Grosse et al. [8]. However, authors have mainly concentrated on adequacy for simple STRIPS-like plan representations. We think its true potential in planning lies in using its full expressiveness. In this paper, we extend linear logic with an appropriate form of induction that allows us to reason about recursive plans.

Section 2 gives a brief introduction to a fragment of intuitionistic linear logic, focussing on its application in planning problems.

S. Biundo and M. Fox (Eds.): ECP-99, LNAI 1809, pp. 252–264, 2000.

Section 3 gives a plan representation which allows a plan to be built directly in the course of a linear logic proof. We then extend this scheme to give deduction rules with plan terms for a larger fragment of linear logic, allowing conditional and conformant plans to be represented.

Section 4 extends the linear logic with an appropriate induction rule for a recursive datatype, allowing proofs to be developed for recursive plans. Section 5 discusses automated proof search (i.e. a planner) using this framework. Sections 6 and 7 relate this work to existing work in the field, and draw conclusions.

## 2    Linear Logic

We briefly describe linear logic here; see [7] for a full exposition, and [15] for an account in relation to a semantics for planning problems.

Linear logic is resource-sensitive. This gives us the ability to model change of state directly using the linear version of implication, written as $\multimap$. The usual example here is that we can model the scenario that we can buy a drink for a pound as follows:

$$have\_pound \multimap have\_drink.$$

The notion of implication here is that if we have a pound, we can have a drink, but (unlike conventional implication) we won't still have the pound. The resource on the left of the implication is used up in producing the resource on the right.

These rules can only be seen as transitions if the logic itself restricts the copying and discarding of resources.

Now if we consider resource-limited versions of familiar connectives such as conjunction, we find that there are two different versions possible, differing in the way the resources are handled.

The *multiplicative conjunction* $A \otimes B$ means that resources $A$ and $B$ are simultaneously present. This is the form of conjunction which is used in STRIPS-like planning problems.

The second form, *additive conjunction*, means that both resources are individually available, but they are exclusive. Only one or the other may be used and the choice is ours.

$$have\_pound \multimap have\_tea \ \& \ have\_coffee$$

Although this looks somewhat more like a disjunction, it is regarded a conjunction, since it would be equivalent to the conventional conjunction if copying and discarding of resources were allowed.

This may be can contrasted with the additive disjunction, $\oplus$.

$$have\_pound \multimap have\_tea \oplus have\_coffee$$

This would correspond to using an erratic drinks machine, which will deliver either tea or coffee, but we cannot choose which.

## 2.1   Planning with Linear Logic

A planning problem can be represented as a sequent of the form $I \vdash G$ where $I$ represents initial conditions and $G$ represents goal conditions. In this intuitionistic version, $I$ is a multiset of formulae (implicitly joined by $\otimes$). The $\vdash$ behaves as linear implication, so we can read this as meaning that the resources $I$ should be consumed in deriving $G$. Similarly, we use transition axioms of the form $P \vdash E$ to represent operators. These axioms can be reused as many times as necessary in the proof, each use corresponding to an action. For instance, we could represent an operator $stack(X, Y)$ for placing a block as follows:

$$hold(X), clr(Y) \vdash empty \otimes clr(X) \otimes on(X, Y)$$

To see the correspondence with STRIPS operators, consider the STRIPS version of $stack(X, Y)$. This can be written as:

```
operator:       stack(X,Y)
preconditions:  hold(X)
                clr(Y)
deletelist:     hold(X)
                clr(Y)
addlist:        clr(X)
                on(X,Y)
                empty
```

Note that the main difference between the two renditions is that there is no equivalent of the delete list in the linear logic description. This is not needed because anything used as a precondition will automatically be consumed by the linear logic version of implication. This can simply represent problems from the STRIPS notation, since any preconditions which are required but not consumed by an action can simply be added back onto the right hand side of the $\vdash$ in the action definition.

Another significant difference is that in linear logic, multiple instances of the same entity are regarded as distinct. For example, we could represent the situation of having two pounds as $have\_pound \otimes have\_pound$.

We can create proofs for solving simple STRIPS-like planning problems using only the $\otimes$ connective and the rules $Ax, cut, l\otimes, r\otimes$.

$$\frac{}{A \vdash A}\ Ax \qquad\qquad \frac{\Gamma \vdash A \quad \Gamma', A \vdash C}{\Gamma, \Gamma' \vdash C}\ cut$$

$$\frac{\Gamma, A, B \vdash C}{\Gamma, A \otimes B \vdash C}\ l\otimes \qquad\qquad \frac{\Gamma \vdash A \quad \Gamma' \vdash B}{\Gamma, \Gamma' \vdash A \otimes B}\ r\otimes$$

The cut rule is crucial here, as it can now be seen as a rule which allows the transition between two states to be made via some intermediate state, and this accounts for the composition of a plan by combining two sub-plans in sequence.

An example of the use of the cut rule is given below. This shows the transition from a state described by $empty, clr(c), on(c, a), clr(b), ontable(b)$ by the application of a *remove* action to a state described by $hold(c), clr(a), clr(b), ontable(b)$.

$$\frac{\overbrace{empty, clr(c), on(c, a) \vdash hold(c) \otimes clr(a)}^{remove(c, a)} \quad \overset{\vdots}{hold(c) \otimes clr(a), clr(b), ontable(b) \vdash G}}{empty, clr(c), on(c, a), clr(b), ontable(b) \vdash G} \; cut$$

If the cut rule is always applied to sequents in which the hypothesis list is a superset of the formulae on the left of a transition axiom, the proof corresponds to a plan built forwards from the initial state.

So we can build a proof tree with a statement of the planning problem at its root, and with instances of transition axioms and $Ax$ at its leaves. If we can build such a proof, we can be satisfied that there is a plan that solves the problem, where applications of the transition axioms correspond to simple actions in the plan. However, some work is still needed to extract the plan from the proof tree. One way to do this is by analysis of the proof, as proposed in [14]. An alternative approach is described in section 3.

# 3   Extraction of Plans

In previous publications, authors have used a procedure for extracting plans from the completed proof. Here, we will make the relationship of the proof to the plan more concrete by attaching proof terms directly to the deduction rules in the style of type theory [16]. This makes the relationship of deduction rules and plan formation clearer, and is easier to extend to deal with a larger subset of linear logic.

A type theory has been defined for linear logic in [1]. Proof terms can be seen as programs in linear lambda calculus — a functional language in which there is a restriction of using each input once.

However, this allows the construction of a more powerful language than we want to deal with in a planning problem. We wish to restrict our consideration to the proof terms which have a simple imperative interpretation. We describe such a system below, in which sequents of the form $A \vdash$ `plan` $: C$ should be interpreted as meaning that *plan* gets us from a state described by resources $A$ to a state described by $C$.

The plan terms are built with the following syntax:

| $plan ::=$ | $step$ | a single-step action |
| | `nop` | do nothing |
| | $plan;plan$ | execute plans sequentially |
| | `par`$(plan, plan)$ | execute plans in parallel |
| | $\lambda v.plan$ | plan parameterised by $v$ |

For notational convenience, we will often omit plan terms on the left side of the sequent.

### 3.1   Transition Axioms

Operator definitions now take the following form:

$$A \vdash \mathsf{step} : C$$

### 3.2   Identity Axiom and Cut Rule

Identity corresponds to a transition to the same state, so the corresponding plan is *nop*. The *cut* rule corresponds to the sequential composition of two plans.

$$\frac{}{A \vdash \mathsf{nop} : A} \; Ax \qquad\qquad \frac{\Gamma \vdash \mathsf{a} : A \quad \Gamma', A \vdash \mathsf{c} : C}{\Gamma, \Gamma' \vdash (\mathsf{a}; \mathsf{c}) : C} \; cut$$

### 3.3   Multiplicative Conjunction

The $\otimes$ rule allows a planning problem to be broken down into two independent sub-problems. Since the sub-problems rely on disjoint sets of resources, the plan term consists of two sub-plans in parallel.

$$\frac{\Gamma, A, B \vdash \mathsf{c} : C}{\Gamma, A \otimes B \vdash \mathsf{c} : C} \; l\otimes \qquad\qquad \frac{\Gamma \vdash \mathsf{a} : A \quad \Gamma' \vdash \mathsf{b} : B}{\Gamma, \Gamma' \vdash \mathsf{par}(\mathsf{a}, \mathsf{b}) : A \otimes B} \; r\otimes$$

### 3.4   Quantifiers

To make recursive plans, we will need to be able to handle quantification in some form. The quantifier rules for linear logic are the same as a version of the standard ones. However, the meaning of $\forall$ is *for any* rather than *for every* — a distinction which is not meaningful in constructive or classical logic. The resultant plan is parameterised, and will provide a plan for a specific instance when supplied with a value for the parameter. We assume $\beta$-reduction is used to compute the plan instances.

$$\frac{\Gamma, \mathsf{j} : P(t) \vdash \mathsf{A} : C}{\Gamma, \mathsf{i} : \forall u.P(u) \vdash \mathsf{A}[\mathsf{i}(\mathsf{t})/\mathsf{j}] : C} \; l\forall \qquad\qquad \frac{\Gamma \vdash \mathsf{c} : C[a/x]}{\Gamma \vdash \lambda \mathsf{a}.\mathsf{c} : \forall x.C} \; r\forall$$

where $a$ is not free in $\Gamma, C$.

### 3.5   Disjunctive Effects and Conditionals

In some planning problems, it may be desirable to represent plans with indeterminate outcomes. These can be represented by actions with disjunctive effects. Here it is appropriate to use the additive disjunction operator $\oplus$. A formula $A \oplus B$ should be interpreted as saying that either resource $A$ or resource $B$ is available, and in a plan, we must cope with both possibilities.

[15] gives an example of a problem in which socks are blindly taken from a drawer. This action is represented by a transition in which a hidden sock ($hs$) becomes either a black sock ($bs$) or a white sock ($ws$).

$$hs \vdash \texttt{pick} : bs \oplus ws$$

In resolving these disjunctions during the planning process, there are two possibilities: the agent which will execute the plan may or may not be capable of performing a test to resolve the disjunction at runtime.

If the agent can perform a test at runtime, it can select between two different plans, i.e. we can build a conditional structure.

$$\frac{\Gamma, A \vdash \texttt{Plan1} : C \quad \Gamma, B \vdash \texttt{Plan2} : C}{\Gamma, A \oplus B \vdash \texttt{if}^{\texttt{A then Plan1}}_{\texttt{B then Plan2}} : C} \; l \oplus (test)$$

This is equivalent to forms used by Brüning et al. [3] and by Abramsky [1].

If the agent cannot perform a test (cannot look at the sock), we must ensure that the same plan will work regardless of which condition holds (black or white sock). This form of planning problem has been called *conformant* or *fail-safe* planning. Here we use the following form of the $l\oplus$ rule, in which the same $Plan$ will work for both $A$ and $B$.

$$\frac{\Gamma, A \vdash \texttt{Plan} : C \quad \Gamma, B \vdash \texttt{Plan} : C}{\Gamma, A \oplus B \vdash \texttt{Plan} : C} \; l\oplus$$

Note that the proof term is here being used to restrict the kind of proof which can be extracted, i.e. to enforce the condition that the plan for each branch is identical. This would not be possible in other linear logic planning schemes, in which the term representing the plan itself is not present in the proof.

For a disjunction of goals, we must simply prove one or the other:

$$\frac{\Gamma \vdash \texttt{p} : B}{\Gamma \vdash \texttt{p} : B \oplus C} \; r1\oplus$$

$$\frac{\Gamma \vdash \texttt{p} : C}{\Gamma \vdash \texttt{p} : B \oplus C} \; r2\oplus$$

## 4  Induction and Formation of Recursive Plans

We can introduce a simple recursive datatype and synthesise recursive plans by performing a proof by induction over that datatype. Note that recursive plans assume the ability to test in the sense of the rule $l \oplus (test)$, to recognise when base and step case actions apply.

## 4.1   Recursive Datatype and Corresponding Induction Rule

We define a datatype for towers:

$$tower := (block :: tower) \mid empty$$

Now we define an induction rule for this datatype. Note that we must be quite careful about how this is defined. Since a given resource may only be consumed once, we must make sure that formulas used in the step case are either dependent on the datatype or are replenished if used. $\Gamma$ is made available only in the base case, which corresponds to a plan that is only executed once.

The syntax for plans is extended by adding the notion of recursive plan, based on the treatment of recursion in type theory [16]. If $t$ is a tower, bp is a plan and $\lambda x.\lambda y.\lambda z.sp$ is a plan with three parameters, then $\texttt{rec\_twr}(t, bp, \lambda x.\lambda y.\lambda z.sp)$ is a recursive plan. Recursive plans are executed as follows. If $t = empty$, then execute bp; otherwise $t$ is of the form $b' :: t'$, so evaluate the plan

$$sp[(\texttt{rec\_twr}(t', bp, \lambda x.\lambda y.\lambda z.sp))/x, b'/y, t'/z]$$

and execute it.

The inference rule that relates induction to the formation of recursive plans is as follows:

$$\frac{\vdash bp : F(empty) \quad r : F(t') \vdash sp : F(b' :: t')}{\vdash \lambda t.\texttt{rec\_twr}(t, bp, \lambda r.\lambda t'.\lambda b'.sp) : \forall t.F(t)}$$

In general, the $F(t)$ will be a plan specification of the form $I(t) \multimap G(t)$. The term $r$ attached to the induction hypothesis will behave as an available action, and its appearance in the plan for the step case signifies the application of the recursive call. It must be used exactly once in the step case plan.

It is helpful to introduce a custom rule in this scenario. This rule is derivable from the standard rules for handling linear implication.

$$\frac{P_1 \vdash c : P_2}{j : P_2 \multimap Q \vdash (c; j) : P_1 \multimap Q} \; c \multimap$$

This is not the most general form, but it will simplify the presentation of the following example significantly.

## 4.2   Example

The following example shows a solution to the problem of inverting a tower of blocks. The specification of the problem itself requires the use of a recursively defined function.

As we shall see in the example below, our proof procedure requires both the application of deduction rules, contributing to the instantiation of a plan term, and the application of rewrite rules, to transform equivalent expressions into the same syntactic form.

For this example, we wish to solve a problem of the form:

$$\vdash \texttt{plan} : \forall t. [twr(t) \multimap twr(rev(t))]$$

which means that a tower $t$ can be reversed by execution of *plan*. $twr(t)$ should be read meaning that a tower $t$ of unknown height is present. *rev* is a reverse function — it allows the relationship between initial and goal states to be described. It does not describe an action.

The following will be used as plan operators.

$$twr(B :: T), hn \ \vdash \texttt{pickup}(B) : \ twr(T) \otimes hold(B)$$
$$twr(T), hold(B) \ \vdash \texttt{put}(B, T) : \ twr(B :: T) \otimes hn$$

where $hn$ stands for "holding nothing". Since we are going to need to pass through intermediate steps in the plan where we have two separate towers, it is necessary to include a reference to a second tower throughout. The problem of finding this generalisation of the original problem is addressed in section 5.

$$\forall t, a. twr(t) \otimes twr(a) \otimes hn \multimap twr(empty) \otimes twr(app(rev(t), a)) \otimes hn$$

The definitions of *rev* and *app* are given by:

$$rev(empty) = empty \tag{1}$$
$$rev(b :: t) = app(rev(t), b :: empty) \tag{2}$$

$$app(empty, u) = u \tag{3}$$
$$app(b :: t, u) = b :: app(t, u) \tag{4}$$

We will also make use of associativity of *app*.

$$app(app(a, b), c) = app(a, app(b, c)) \tag{5}$$

We can begin the proof by using an induction on $t$ (for clarity, we will omit plan terms).

For the base case, we must prove:

$$\vdash \forall a. twr(empty) \otimes twr(a) \otimes hn \multimap twr(empty) \otimes twr(app(rev(empty), a)) \otimes hn$$

After application of rewrites 1,3 we get:

$$\vdash \forall a. twr(empty) \otimes twr(a) \otimes hn \multimap twr(empty) \otimes twr(a) \otimes hn$$

which can be proved trivially.

For the step case we will walk through the proof in the direction of its construction i.e. from bottom to top. The full sequent proof is given as an appendix. The initial goal is:

$$\begin{array}{c} \forall a. twr(t) \otimes twr(a) \otimes hn \multimap \\ twr(empty) \otimes twr(app(rev(t), a)) \otimes hn \end{array} \vdash \begin{array}{c} \forall a. twr(b :: t) \otimes twr(a) \otimes hn \multimap \\ twr(empty) \otimes twr(app(rev(b :: t), a)) \otimes hn \end{array}$$

We can eliminate the universal quantifiers: first on the right, then on the left. We allow meta-variables ($A$ in this proof) so as to delay the choice of witness term $t$ in the use of the rule $l\forall$

$$\frac{twr(t) \otimes twr(A) \otimes hn \multimap}{twr(empty) \otimes twr(app(rev(t), A)) \otimes hn} \vdash \frac{twr(b :: t) \otimes twr(a') \otimes hn \multimap}{twr(empty) \otimes twr(app(rev(b :: t), a')) \otimes hn}$$

Now we apply rewrite rules 2,5,4,3 to the R.H.S.

$$\frac{twr(t) \otimes twr(A) \otimes hn \multimap}{twr(empty) \otimes twr(app(rev(t), A)) \otimes hn} \vdash \frac{twr(b :: t) \otimes twr(a') \otimes hn \multimap}{twr(empty) \otimes twr(app(rev(t), b :: a')) \otimes hn}$$

By instantiating the meta-variable to $b :: a'$, we can now apply the $c \multimap$ rule, which leads to:

$$twr(b :: t), twr(a'), hn \vdash twr(t) \otimes twr(b :: a') \otimes hn$$

Now, we can complete the proof by cutting in applications of the *pickup* and *put* actions (see appendix for complete proof of step case).
The plan can now be given as:

$\lambda\mathtt{t.rec\_twr(t, nop,}$
$\quad\quad\quad \lambda\mathtt{r.\lambda t'.\lambda b'.\lambda a'.(par(nop, pickup(b')); par(nop, put(b', a')); r(t', a')))}$

which can be simplified to:

$\quad\quad \lambda\mathtt{t.rec\_twr(t, nop, \lambda r.\lambda t'.\lambda b'.\lambda a'.(pickup(b'); put(b', a'); r(t', a')))}$

As we are synthesising a plan with two parameters (corresponding to the two universal quantifiers in the original specification), the recursive call $\mathtt{r(t', a')}$ requires values $t', a'$ for the parameters.

## 5   Automated Proof Search

An attractive approach to proof search in linear logic is to use a form of logic programming based on linear logic rather than on classical (or constructive) logic. We have used the language Lolli [10] as a theorem prover in our experiments. These languages are based on intuitionistic linear logic, with some restrictions to permit a uniform proof procedure to be carried out (for example, expressions of the form $A \otimes B$ are not allowed as the head of a clause). These restrictions prevent us from writing operators directly as clauses in Lolli, and finding a plan by executing the query that asks whether the initial resources can be transformed into the goal resources. However, a plan-forming routine can be written to interpret declared operators and queries in the style of a meta-interpreter [18], while building up the extracted plan, and allowing an induction rule to be used. This provides a tidy representation of the planning problem, and allows bookkeeping of the resources to be handled by the logic programming machine, but is not particularly efficient for finding a solution.

In the recursive planning problems, the process of synthesising the plan (and the proof) now consists not only of manipulations which correspond to plan steps, but also rewrites of terms to equivalent forms. Uniform search is not enough in this situation, and we lean on work in guiding inductive proof in standard logic, described in [4]. Heuristics are available to help at the difficult choice points, namely which induction to use (there may be several schemes possible, and variables to choose from), and choice of rewrites to apply.

The generalisation step in the example in section 4.2 is amenable to automation. A similar problem is discussed in [9,11]: when the problem cannot be solved in its given form, the failed proof attempt gives rise to information about the form of generalisation that would allow the proof to be completed. In program synthesis, this may suggest the need for an accumulator variable, and an appropriate specification of the generalised procedure derived in the course of the proof. A similar analysis applies for plan synthesis.

# 6   Related Work

## 6.1   Linear Logic Planners

The work of Masseron et al. [15,14] has been the starting point for this work. Masseron showed the adequacy of linear logic for simple conjunctive planning problems, for which a procedure for extracting plans in the form of directed graphs was given. Their representation for partially-ordered plans is less redundant than ours, but more awkward to extend to a larger fragment of the logic. Problems involving disjunction are covered by the representation but not by the geometric argument relating proofs to plans. A procedure for proof search using the conjunctive fragment is given in [12].

[8] makes clear the close relationship between their technique of planning with equational resolution, Bibel's linear connection method [2], and linear logic. Techniques based on the connection method have been used both as deductive planners and as linear logic theorem provers.

## 6.2   Recursive Planning

The approach of Manna and Waldinger [13] uses a version of the situation calculus. Their formalism accounts for the derivation of conditional and recursive plans. Their work identifies the need for generalisation in performing inductive proof as being a major problem.

The RNP planner of Ghassem-Sani and Steel [6] described RNP, an implementation of a recursive planner based on goal-driven partial order planning. This allowed for a principled method for deciding when to introduce a recursive construct. The plans generated by this planner used a conventional representation of a plan as a directed graph, and conditional and recursive constructs were considered as special types of node containing sub-plans. An interesting feature of their approach is that induction rules are constructed dynamically from a well-founded order.

The approach described in this paper differs from these works in that our induction is based on a given rule for an inductive datatype, rather than constructing rules from a well-founded order. We believe our formulation allows stronger control of the proof search, while obviating the need to reason about partially defined functions.

The deductive planning work of Stephan and Biundo [17] uses a temporal logic to model problems and plans. There is a sophisticated theorem-proving environment for domain modelling and plan formation. In their approach, formation of recursive procedures in considered part of the domain modelling, where such procedures can be developed and proved interactively with the theorem prover. After this stage, plans can then be constructed automatically by refinement of specifications.

Unlike Stephan and Biundo, we are interested in automating the recursive plan formation process.

## 7    Conclusion

We believe the combination of linear logic and induction is a promising avenue for the formation of recursive plans in a declarative way. It allows us to bring experience in the control of proof search in inductive proofs to bear on this problem. Furthermore the expressive nature of linear logic allows a natural and concise formulation of several interesting extensions to conventional planning that can be incorporated in our framework.

In order to demonstrate the usefulness of this approach, we are currently devising a set of examples using different inductively defined datatypes, for example, trees.

As other authors have noted [13], the generalisation problem is central to full automation. As was mentioned in section 5, this problem is amenable to automation, but we have not yet explored the applicability of that approach in this context.

**Acknowledgements.** Thanks to Louise Pryor for discussions on this and related material. Stephen Cresswell was supported by an EPSRC studentship. Alan Smaill and Julian Richardson are supported by EPSRC grant GR/M45030.

## References

1. S. Abramsky. Computational interpretations of linear logic. *Theoretical Computer Science*, 111:3–57, 1993. (Revised version of Imperial College Technical Report DoC 90/20).
2. W. Bibel. A deductive solution for plan generation. *New Generation Computing*, 4:115–132, 1986.
3. S. Brüning, S. Hölldobler, J. Schneeberger, U. Sigmund, and M. Thielscher. Disjunction in resource-oriented deductive planning. In *Proceedings of the International Symposium on Logic Programming*, page 670, 1993.

4. A. Bundy, A. Stevens, F. van Harmelen, A. Ireland, and A. Smaill. Rippling: A heuristic for guiding inductive proofs. *Artificial Intelligence*, 62:185–253, 1993. Also available from Edinburgh as DAI Research Paper No. 567.

5. R. E. Fikes and N. J. Nilsson. STRIPS: A new approach to the application of theorem proving to problem solving. *Artificial Intelligence*, 2:189–208, 1971.

6. G. R. Ghassem-Sani and S. W. D. Steel. Recursive plans. In *Proc. of the European Workshop on Planning EWSP-91*, pages 53–63, St. Augustin, Germany, 1991.

7. J. -Y. Girard. Linear logic: its syntax and semantics. In J. -Y. Girard, Y. Lafont, and L. Regnier, editors, *Advances in Linear Logic*, number 222 in London Mathematical Society Lecture Notes Series. Cambridge University Press, 1995.

8. G. Große, S. Hölldobler, and J. Schneeberger. Linear deductive planning. *Journal of Logic and Computation*, 6(2):233–262, 1996.

9. J. Hesketh, A. Bundy, and A. Smaill. Using middle-out reasoning to control the synthesis of tail-recursive programs. In Deepak Kapur, editor, *11th International Conference on Automated Deduction*, pages 310–324, Saratoga Springs, NY, USA, June 1992. Published as Springer Lecture Notes in Artificial Intelligence, No 607.

10. J. S. Hodas and D. Miller. Logic programming in a fragment of intuitionistic linear logic. *Information and Computation*, 110(2):327–365, 1994. Extended abstract in the Proceedings of the Sixth Annual Symposium on Logic in Computer Science, Amsterdam, July 15–18, 1991.

11. A. Ireland and A. Bundy. Extensions to a Generalization Critic for Inductive Proof. In M. A. McRobbie and J. K. Slaney, editors, *13th International Conference on Automated Deduction*, pages 47–61. Springer-Verlag, 1996. Springer Lecture Notes in Artificial Intelligence No. 1104. Also available from Edinburgh as DAI Research Paper 786.

12. E. Jacopin. Classical AI planning as theorem proving: The case of a fragment of linear logic. In *AAAI Fall Symposium on Automated Deduction in Nonstandard Logics, Technical Report FS-93-01*, pages 62–66. AAAI Press Publications, Palo Alto, 1993.

13. Z. Manna and R. Waldinger. How to clear a block: a theory of plans. *Journal of Automated Reasoning*, 3(4):343–377, 1986.

14. M. Masseron. Generating plans in linear logic II: A geometry of conjunctive actions. *Theoretical Computer Science*, 113:371–375, 1993.

15. M. Masseron, C. Tollu, and J. Vauzeilles. Generating plans in linear logic I: Actions and proofs. *Theoretical Computer Science*, 113(2):349–371, 1993.

16. B. Nordström, K. Petersson, and J. Smith. *Programming in Martin-Löf Type Theory*. Oxford University Press, 1990.

17. W. Stephan and S. Biundo. Deduction based refinement planning. In B. Drabble, editor, *Proceedings of the 3rd International Conference on Artificial Intelligence Planning Systems (AIPS-96)*, pages 213–220. AAAI Press, 1996.

18. L. Sterling and E. Shapiro. *The Art of Prolog*. MIT Press, Cambridge, MA, second edition, 1994.

# Appendix: Step Case Proof

```
                                                                          ──────────────────── Ax    ──────────────────────────────────────── (put)
                                                                          twr(t) ⊢ twr(t)            twr(a'), hold(b) ⊢ twr(b :: a') ⊗ hn
──────────────────── Ax    ──────────────────────────────────── (pickup)  ──────────────────────────────────────────────────────────────── r⊗
twr(a') ⊢ twr(a')          twr(b :: t), hn ⊢ twr(t) ⊗ hold(b)             twr(t), twr(a'), hold(b) ⊢ twr(t) ⊗ twr(b :: a') ⊗ hn
─────────────────────────────────────────────────────────── r⊗            ──────────────────────────────────────────────────────────────── l⊗
twr(b :: t), twr(a'), hn ⊢ twr(t) ⊗ twr(a') ⊗ hold(b)                     twr(t) ⊗ twr(a') ⊗ hold(b) ⊢ twr(t) ⊗ twr(b :: a') ⊗ hn
─────────────────────────────────────────────────────────── l⊗            ──────────────────────────────────────────────────────────────── cut
twr(b :: t) ⊗ twr(a') ⊗ hn ⊢ twr(t) ⊗ twr(a') ⊗ hold(b)
                                                            ────────────────────────────────────────────────────────────────────
                                                            twr(b :: t), twr(a'), hn ⊢ twr(t) ⊗ twr(b :: a') ⊗ hn
                                                            ──────────────────────────────────────────────────────── 1⊗, l⊗
                                                            twr(b :: t) ⊗ twr(a') ⊗ hn ⊢ twr(t) ⊗ twr(b :: a') ⊗ hn

                      twr(t) ⊗ twr(A) ⊗ hn ⊸                          twr(b :: t) ⊗ twr(a') ⊗ hn ⊸                   c⊸,
                      ────────────────────────────────────────────────────────────────────────────────────────      A = b :: a'
                      twr(empty) ⊗ twr(app(rev(t), A)) ⊗ hn ⊢ twr(empty) ⊗ twr(app(rev(t), b :: a')) ⊗ hn
                      ────────────────────────────────────────────────────────────────────────────────────── 2, 5, 4, 3
                      twr(t) ⊗ twr(A) ⊗ hn ⊸
                      twr(empty) ⊗ twr(app(rev(t), A)) ⊗ hn ⊢ twr(empty) ⊗ twr(app(rev(b :: t), a')) ⊗ hn
                      ────────────────────────────────────────────────────────────────────────────────────── r∀, l∀
                      ∀a.twr(b :: t) ⊗ twr(a) ⊗ hn ⊸
                      twr(empty) ⊗ twr(app(rev(b :: t), a)) ⊗ hn
```

# Sensor Planning
# with Non-linear Utility Functions

Sven Koenig and Yaxin Liu

College of Computing
Georgia Institute of Technology
Atlanta, GA 30332-0280, USA
{skoenig,yxliu}@cc.gatech.edu

**Abstract.** Sensor planning is concerned with when to sense and what to sense. We study sensor planning in the context of planning objectives that trade-off between minimizing the worst-case, expected, and best-case plan-execution costs. Sensor planning with these planning objectives is interesting because they are realistic and the frequency of sensing changes with the planning objective: more pessimistic decision makers tend to sense more frequently. We perform sensor planning by combining one of our techniques for planning with non-linear utility functions with an existing sensor-planning method. The resulting sensor-planning method is not only as easy to implement as the sensor-planning method that it extends but also (almost) as efficient. We demonstrate empirically how sensor plans change as the planning objective changes, using a common testbed for sensor planning.

## 1 Introduction

Sensor planning [4,1,11,12,19,10] is concerned with when to sense and what to sense. Sensor planning problems are interesting because sensing provides information that can reduce the plan-execution costs but comes at a cost itself. Although sensor planning can be used in conjunction with different planning objectives, sensor planners traditionally minimize the expected plan-execution costs. However, decision makers often trade-off between minimizing the worst-case, expected, and best-case plan-execution costs. Some decision makers are even so pessimistic (conservative) that they prefer plans with minimal worst-case plan-execution costs, a planning objective often used in robotics [13]. Other decision makers are more optimistic. This is often necessary even for pessimistic decision makers because often the worst-case plan-execution costs are infinite for all plans. Then, pessimistic decision makers need to adopt a more optimistic planning objective than minimizing the worst-case plan-execution costs but still prefer a more pessimistic planning objective than minimizing the expected plan-execution costs. Planners should be able to adopt this planning objective and similar planning objectives. Studying sensor-planning in the context of these planning objectives is interesting not only because they are realistic but also because the frequency of sensing changes with the planning objective. Sensing often provides additional information that can be used to avoid catastrophes, which is important for pessimistic decision makers. Consequently, more pessimistic decision makers tend to sense more frequently.

S. Biundo and M. Fox (Eds.): ECP-99, LNAI 1809, pp. 265–277, 2000.
© Springer-Verlag Berlin Heidelberg 2000

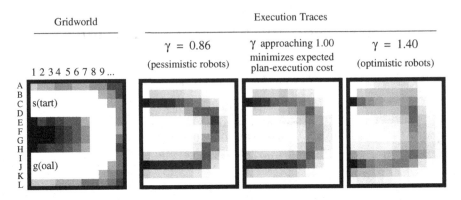

**Fig. 1.** Gridworld and Execution Traces

In this paper, we describe how one can generalize an existing sensor-planning method by Hansen [5] that minimizes the expected plan-execution costs to these more realistic planning objectives, thereby extending its applicability while maintaining its efficiency. We achieve this by combining one of our techniques for planning with nonlinear utility functions [8] with this efficient sensor-planning method, demonstrating the ease with which our technique can be combined with existing planning methods that minimize the expected plan-execution costs. The resulting sensor-planning method is not only as easy to implement as the sensor-planning method that it extends but is also almost as efficient. To gain insight into how sensor plans change as the planning objective changes, we use a simple robot-navigation domain that is easy to visualize. In particular, we study robot-navigation problems with actuator uncertainty. The robot can always decide to sense its current location, which is costly but prevents it from deviating from the nominal path into muddy terrain that is costly to traverse. Problems of this kind have been studied before [5], traditionally with the planning objective of minimizing the expected plan-execution costs, and can therefore be considered good test problems for new sensor-planning methods. We demonstrate experimentally how the sensor plans change for these problems as the planning objective changes.

## 2    The Sensor-Planning Problem

We consider robot-navigation problems with the following properties [5]. The robot has to navigate from a given start location to a given goal location in a known environment. Since motion is noisy, the robot can deviate from the nominal path but it can always opt to sense its current location. Sensing provides certainty about its current location but is costly (for example, consumes energy). We assume that there is a finite set of locations $L$. The robot knows that it starts at location $l_{start} \in L$ and its task is to navigate to location $l_{goal} \in L$ and be sure that it stops at exactly that location. There is a finite set $M$ of movement actions, all of which can be executed at all locations. Motion uncertainty is modeled with conditional probability distributions. Executing movement action $m \in M$ results with cost $c(l, m) > 0$ and probability $p(l'|l, m)$ in location $l'$. The robot receives no feedback as to what its new

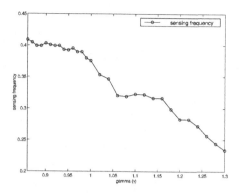

**Fig. 2.** Sensing Frequency

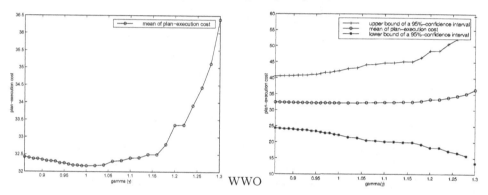

**Fig. 3.** Mean of Plan-Execution Costs (left) and Mean with Confidence Interval (right)

location is (which makes the simplifying assumption that the costs of the executed actions cannot be observed directly) but there is one sensing action $o$ that can be executed at all locations. Executing it incurs cost $c(l, o) > 0$ and reports the current location of the robot with certainty. We assume that it is possible to reach every location from every other location.

## 3  Example

We illustrate the sensor-planning problem using simple gridworlds, similar to those used on real robots [18]. Figure 1 (left) shows the gridworld that we use in our experiments, where the locations are squares. The start location is C1 and the goal location is J1. The robot can always sense its current location (O) or move north (N), east (E), south (S), or west (W) to an adjacent square. If the robot attempts to move in a certain direction (say, move east in square C1), then it either moves as intended (C2, with probability 0.6) or strays off by one square to the left (B2, with probability 0.2) or right (D2, with probability 0.2) due to actuator noise and not facing precisely in the right direction. The robot does not move when it bumps into the border of the gridworld. The movement costs range from 1.0 to 10.0. They

$\gamma = 0.86$ (pessimistic robots)

| | 1 | 2 | 3 | 4 | 5 | 6 | 7 | 8 | 9 | 10 | 11 |
|---|---|---|---|---|---|---|---|---|---|---|---|
| A | | | | | | | | SSSO | | | |
| B | | SEO | SEO | SEO | SEO | SEO | SO | SSO | | | |
| C | EO | EO | EO | EO | EO | EO | SEO | SO | SO | WSO | |
| D | | NEO | NEO | NEO | EEEO | EEO | EO | ESO | SO | WSO | WWSO |
| E | | | | | | | EESO | ESO | SO | WSO | |
| F | | | | | | | | SSSO | SO | WSO | |
| G | | | | | | | | SSO | SO | WSO | |
| H | | | | | | SSWO | SSWO | SO | WSO | WO | |
| I | SO | SWO | SWO | SWO | SWO | SWO | SWO | SWO | WO | WWO | WWWO |
| J | goal | WO | WO | WO | WO | WO | WO | WO | NWO | | |
| K | NO | NWO | NWO | NWO | NWO | NWO | NWO | NWO | NWWO | | |
| L | | | | | | | NNWO | NNWO | | | |

$\gamma = 1.40$ (optimistic robots)

| | 1 | 2 | 3 | 4 | 5 | 6 | 7 | 8 | 9 | 10 | 11 |
|---|---|---|---|---|---|---|---|---|---|---|---|
| A | SSEEEO | | | SSEO | SSEO | SSSO | SSSO | | | | |
| B | SEEEEO | | | SEEO | SEEO | SEO | SSO | SSO | SSSSSSSWO | | |
| C | EEEEEESO | | | EEEESO | EEESO | EESO | SO | SO | SSSSSSWO | | |
| D | EEEEEESO | | | EEEESO | EEESO | EESO | ESO | SSSSSWO | SSSSSWO | WSSO | WWSSO |
| E | NNEEEEO | NEEEEO | NEEEEO | NEEEO | NEEO | EESSO | EESSO | SSSSWO | SSSSWO | WSSO | WWSSO |
| F | SSSSO | SSSSO | SSSO | SSSSWO | SSSSWWO | EESSO | ESSSO | SSSWO | SSSWO | WSSO | WWSO |
| G | SSSO | SSSO | SSSO | SSSWO | SSSWWO | SSSWWWO | SSWWWWO | SSWWWWO | SSWO | WSO | WWSO |
| H | SSO | SSO | SSWO | SSWWO | SSWWO | SSWWWO | SSWWWWO | SSWWWWWO | SWO | WWO | WWWO |
| I | SO | SO | SWO | SWWO | SWWWO | SWWWWO | SWWWWWO | SWWWWWWO | WO | WWO | WWWO |
| J | goal | WO | WWO | WWWO | WWWWO | WWWWWO | WWWWWWO | WWWWWWWO | WWWWWWWO | WWWWWWWWO | WWWO |
| K | NO | NO | NWO | WWWO | WWWWO | WWWWWO | WWWWWWO | WWWWWWWO | WWWWWWWWO | WWWWWWWWO | WWWWWWWWWO |
| L | NNO | NNO | NNWO | NWWO | NWWWO | NWWWWO | | NWWWWWWO | | | |

**Fig. 4.** Optimal Sensor Plans

are low for roads (white) and high for muddy terrain (darker colors). The sensing costs are always 0.2. We express the trade-off between minimizing the worst-case, expected, and best-case plan-execution costs using a parameter $\gamma$ with $0 < \gamma < \infty$. As $\gamma$ approaches infinity, the robots become more optimistic and thus more interested in plans with small best-case plan-executions costs. As $\gamma$ approaches one, the robots become more interested in plans with small expected plan-execution costs. Finally, as $\gamma$ approaches zero, the robots become more pessimistic and thus more interested in plans with small worst-case plan-execution costs. Notice that pessimistic robots have to trade off between minimizing the worst-case and expected plan-execution costs in our example domain. It is not possible to minimize the worst-case plan-execution costs directly because all plans cycle with some probability and thus have worst-case plan-execution costs that are infinite. Figure 2 shows that the sensing frequency (that is, the percentage of sensing actions among all executed actions) increases as the robots become more pessimistic and $\gamma$ decreases. This is also illustrated in Figure 4, that shows optimal sensor plans for two different values of $\gamma$.[1] The sensor plans are depicted as gridworlds, each location of which is annotated with an action sequence.

---

[1] There are some exceptions to this trend, for example, in the vicinity of the goal. This can be explained as follows: Optimistic robots assume that short action sequences that have a chance of reaching the goal location will indeed reach it. Thus, they execute these action sequences followed by a sensing action to confirm that they have reached the goal location. For example, for $\gamma = 1.40$, the action sequence of location I2 is SO. The robot hopes that it will drift to the goal location J1 as it moves south, although this is less likely than moving to location J2. More pessimistic robots are more cautious and execute longer action sequences. For example, for $\gamma = 1.86$, the action sequence of location I2 is SWO, which reaches the goal location with higher probability than the action sequence SO. This phenomenon and similar phenomena contribute to the small local minima in the graph of Figure 2.

These action sequences are used as follows: After the robots have executed a sensing action, they look up the action sequence that corresponds to the sensed location, execute it, and repeat the process, until they sense that they are at the goal location. For example, for $\gamma = 0.86$, the action sequence of location B2 is SEO. Consequently, after the robot has sensed that it is at location B2, it first moves south (S), then moves east (E), and finally senses again (O). Locations whose action sequences are not used for getting from the start location (C1) to the goal location (J1) are left blank. Figure 1 (right) shows how often the robots visit each grid square during two million runs for three different values of $\gamma$. Darker colors indicate a larger number of visits. Thus, more pessimistic robots are more likely to stay on the road and close to the nominal path, which is possible due to the increased sensing frequency. That more pessimistic robots are more likely to stay on the road can be explained as follows: By staying on the road, the robots are likely able to avoid the large costs necessary for getting out of the mud, which pessimistic robots consider to be important. On the other hand, smaller sensing frequencies and attempts to cut the corners decrease the probability that the robots stay on the road but also decrease the plan-execution costs in the best case, which optimistic robots consider to be important. (In fact, the action sequences of the start location get longer and longer as the robots become more optimistic until the action sequences are able to move the robots to the goal location if they are lucky.) This explanation suggests that there is a mean-variance trade-off in our example domain. Mean-variance trade-offs are often used as crude but easy-to-understand explanations for trade-offs between minimizing the worst-case, expected, and best-case plan-execution costs. For example, the graph in Figure 3 (left) shows the mean of the plan-execution costs, and the difference of the upper and lower graphs in Figure 3 (right) corresponds to four times the standard deviation of the plan-execution costs. The mean of the plan-execution costs increases but the variance decreases as the robots become more pessimistic and $\gamma$ decreases from one to zero. The variance decreases because more pessimistic robots stay on the road and close to the nominal path. The mean-variance trade-off can be explained as follows: More pessimistic robots are willing to accept a larger mean of the plan-execution costs for a decrease in variance because they expect to be unlucky and fear for the worst case. A small variance avoids plan-execution costs that are much larger than average. Figure 3 (right) illustrates this using the upper bound of a 95-percent-confidence interval (that is, mean plus twice the standard deviation) as an approximation of the worst-case plan-execution costs. The upper bound indeed decreases as the robots become more pessimistic since the decrease of the variance outweighs the increase of the mean. On the other hand, more optimistic robots are willing to accept a larger mean for an increase in variance. They expect to be lucky and hope for the best case since a larger variance promises a chance to realize plan-execution costs that are much smaller than the expected plan-execution costs. Figure 3 (right) illustrates this using the lower bound of a 95-percent-confidence interval (that is, mean minus twice the standard deviation) as an approximation of the best-case plan-execution costs. The lower bound indeed decreases as the robots become more optimistic since the increase of the variance outweighs the increase of the mean. The optimal sensor plans can be calculated efficiently with our sensor-planning method that we discuss in the following. For $\gamma = 0.86$, our sensor-planning method expands 2,071 nodes and needs an average of 4.6 milliseconds per node expansion on a Sun Ultra 1 running Solaris 7. The original sensor-planning method by Hansen, that our sensor-planning method extends, corresponds to the case where $\gamma$ approaches one. It needs 2,815 node expansions and 2.0 milliseconds per node expansion. For $\gamma = 1.40$, our sensor-planning method needs 5,808 node expansions and 5.2 milliseconds per node expansion. The number of node expansions depends on the sensing frequency. It increases as the sensing frequency of the optimal plans decreases. Our sensor-planning method has a slight run-time disadvantage per node expansion compared to the original sensor-planning method by Hansen because it has to calculate exponentials and logarithms, and its heuristic search method cannot calculate the heuristic values quite as efficiently as the original sensor-planning method.

## 4    Formalizing the Sensor-Planning Problem

The robot-navigation problems are special cases of partially observable Markov decision process models that can be translated into goal-directed Markov decision process models. Goal-directed Markov decision process models are often used for probabilistic planning in artificial intelligence [2]. They are totally observable Markov decision process models and consist of a finite set of states, including a start state and a goal state in which execution stops. Each non-goal state has a finite set of actions associated with it that can be executed in that state. Executing an action incurs a given finite cost and results in a successor state that is determined by a given probability distribution that depends only on the action and the state that it is executed in. A goal-directed Markov decision process is a stream of ⟨state, action, cost⟩ triples. The process is always in exactly one state and makes state transitions at discrete time steps. It starts in the start state. A policy determines which actions to execute. A (stationary, deterministic) policy is a mapping from non-goal states to actions, and the goal-directed Markov decision process always executes the action that the policy assigns to its current state. Execution stops only in goal states, in which case the goal-directed Markov decision process does not incur any further costs. In case of our robot-navigation problems, the states of the goal-directed Markov decision process model correspond to the locations. The start state is the start location, and the goal state is the goal location. The actions of the goal-directed Markov decision process correspond to sequences of one or more movement actions followed by the sensing action. (This assumes that the robot always needs to execute the sensing action before stopping to ensure that it is at the goal location.) In the remainder of the paper, we first explain a special case that is easy to understand but makes several simplifying assumptions. First, we assume that $\gamma > 1$. Second, we assume that the robot has to execute the sensing action after at most $b \geq 1$ movement actions, which makes the number of action sequences finite. We use $M^b$ to denote all sequences of movement actions whose lengths are between one and $b$. Third, we assume that the costs of the movement and sensing actions do not depend on the location. Executing movement actions $m \in M$ incurs costs $c(m) > 0$ and executing sensing actions $o$ incurs costs $c(o) > 0$. A policy $\delta$ then assigns action sequences to locations. Executing action sequence $m_1 \ldots m_j o$ at location $l$ results with cost $\sum_{i=1}^{j} c(m_i) + c(o)$ and probability $P(l'|lm_1 \ldots m_j)$ in location $l'$, where the value $P(l'|lm_1 \ldots m_j)$ can be calculated recursively as follows:

$$P(l'|l) = \begin{cases} 1 \text{ if } l = l' \\ 0 \text{ otherwise} \end{cases}$$

$$P(l'|lm_1 \ldots m_i) = \sum_{l''} p(l'|l''m_i)P(l''|lm_1 \ldots m_{i-1}) \qquad \text{for all } 1 \leq i \leq j.$$

We will relax all of these assumptions in Section 6. The example in Section 3 did not include them.

## 5   Planning Objectives

The plan-execution costs of plans are the sum of the costs of all executed actions. If a plan has plan-execution costs $c_i$ with probabilities $p_i$, then its expected plan-execution cost is $\sum_i p_i c_i$. Minimizing the expected plan-execution costs is a traditional planning objective. An optimal policy for this planning objective satisfies the following system of $|L|$ Bellman equations, where the $|L|$ unknowns $V(l)$ are the expected plan-execution costs if the start location is known to be $l$.

$$V(l) = 0 \qquad\qquad\qquad\qquad \text{for all } l \in L \text{ with } l = l_{goal}$$

$$V(l) = \min_{m_1 \ldots m_j o \in M^b o} \sum_{l'} P(l'|lm_1 \ldots m_j)(\sum_{i=1}^{j} c(m_i) + c(o) + V(l')) \text{ for all } l \in L \text{ with } l \neq l_{goal}.$$

A policy that minimizes the expected plan-execution costs can be found with dynamic programming methods from operations research, including policy iteration [6], as well as combinations of these methods and heuristic search methods from artificial intelligence [5]. We modify these methods for a different planning objective, namely to maximize the expected utilities of the plan-execution costs for given exponential utility functions, which allows one to trade-off between minimizing the worst-case, expected, and best-case plan-execution costs. Markov decision process models with this planning objective are studied in operations research in the context of risk-sensitive Markov decision process models [14]. A utility function $u$ maps plan-execution costs $c$ into the corresponding real-valued utilities $u(c)$. If a plan has plan-execution costs $c_i$ with probabilities $p_i$, then its expected utility is $\sum_i p_i u(c_i)$ and its certainty equivalent is $u^{-1}(\sum_i p_i u(c_i))$ for the given utility function $u$. Our approach uses exponential utility functions, both convex exponential utility functions: $u(c) = \gamma^{-c}$ for parameter $\gamma > 1$, and concave exponential utility functions: $u(c) = -\gamma^{-c}$ for parameter $\gamma$ with $0 < \gamma < 1$. Exponential utility functions are perhaps the most often used utility functions in utility theory [20]. Their parameter $\gamma$ can be used to trade-off between minimizing the best-case, expected, and worst-case plan-execution costs [8]. For example, the plan with maximal expected utility minimizes the best-case plan-execution costs as gamma approaches infinity (under appropriate assumptions). The plan with maximal expected utility minimizes the expected plan-execution costs as gamma approaches one. Finally, the plan with maximal expected utility minimizes the worst-case plan-execution costs as gamma approaches zero. As mentioned earlier, we assume for now that the exponential utility functions are of the form $u(c) = \gamma^{-c}$ for parameter $\gamma > 1$. An optimal policy for this planning objective satisfies the following system of $|L|$ Bellman equations, where the $|L|$ unknowns $V(l)$ are the expected utilities if the start location is known to be $l$.

$$V(l) = u(0) = \gamma^{-0} = 1 \qquad\qquad \text{for all } l \in L \text{ with } l = l_{goal} \quad (1a)$$

$$V(l) = \max_{m_1 \ldots m_j o \in M^{bo}} \sum_{l'} P(l'|lm_1 \ldots m_j) u(\sum_{i=1}^{j} c(m_i) + c(o) + u^{-1}(V(l'))) \qquad\qquad (1b)$$

$$= \max_{m_1 \ldots m_j o \in M^{bo}} \sum_{l'} P(l'|lm_1 \ldots m_j) \gamma^{-(\sum_{i=1}^{j} c(m_i) + c(o) - \log_\gamma(V(l')))}$$

$$= \max_{m_1 \ldots m_j o \in M^{bo}} \gamma^{-\sum_{i=1}^{j} c(m_i) - c(o)} \sum_{l'} P(l'|lm_1 \ldots m_j) V(l') \qquad \text{for all } l \in L \text{ with } l \neq l_{goal}.$$

That this set of equations has a unique solution follows directly from [15], and that the optimal policy reaches the goal with probability one under the assumption that it is possible to reach every location from every other location follows from [9].

*Explanation of the Formulae:* If the current location $l$ of the robot is known to be the goal location then plan execution stops right away with plan-execution cost zero and thus expected utility one, which explains equation (1a). To derive the other equation we utilize a property of exponential utility functions called "delta property" [7] or, equivalently, "constant local risk aversion" [17], namely that the certainty equivalent of a plan increases by $c$ if we add the same costs $c$ to all of its plan-execution costs. To see this, assume that a plan has plan-execution costs $c_i$ with probabilities $p_i$. Then,

$$u^{-1}(\sum_i p_i u(c + c_i)) = -\log_\gamma(\sum_i p_i \gamma^{-(c+c_i)}) = -\log_\gamma(\gamma^{-c} \sum_i p_i \gamma^{-c_i})$$

$$= c - \log_\gamma(\sum_i p_i \gamma^{-c_i}) = c + u^{-1}(\sum_i p_i u(c_i)).$$

If the current location $l$ of the robot is known not to be the goal location then the execution of action sequence $m_1 \ldots m_j o$ at location $l$ results with cost $\sum_{i=1}^{j} c(m_i) + c(o)$ and probability $P(l'|lm_1 \ldots m_j)$ in location $l'$, where the robot proceeds to execute a plan with expected utility $V(l')$ and thus certainty equivalent $u^{-1}(V(l'))$. The previously executed actions increase this certainty equivalent by their costs (due to the delta property), resulting in a certainty equivalent of $\sum_{i=1}^{j} c(m_i) + c(o) + u^{-1}(V(l'))$ and thus an expected utility of $u(\sum_{i=1}^{j} c(m_i) + c(o) + u^{-1}(V(l')))$. Since this utility occurs with probability $P(l'|lm_1 \ldots m_j)$, the expected utility is $\sum_{l'} P(l'|lm_1 \ldots m_j) u(\sum_{i=1}^{j} c(m_i) + c(o) + u^{-1}(V(l')))$, which explains equation (??b).

A policy that satisfies equations (1a) and (1b) and thus maximizes the expected utility for the given exponential utility function can be found with dynamic programming methods from operations research. For example, Figure 5 shows how to apply policy iteration to this problem. Step 3 corresponds to a search problem that is essentially deterministic. Thus, heuristic search methods from artificial intelligence can be used to implement it efficiently. For example, Figure 6 shows

1. Start with an arbitrary initial policy $\delta$.
2. Compute the value function $V^\delta$ for policy $\delta$ by solving the following system of $|L|$ equations, where the $|L|$ unknowns $V^\delta(l)$ are the expected utilities if policy $\delta$ is followed and the start location is known to be $l$. Assume that $\delta(l) = m_1 \ldots m_j o$. (To keep the notation simple, we do not index the action sequence with $\delta(l)$.). Then,

$$V^\delta(l) = 1 \qquad\qquad\qquad \text{for all } l \in L \text{ with } l = l_{goal}$$
$$V^\delta(l) = \gamma^{-\sum_{i=1}^{j} c(m_i) - c(o)} \sum_{l'} P(l'|lm_1 \ldots m_j)V^\delta(l') \text{ for all } l \in L \text{ with } l \neq l_{goal}.$$

3. For each location $l \in L$ with $l \neq l_{goal}$, attempt to find an action sequence $m_1 \ldots m_j o \in M^b o$ such that

$$V^\delta(l) < \gamma^{-\sum_{i=1}^{j} c(m_i) - c(o)} \sum_{l'} P(l'|lm_1 \ldots m_j)V^\delta(l').$$

If such an action sequence is found, set $\delta(l) = m_1 \ldots m_j o$ otherwise do not change $\delta(l)$. The search procedure has to find such an action sequence for at least one location if such action sequences exist.
4. If no $\delta(l)$ has changed in the previous step, then stop. Otherwise go to step 2.

**Fig. 5.** Policy Iteration

---

1. Start with the empty OPEN list.
2. Create a search node $lm$ for each movement action $m \in M$ and add them to the OPEN list.
3. Remove the search node with the smallest f-value. If the search node is of the form $lm_1 \ldots m_j o \in lM^b o$ (that is, ends in the sensing action) then go to step 5.
4. *(The search node is of the form $lm_1 \ldots m_j \in lM^b$:)* Create a search node $lm_1 \ldots m_j o$ and add it to the OPEN list. If $j < b$ then also create search nodes $lm_1 \ldots m_j m$ for each movement action $m \in M$ and add them to the OPEN list. Go to step 3.
5. *(The search node is of the form $lm_1 \ldots m_j o \in lM^b o$:)* If the f-value of $lm_1 \ldots m_j o$ is smaller than $-\log_\gamma V^\delta(l)$, that is, if

$$\sum_{i=1}^{j} c(m_i) + c(o) - \log_\gamma\left(\sum_{l'} P(l'|lm_1 \ldots m_j)V^\delta(l')\right) < -\log_\gamma V^\delta(l),$$

then set $\delta(l) = m_1 \ldots m_j o$. Return.

**Fig. 6.** A* Implementation of Step 3 of Policy Iteration for a Given Location $l$

---

how to use A* [16] to implement it for a given location $l \in L$, adapting a method developed by Hansen in the context of minimizing the expected plan-execution costs [5]. The states of the search problem are of the form $lm_1 \ldots m_j \in l \cup lM^b$

or $lm_1 \ldots m_j o \in lM^b o.^2$ The start state is $l$. The goal states are of the form $lm_1 \ldots m_j o$, that is, the states ending in the sensing action. The actions of the search problem are the movement actions and the observation action. The movement actions $m \in M$ apply to all states of the form $lm_1 \ldots m_j$ with $j < b$. We assume that they result with costs $c(m)$ in the state $lm_1 \ldots m_j m$. The observation actions $o$ apply to all states of the form $lm_1 \ldots m_j$ with $j \leq b$. We assume that they result with costs $c(o) - \log_\gamma(\sum_{l'} P(l'|lm_1 \ldots m_j) V^\delta(l'))$ in the state $lm_1 \ldots m_j o$. The h-value (heuristic estimate of the goal distance) of state $lm_1 \ldots m_j$ is $-\log_\gamma(\sum_{l'} P(l'|lm_1 \ldots m_j) V^\delta(l'))$, resulting in an f-value (heuristic estimate of the smallest cost of a path from the start state to a goal state via state $lm_1 \ldots m_j$) of $\sum_{i=1}^j c(m_i) - \log_\gamma(\sum_{l'} P(l'|lm_1 \ldots m_j) V^\delta(l'))$. The h-value of state $lm_1 \ldots m_j o$ is zero, resulting in an f-value of $\sum_{i=1}^j c(m_i) + c(o) - \log_\gamma(\sum_{l'} P(l'|lm_1 \ldots m_j) V^\delta(l'))$. Although the h-values are not admissible, A* finds an action sequence $m_1 \ldots m_j o$ for at least one location $l$ such that

$$V^\delta(l) < \gamma^{-\sum_{i=1}^j c(m_i) - c(o)} \sum_{l'} P(l'|lm_1 \ldots m_j) V^\delta(l')$$

$$-\log_\gamma V^\delta(l) > -\log_\gamma(\gamma^{-\sum_{i=1}^j c(m_i) - c(o)} \sum_{l'} P(l'|lm_1 \ldots m_j) V^\delta(l'))$$

$$-\log_\gamma V^\delta(l) > \sum_{i=1}^j c(m_i) + c(o) - \log_\gamma(\sum_{l'} P(l'|lm_1 \ldots m_j) V^\delta(l')) = f(m_1 \ldots m_j o) \qquad (1)$$

if such action sequences exist, as required by policy iteration.

*Sketch of the Proof:* The proof is similar to Hansen's proof for the planning objective of minimizing the expected plan-execution costs [5]. It is by contradiction. A* terminates because the search space is finite. Assume that there exists an action sequence for some location with property (1) but that A* does not find any such action sequence for any location and consequently does not improve policy $\delta$. Let the action sequence $m_1 \ldots m_j o$ at location $l$ be a shortest action sequence with property (1). When A* terminates for location $l$ there is some prefix of $lm_1 \ldots m_j o$ on the OPEN list, since initially $lm_1$ is put on the OPEN list. This prefix cannot be $lm_1 \ldots m_j$ or $lm_1 \ldots m_j o$ because $f(m_1 \ldots m_j) < f(m_1 \ldots m_j o) < -\log_\gamma V^\delta(l)$ and thus A* would have been able to improve policy $\delta$. Thus, the prefix must be of the form $lm_1 \ldots m_k$ for some $1 \leq k < j$. This state has an f-value that is no smaller than the f-value of the state that terminated the search, otherwise A* would have chosen $lm_1 \ldots m_k$ instead. The f-value of the state that terminated the search, in turn, is no smaller than $-\log_\gamma V^\delta(l)$, otherwise A* would have been able to improve policy $\delta$. Finally, we assumed that $-\log_\gamma V^\delta(l)$ was strictly larger than the f-value of $lm_1 \ldots m_j o$. Put together,

---

$^2$ Instead of using $lm_1 \ldots m_j$ as states, we could also use the corresponding probability distributions $P(\cdot|lm_1 \ldots m_j)$. This can result in smaller state spaces at the expense of more extensive bookkeeping.

$$f(m_1 \ldots m_k) > f(m_1 \ldots m_j o)$$

$$\sum_{i=1}^{k} c(m_i) - \log_\gamma(\sum_{l'} P(l'|lm_1 \ldots m_k)V^\delta(l')) > \sum_{i=1}^{j} c(m_i) + c(o) - \log_\gamma(\sum_{l''} P(l''|lm_1 \ldots m_j)V^\delta(l''))$$

$$-\log_\gamma(\sum_{l'} P(l'|lm_1 \ldots m_k)V^\delta(l')) > \sum_{i=k+1}^{j} c(m_i) + c(o) - \log_\gamma(\sum_{l''} P(l''|lm_1 \ldots m_j)V^\delta(l''))$$

$$-\log_\gamma(\sum_{l'} P(l'|lm_1 \ldots m_k)V^\delta(l')) > -\log_\gamma(\sum_{l''} P(l''|lm_1 \ldots m_j)\gamma^{-\sum_{i=k+1}^{j} c(m_i)-c(o)}V^\delta(l''))$$

$$\sum_{l'} P(l'|lm_1 \ldots m_k)V^\delta(l') < \sum_{l''} P(l''|lm_1 \ldots m_j)\gamma^{-\sum_{i=k+1}^{j} c(m_i)-c(o)}V^\delta(l'')$$

$$\sum_{l'} P(l'|lm_1 \ldots m_k)V^\delta(l') < \sum_{l''}(\sum_{l'} P(l'|lm_1 \ldots m_k)P(l''|l'm_{k+1} \ldots m_j))\gamma^{-\sum_{i=k+1}^{j} c(m_i)-c(o)}V^\delta(l'')$$

$$\sum_{l'} P(l'|lm_1 \ldots m_k)V^\delta(l') < \sum_{l'} P(l'|lm_1 \ldots m_k)(\sum_{l''} P(l''|l'm_{k+1} \ldots m_j)\gamma^{-\sum_{i=k+1}^{j} c(m_i)-c(o)}V^\delta(l'')).$$

Thus, there exists a location $l'$ such that

$$V^\delta(l') < \sum_{l''} P(l''|l'm_{k+1} \ldots m_j)\gamma^{-\sum_{i=k+1}^{j} c(m_i)-c(o)}V^\delta(l'')$$

$$-\log_\gamma V^\delta(l') > \sum_{i=k+1}^{j} c(m_i) + c(o) - \log_\gamma \sum_{l''} P(l''|l'm_{k+1} \ldots m_j)V^\delta(l'')$$

$$-\log_\gamma V^\delta(l') > f(l'm_{k+1} \ldots m_j o).$$

But this is in contradiction to our assumption that the action sequence $m_1 \ldots m_j o$ at location $l$ was the shortest action sequence with property (1) since the action sequence $m_{k+1} \ldots m_j o$ at location $l'$ also has property (1) but is strictly smaller than the action sequence $m_1 \ldots m_j o$. Consequently, A* implements step 3 of policy iteration correctly.

This combination of policy iteration and A* finds a policy that maximizes the expected utility for the given exponential utility function under the assumption that the robot always executes the sensing action after at most $b$ movement actions.

# 6  Extensions

In the following, we briefly explain (without proofs) how to relax the assumptions that we made earlier for didactic reasons. First, we assumed that $\gamma > 1$. If this is not the case and $0 < \gamma < 1$, then the robots are pessimistic. Our sensor-planning method still applies, with the following changes: The value of the goal location is now minus one instead of one, and the initial policy now has to guarantee that the robot reaches a goal location with probability one from every start location. Such a policy can easily be found by assuming that the robot senses after every movement action. Second, we assumed that the robot had to execute the sensing action after at most $b$ movement actions. If this is not the case, our sensor-planning method can increase $b$ iteratively as follows, as long as it is optimal to sense in finite intervals (which is always the case for pessimistic robots but might not be the case for optimistic robots): If the length bound $b$ is not used to prune any search path in step 4 of Figure 6 during the last iteration of policy iteration before termination, the policy to which it converges maximizes the expected utility even if no length bound exists. If the length bound is used to prune some search path during the last iteration of policy iteration, then one can increase the length bound and continue to execute policy iteration, until eventually the policy does not change from one iteration to the next and the length bound is not used to prune any search path during the last iteration of policy iteration. Third, we assumed that the costs of the movement and sensing actions did not depend on the location. If this is not the case, then the f-values have to be calculated as follows during the heuristic search, making their calculation slightly less efficient:

$$f(lm_1 \ldots m_j) = \sum_{l'} \bar{P}(l'|lm_1 \cdots m_j)V(l')$$

$$f(lm_1 \ldots m_j o) = \sum_{l'} \bar{P}(l'|lm_1 \cdots m_j)\gamma^{-c(l',o)}V(l'),$$

where

$$\bar{P}(l'|l) = \begin{cases} 1 \text{ if } l = l' \\ 0 \text{ otherwise} \end{cases}$$

$$\bar{P}(l'|lm_1 \ldots m_i) = \sum_{l''} p(l'|l''m_i)\gamma^{-c(l'',m_i)}\bar{P}(l''|lm_1 \ldots m_{i-1}) \qquad \text{for all } 1 \leq i \leq j,$$

which can be calculated when a node is expanded (at the cost of an additional amount of bookkeeping).

## 7  Conclusions

We described a sensor-planning method that seamlessly trades off between minimizing the best-case, expected, and worst-case plan-execution costs, and thus generalizes the planning objectives of traditional sensor planners that often either minimize the expected plan-execution costs or the worst-case plan-execution costs. Sensor planning with our more realistic planning objectives is interesting because the frequency of sensing depends on the trade-off between minimizing the best-case, expected, and worst-case plan-execution costs: more pessimistic decision makers tend to sense more frequently. We derived our sensor-planning method by combining one of our techniques for planning with non-linear utility functions with an existing sensor-planning method. The resulting sensor-planning method is not only as easy to implement as the sensor-planning method that it extends but also as almost efficient. It assumes that sensing provides perfect information about the state of the world. We are currently working on more general sensor-planning methods that do not make this assumption, using results from the theory of risk-sensitive partially observable Markov decision process models [3].

**Acknowledgments.** The Intelligent Decision-Making Group is partly supported by an NSF Career Award to Sven Koenig under contract IIS-9984827. The views and conclusions contained in this document are those of the author and should not be interpreted as representing the official policies, either expressed or implied, of the sponsoring organizations and agencies or the U.S. government.

# References

1. B. Abramson. A decision-theoretic framework for integrating sensors in AI plans. *IEEE Transactions on Systems, Man, and Cybernetics*, 23:366–373, 1993.
2. C. Boutilier, T. Dean, and S. Hanks. Decision-theoretic planning: Structural assumptions and computational leverage. *Journal of Artificial Intelligence Research*, 1999.
3. E. Fernandez-Gaucherand and S. Marcus. Risk-sensitive optimal control of hidden Markov models: Structural results. *IEEE Transactions on Automatic Control*, 1997.
4. G. Hager. *Task-Directed Sensor Fusion and Planning: A Computational Approach.* Kluwer Academic Publishers, 1990.
5. E. Hansen. Markov decision processes with observation costs. Technical Report CMPSCI 97–01, Department of Computer Science, University of Massachusetts, Amherst (Massachusetts), 1997.
6. R. Howard. *Dynamic Programming and Markov Processes.* MIT Press, third edition, 1964.
7. R. Howard and J. Matheson. Risk-sensitive Markov decision processes. *Management Science*, 18(7):356–369, 1972.
8. S. Koenig and R.G. Simmons. How to make reactive planners risk-sensitive. In *Proceedings of the International Conference on Artificial Intelligence Planning Systems*, pages 293–298, 1994.
9. S. Koenig and R.G. Simmons. Risk-sensitive planning with probabilistic decision graphs. In *Proceedings of the International Conference on Principles of Knowledge Representation and Reasoning*, pages 2301–2308, 1994.
10. S. Kristensen. Sensor planning with Bayesian decision theory. In L. Dorst, M. van Lambalgen, and R. Voorbraak, editors, *Reasoning with Uncertainty in Robotics*, volume 1093 of *Lecture Notes in Artificial Intelligence*, pages 353–367. Springer, 1996.
11. P. Langley, W. Iba, and J. Shrager. Reactive and automatic behavior in plan execution. In *Proceedings of the International Conference on Planning Systems*, pages 299–304, 1994.
12. S. Lee and X. Zhao. Sensor planning with hierarchically distributed perception net. In *Proceedings of the International Conference on Multisensor Fusion and Integration for Intelligent Systems*, pages 591–598, 1994.
13. T. Lozano-Perez, M. Mason, and R. Taylor. Automatic synthesis of fine-motion strategies for robots. *International Journal of Robotics Research*, 3(1):3–24, 1984.
14. S. Marcus, E. Fernàndez-Gaucherand, D. Hernàndez-Hernàndez, S. Colaruppi, and P. Fard. Risk-sensitive Markov decision processes. In C. Byrnes et. al., editor, *Systems and Control in the Twenty-First Century*, pages 263–279. Birkhauser, 1997.
15. H. Mine and S. Osaki. *Markovian Decision Processes.* Elsevier, 1970.
16. N. Nilsson. *Problem-Solving Methods in Artificial Intelligence.* McGraw-Hill, 1971.
17. J. Pratt. Risk aversion in the small and in the large. *Econometrica*, 32(1-2):122–136, 1964.
18. A. Stentz and M. Hebert. A complete navigation system for goal acquisition in unknown environments. *Autonomous Robots*, 2(2):127–145, 1995.
19. K. Tarabanis, P. Allen, and R. Tsai. A survey of sensor planning in computer vision. *IEEE Transactions on Robotics and Automation*, 11(1):86–104, 1995.
20. S. Watson and D. Buede. *Decision Synthesis.* Cambridge University Press, 1987.

# Propice-Plan: Toward a Unified Framework for Planning and Execution

Olivier Despouys* and François Félix Ingrand

LAAS/CNRS, 7 avenue du Colonel Roche,
F-31077 Toulouse Cedex 04, France
{despouys,felix}@laas.fr

**Abstract.** In this paper, we investigate the links between planning and plans execution. We propose a new approach (Propice-Plan) which integrates both activities. It implements supervision and execution capabilities, combined with different planning techniques:

- *plan synthesis* to complement existing operational plans; and
- *anticipation planning* to advise the execution for the best option to take when facing choices (by anticipating plans execution), and to forecast problems that may arise due to unforeseen situations.

This approach relies on a common language to represent plans, actions, operational procedures and constraints. In particular, the description we propose makes transitions between planning activities and execution seamless.

This work is used in two complex real-world problems: planning and control for autonomous mobile robots, and for the transition phases of a blast furnace.

## 1 Introduction

For years, the AI planning community has developed a wide range of techniques for *plan generation*. With few exceptions, this great amount of work did not take into account the issue of *plan execution* nor the impact execution may have on the planning process itself. Nevertheless, recent works study the links between plan synthesis and their execution. This point is indeed critical when tackling real-world problems, since several simplifying assumptions used in off-line planning reveal to be far too optimistic:

- the environment cannot be considered as *static*: unexpected events may occur during plan synthesis or its execution, thus invalidating parts of it, if not all.
- when performing planned actions, execution failures may occur; thus, the planner should take them into account to adapt its plan.

These observations motivated important research studies in order to relax some of these assumptions: probabilistic planning [Kushmeric *et al.*, 1995], possibilistic planning [Guéré and Alami, 1999], conditional planning [Pryor and Collins, 1996], Markov decision processes [Geffner and Bonet, 1998] and transformational planning [Beetz and McDermott, 1994], among others, can handle

---

* Part of this work is funded under contract with Usinor/Sollac, SACHEM Project.

S. Biundo and M. Fox (Eds.): ECP-99, LNAI 1809, pp. 278–293, 2000.
© Springer-Verlag Berlin Heidelberg 2000

uncertainty about the environment states and possible actions outcomes. Nevertheless, these approaches still concentrate mostly on the planning process.

On the opposite side, systems and techniques dealing with plans execution have also been developed: PRS [Ingrand et al., 1996], RAPS [Firby, 1994]. However, these approaches do not provide any planning activity, and fail when encountering new goals for which no explicit method is provided.

Some work has already been done to combine planning and execution. In [Wilkins et al., 1994; Myers, 1998], the authors define the ACT language, a superset of the languages used in SIPE and PRS. But the resulting system is more a concatenation of SIPE and PRS algorithms than a combined approached. 3T [Bonasso et al., 1997], based on RAPS and the Adversarial Planner (AP), provides similar features in addition to a reactive skills manager. In [Levinson, 1994], the author presents Propel, which provides a unified representation for anticipation planning and execution, but here also, the operators appear to be used by either planning or execution, but not both.

Propice-Plan, the approach we propose, combines an execution model based on PRS, and various planning techniques (plan synthesis and anticipation planning) in a unified framework. In this paper, we use the term *OP* (operational plan) as the result of a planning activity, but also an operational procedure defined by a domain expert.

## 1.1   Application Domains

Propice-Plan is intended to be applied to perform dynamic planning and plan execution in real-world applications. Our study is motivated by an industrial collaboration for aided transition operations for a blast furnace (such as shutting it down for maintenance). These transition phases last several hours (eight hours to stop a furnace when no problem occurs). Such a process represents a dynamic continuous system for which action execution effects are poorly modelled. Nevertheless, human operators make use of a large body of operational plans which guide them through these transitional phases. They include conditional branching, loops, sub-goal posting and refinement. The set of all empirical plans represent the known paths to reach an objective. Our participation to real shutdown operations revealed that following operational plans is not sufficient and planning new sequences of actions or forecasting particular situations may help to produce better plans.

Beyond this particular application, we also use Propice-Plan on our autonomous mobile robots. Indeed, we have been using PRS for supervision and execution control [Ingrand et al., 1996]. However, as pointed out by the authors in [Alami et al., 1998], robotics applications also require planning activities to synthesise a particular sequence of actions to achieve certain goals. But here also, we would like to use the large body of operational plans already available for execution and control to examine and plan decision choices in advance.

## 1.2   Issues

Complex applications such as autonomous robots or the furnace domain require different kinds of planning techniques. In case of situations for which the expert did not specify operational plans or for which they failed, the system should try to synthesise a new one, based on the declarative information of existing operational plans (we will refer to this kind of planning as **plan synthesis**).

Besides, the execution of some operational plans may lead to critical states or may be inefficient to reach a goal in some cases. We want our system to provide some kind of look-ahead capabilities, based on these plans and their execution model, in order to advise the execution of the best option ahead, but also to opportunistically adapt plans. On contrary to XFRM [Beetz and McDermott, 1994], adaptations are not provided by a user-defined set of transformation rules, but directly derived from the model of operational plans. We will refer to this kind of planning as **anticipation planning**.

These planning activities must take into account changes in the world during the planning itself, and also produce robust plans.

Planning and execution must be integrated seamlessly. For example, from a reactivity point of view, planning should be interruptible and execution should remain the highest priority activity. From the data structure point of view, planning and execution should use a common language for plans, operators and also constraints.

The rest of this paper is organised as follows. Section 2 presents the overall organisation of Propice-Plan: the data structures and functional modules. We introduce the plan, operator and constraint representation in Section 3, followed by a brief presentation of the execution module in Section 4. Section 5 describes the planning module and Section 6 the anticipation module, which are the two planning approaches we have integrated and implemented. Section 7 concludes the paper and proposes some future work.

## 2    **Propice-Plan** Organisation

The overall architecture of our applications corresponds to the one presented in [Alami et al., 1998]. This section focuses on the organisation of the decisional level and sketches the various data structures and functional components (see Fig. 1).

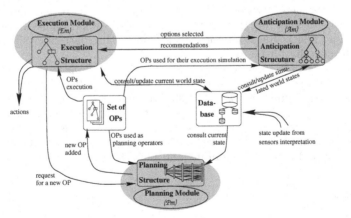

**Fig. 1.** Organisation of the various components.

The data structures used by the functional components follows:

• The database represents the current state of the world, as sensed by the system. It is automatically updated as new events occur. Moreover, this database provides a mechanism to handle multiple world developments, in order to enable the anticipation modules to simulate possible plan executions.

• The OPs set is initialised with the operational plans and operators which are interpreted by the execution module when receiving particular events or goals. This set can be supplemented with new OPs produced by the planning module on request.

The functional components are:

• The execution module ($\mathcal{E}m$) executes the OPs in a structure containing their execution state in response to events or to explicit goals given by the user. It is also able to recognise failures when achieving a particular goal, and then may request a new plan from the planning module.

• The anticipation module ($\mathcal{A}m$) is based on the current state of the execution structure. It simulates the execution of the OPs and evaluates the various outcomes of the different possible simulation paths to guide the $\mathcal{E}m$ if possible. These outcomes are stored and ordered according to their estimated adequacy to the current situation in the *anticipation structure* which is also consulted and updated by the $\mathcal{E}m$.

• The planning module ($\mathcal{P}m$) is currently composed of the IPP planner [Koehler *et. al.*, 1997]. When asked to achieve a particular goal state, it uses the database and the set of OPs to produce a new one.

# 3    Plans and Operators Representation

Like Propel [Levinson, 1994], one of the goals of this work is to provide a unified representation, OPs, for planning and execution control, i.e. for operational plans and operators. We illustrate this representation, inspired to some extend by the PDDL formalism [McDermott *et al.,* 1998], with an OP (see Fig. 2) from the blast furnace application[1].

▶ **Notation**
Symbols prefixed with $ denote logical variables (@@ denotes global variables). Variable typing is allowed: (syntax `<variable>:<type>`).
Modal operators may be used for goals to achieve (!), tests (?) or waiting ($\wedge$).
=>, ~> (respectively +>, −>) assert or retract facts systematically (respectively under some condition indicated by the first expression).

▶ **Declarative Information**
The following fields represent the context of use of an OP and its known effects:
**Invocation field.** This corresponds to the main effect of an OP. An OP may be goal-triggered (its invocation part is prefixed with '!', like in the example above) or event-triggered (without '!': see Fig. 3).
**Call field.** This field is available to uniquely identify the OP and to bind all variables upon calling (indeed, different OPs may have the same invocation part).

---

[1] In this application, one has to manage a complex pipes network, in which various gas and liquids flow. This particular OP specifies how to transfer purge fluid from a pipe section ($up-sec) to the next downstream one ($down-sec).

```
(defop |Purge Fluid Transfer|
    :invocation (! (in $down-sec purge-fluid))
    :call (<> (Purge-Fluid-Transfer $up-sec:section $down-sec:section))
    :context ((? (upstream-section $down-sec $up-sec))
              (! (in $up-sec purge-fluid)) (! (in $down-sec nil)))
    :effects ((=> (status $up-sec purged)) (=> (in $up-sec nil))
              (~> (in $down-sec nil)) (~> (in $up-sec purge-fluid))
              (-> (status $down-sec purged) (status $down-sec purged)))
    :properties ()
    :body ((? (upstream-valve $down-sec $up-valve))
           (? (downstream-valve $down-sec $down-valve))
           (! (valve-position $down-valve cl)) ; close downstream valve
           (if (? (drain-valve $down-sec $drain-valve))
               (! (valve-position $drain-valve cl))) ; close drain valve
           (! (valve-position $up-valve op)) ; open upstream valve
           (∧ (elapsed-time (time) @@fill-in-delay))
           (! (valve-position $up-valve cl)))) ; close upstream valve
```

**Fig. 2.** Example of OP.

It allows the system to call the OP directly (thus avoiding the standard trigge-ring/filtering mechanism).

**Context field.** This field gathers all preconditions required to apply the OP. As suggested in the PDDL formalism, we provide *filtering conditions* (prefixed with ?), *feasible conditions* (prefixed with !) which may be made true by an other OP, and also *maintain conditions* (prefixed with #), which must remain true during the OP execution.

**Effect field.** This field contains the expected effects (apart from the one ex-pressed in the `invocation` field) of a successful OP execution. Note that it is possible to use conditional effects.

**Properties field.** Some OPs may not be suitable for a particular module ($Am$, $Pm$); this can be specified by setting a particular property in this field.

▶ **Terminology**

Based on the field definitions above, we define some terms used in the following sections:

An OP is *relevant* for a goal (or a fact) if its `invocation` part unifies with it.

An OP is *applicable* if it is *relevant* and there exists a valid unification of the `context`.

An OP is *potentially applicable* if it is *relevant* and there exists a unification satisfying all the *filtering conditions* of its `context`.

An OP is *non applicable* if it is *relevant* and there is an unsatisfied *filtering conditions* in its `context`.

▶ **Executive Information**

An OP also contains informations describing what has to be done to execute it. It is either an *action* field (linked to an external execution code), or a `body` field; the `body` is a sequence of subgoals to satisfy (test, achieve, wait) if they are not already established in the current situation, combined with conditional constructs and loops (if-then-else, while, repeat), and also parallelism (expressed with //). Execution heuristics may be specified to favour an OP among various applicable ones.

The **body** described in the example can therefore be interpreted as follows: "Determine the upstream and the downstream valves of the section \$down-sec; close the downstream one; close the possibly existing drain valve; open the upstream one; wait for a given duration (corresponding to the time it usually takes to fill in a section); at last, close the upstream valve."

▶ **Discussion**

Unlike HTN representation [Nau *et al.*, 1999; Currie and Tate, 1991], OPs emphasise the dynamic and execution aspects of operational plans. For example one can explicitly write an OP which loops over a sequence of subgoals until a condition is true, or which follows one execution path or another according to a condition.

The OP description is limited to its abstraction level, regardless of lower level information; for example, the OP above does not clarify the valves positions in the **context** or **effects** fields. In this sense, the OP representation is *incomplete*, and corresponds to the *relaxed models* [Yang, 1997] used in hierarchical planning. Even if a sequence of OPs seems correct for some level of abstraction, interaction problems between effects and expected preconditions may arise at lower ones. Although it may be possible for each OP to compile off-line its extended **effects** and **context** fields, with respect to OPs used in its **body**, this technique would lead to a prohibiting number of highly specific OPs. Another solution is developed in § 6.

At last, as shown on Fig. 3, the representation proposed here makes it possible to define domain-specific constraints (ex: for security).

```
(defconst |Upper Bound For Surface Temperature|
   :invocation (surface-temperature $bf:furnace $temp:temperature-threshold)
   :context ((? (greater $temp @@max-surf-temp)))
   :body (// (! (coal-load-policy $bf suspend))
            (! (vapor-injection $bf (* @@u-vap (val-temp $temp))))))
```

**Fig. 3.** Example of constraint.

## 4   Supervision and Execution Control

The $\mathcal{E}m$ uses all the data structures presented in Fig. 1. Its algorithms are heavily inspired from the PRS and we invite the reader to check [Ingrand *et al.*, 1992; Ingrand *et al.*, 1996] for a more detailed account of how it works. Roughly, its main loop runs as follows: it takes into account new events in the database and new goals. It checks sleeping and maintained conditions in the execution structure, selects appropriate OPs and constraints according to these new events or goals, and the recommendations given by the anticipation structure. Then, it places the selected OPs in its execution structure to execute them (step by step to make the execution loop tight). It results in a primitive action, the assertion of new facts in the database, or the establishment of a new goal.

The $\mathcal{E}m$ differs from PRS in various ways. It explicitly takes into account the recommendations made by the $\mathcal{A}m$ in the anticipation structure (either OP , unification choice) and updates the anticipation structure accordingly. Moreover, it may request a new OP from the $\mathcal{P}m$ when it has to achieve a goal or a conjunction such that no corresponding OP is successfully applicable.

# 5    Plan Synthesis

As mentioned in Section 1, one of our goal is to use and integrate, when possible, existing efficient planning paradigms. We choose one of Graphplan [Blum and Furst, 1997] successors: IPP [Koehler et al., 1997], which provides some interesting extensions for real world applications.

## 5.1    Principle

When Propice-Plan has to face a situation for which there is no applicable OP, it may request help from the $\mathcal{P}m$. This planner takes as inputs the current contents of the database (initial state), the desired situation (goal), and the OPs available for planning.

For each OP, the $\mathcal{P}m$ only uses the *declarative informations* (see Section 1). If no solution is found, it reports a failure. Otherwise, a plan is sent as a new OP (in the OPs set) with a **body** corresponding to the sequence of OPs to execute (referenced by their **call** fields), and with its **context** set to the conjunction of the facts which entail the plan (such conjunction is computed during the final backchaining process in IPP). Then, the $\mathcal{E}m$ will check the returned plan **context** (i.e. its applicability) before executing it, since the environment state may have evolved since the plan was produced. Fig. 4 shows an OP synthesised by the $\mathcal{P}m$.

```
(defop |IPP air-regulation_bf1_thres-20000 (0)|
    :invocation (! (air-regulation bf1 thres-20000))
    :context ((? (status reg12 ok)) (? (air-feeder reg12))
            ... (? (bf-regulation reg12 bf1)))
    :body ((// (<> (Downstream-Section-Connection c12 reg12 tc1))
            (<> (Upstream-Section-Connection c11 reg11 tc1)))
        (<> (Regulation-Feeder reg12 reg11 sw1))
        ...
        (<> (Air-Regulation bf1 reg12 thres-20000))))
    :properties ((op-type synthesised)))
```

**Fig. 4.** Example of a plan returned by the $\mathcal{P}m$.

## 5.2   Selecting Relevant Information

Planners based on Graphplan algorithm suffer from a drastic drop in performance when the number of operators and the size of the initial state increase [Nebel *et al.*, 1997]. Contrary to most "toy" examples where planning problems are well-conditioned, our applications have a wide range of OPs available for planning and a large number of facts used to describe the current state. As a consequence, we need an effective selection method to preserve IPP efficiency.

Our first attempt consisted in using the RIFO technique in IPP [Nebel et al., 1997] to determine all operators and initial facts that seem relevant to the goal. Although this is very effective in most cases, it does not preserve the solution existence for some problems (which we encountered); this leads us to consider another method: off-line preprocessing is used to cluster the set of OPs according to their **effects** and **context** fields. As OPs are loaded in Propice-Plan, various clusters can be determined by analysing the attributes and the types of the variables in those fields. For the sake of concision, this technique will not be developed in this paper.

## 5.3   Monitoring the Initial State

The environment being dynamic, the $\mathcal{P}m$ has to monitor changes occurring in the database which may invalidate the initial state on which the planning graph is being built. This state is in fact restricted to those attributes described in the declarative informations of the selected OPs. We have been studying a technique to adapt the $\mathcal{P}m$ planning graph *on the fly*; the underlying idea is to stop the planning process in a stable state (ie. in the latest proposition level developed so far, once it has been complemented with mutual exclusive relations among its node, see [Blum and Furst, 1997] for more details) and clean off the rest of the graph. Then, starting from the initial proposition level, we remove successively all outdated facts, with the mutex relations they were possibly linked to, all operators they were precondition of (including noop), and restart the process for the next proposition level to remove all effects of deleted operators. This will stop when an empty level or the very last propositional one has been reached. The next step consists in completing the first proposition level with the missing facts described in the initial state, and restart IPP's forward-chaining process to complete its successors until the goal is found. At last, the plan search in the graph is performed as in IPP original algorithm. Unfortunately, this method does not guarantee the plan synthesis ending. In such a situation, we claim that the knowledge elements used for planning are inadequate with respect to the application environment dynamic.

## 5.4   Constraints Management

As illustrated on Fig. 3, Propice-Plan provides means to express the constraints of the domain. Thus, the $\mathcal{P}m$ provides basic constraint management mechanisms. Constraints (for instance (**surface-temperature $temp**) with $temp higher than @@max-surf-temp) are checked successively for each proposition level, when building the planning graph. If one is violated, the undesired facts are discarded in the current propositional level by use of IPP's Mutex relations (inherited from Graphplan's).

Of course, this technique is inadequate for constraints between more than two facts, and only unary and binary constraints are taken into account in Propice-Plan. However, this did not reveal too restrictive for the applicative domains we encoded so far. Moreover, this approach could be complemented with recent works using dynamic CSP techniques during the plan extraction phase [Kambhampati, 1999].

# 6    Anticipation Planning

The $\mathcal{Am}$ uses the OPs set and the database to simulate and examine in advance a number of possible options available to the system ahead of the execution. It provides two types of information useful to the $\mathcal{Em}$:

- it can evaluate in advance choices in order to advise the $\mathcal{Em}$ for the best option with respect to the current state of the system, and the set of projections developed so far (cf. 6.1).
- it can anticipate some unsatisfied preconditions to come, and try to establish them with an adequate *opportunistic* strategy (cf. 6.2).

Supervision and execution systems such as those mentioned in Section 1 do not perform explicit planning. They merely choose among possible available options at run-time (such as different applicable plans for a specific goal) without any projection of the current situation in the future. Our goal is to take advantage of the large number of OPs available and of the spare time left while performing control execution and supervision.[2]

The whole idea of anticipation planning is thus based on the simulation of OPs executive information. However, one has to keep in mind that some actions, mostly those related to sensing the value of an attribute, are non deterministic. As a consequence, we need to treat these attributes accordingly.

## 6.1    Guiding the $\mathcal{Em}$ through Choice Points

**Choices and Preferences.** For a given goal, the $\mathcal{Em}$ uses execution heuristics to choose among various applicable and relevant OPs for the most trustworthy one; though suited for most situations, the selected operational plan is generally more expensive than others (in terms of execution duration, resources,...). The first role of the $\mathcal{Am}$ is thus to evaluate such choice points before the $\mathcal{Em}$, with anticipation heuristics indicating less reliable but possibly cost-effective OPs. The corresponding results are gathered and regularly updated in the anticipation structure which contains all the projections examined so far, which will be consulted later by the $\mathcal{Em}$ when it faces a choice point. If the anticipation structure contains an option that led to a successful execution simulation, the $\mathcal{Em}$ uses it. Otherwise, the $\mathcal{Em}$ uses execution heuristics to select an adequate OP.

A specific processing is required when loops and conditional branching are simulated. Indeed, the condition value may be unpredictable due to non determinism (if (? (& (nitrogen-flow \$nf) (< \$nf 12))) ... ). Then, there is no mean

---

[2] For example, a blast furnace shutdown lasts eight hours and the supervision/execution process is idle most of the time. However, when a decision has to be taken, the response time must not exceed one minute.

to foresee which execution branch will be performed by the $\mathcal{E}m$. Both branches are then simulated and labeled with the corresponding constraints (here: one with a nitrogen flow value less than 12, and one for the opposite).

$\mathcal{E}m$ and $\mathcal{A}m$ Interactions. The initial state of the anticipation process is based on the current environment state. Then, when evaluating all possible outcomes for a choice point, the $\mathcal{A}m$ will develop different states accordingly; these will in turn lead to different options, when the next choice point will be processed. Therefore, the projections examined by the $\mathcal{A}m$ is structured as a *projection tree* (see Fig. 5). Then, assigning instantiated OPs to choice points is equivalent to searching a path in this tree, and search control is performed by anticipation heuristics.

In return, the $\mathcal{E}m$ synchronises the anticipation process with the actual environment state by updating the database (and therefore the projection tree root). Obsolete paths are automatically cut off by the anticipation structure; indeed, edges corresponding to a choice option are systematically labeled with the condition enabling this option (for example, when various OP instances are relevant for a goal, each outgoing edge from the current choice point is labeled with the corresponding context fields; see Fig. 5). Updates are processed every time the $\mathcal{E}m$ steps across a choice point even if it followed the $\mathcal{A}m$'s advise to take unexpected events into account.

## 6.2  Opportunistic Preconditions Achievement

The operational plans in the OPs set cover a wide range of expected situations including emergency ones (see Fig. 3). Yet, if the $\mathcal{E}m$ keeps executing and refining OPs without any anticipation, some preconditions required at a lower execution level may lead to inefficient plan, as they were unforeseen at the higher level.

• *Detecting inadequacy.* The $\mathcal{A}m$ detects an OP inadequacy if its execution simulation failed, eg. one of the goal in its body field cannot be reached. If it occurs that the relevant OPs for this goal are only potentially applicable (as defined in Section 1), the $\mathcal{A}m$ will try to adapt the current plan.

• *Fitting the current plan.* This is done by inserting, at the best place in the current plan, the proper OP which establishes the missing precondition(s). This adaptation should occur according to an *opportunistic* and *conservative* strategy, to modify the original plan as little as possible and minimise harmful interactions due to these modifications. The following example will illustrate these ideas.

**Example.** This example was inspired by a real problem that occurred during a furnace shutdown, while purging a pipes network. A known defective section valve could not be closed in time because the operational plan dealing with such valves requires an operator with a special tool. Without anticipation, such precondition could not be foreseen and an inefficient plan execution resulted (indeed, the shutdown was delayed by an half hour to send an operator on site).

▶ **OPs and Initial State**

To illustrate this, we introduce new OPs which body (not figured) will not be simulated for the sake of concision. Two OPs set valves positions: either manually (for defective valves that require assistance from an operator equipped with a toolbox), or automatically from a console, if the valve is not defective. The two others manage the operator equipment and location.

```
(defop |Automatic Valve Control|
  :invocation (! (valve-position $v $p))
  :call (<> (Automatic-Valve-Ctl $v:valve
                     $p:pos $old-p:pos))
  :context ((? (status $v OK))
            (? (valve-position $v $old-p))
            (! (ope-loc CO))) ;at the console
  :effects ((~> (valve-position $v $old-p))))
```

```
(defop |Get Toolbox|
  :invocation (! (has-toolbox))
  :call (<> (Get-Toolbox $l:location))
  :context ((? (toolbox-loc $l))
            (! (ope-loc $l)))
  :effects ())
```

```
(defop |Manual Valve Control|
  :invocation (! (valve-position $v $p))
  :call (<> (Manual-Valve-Ctl $v:valve
                     $p:pos $old-p:pos))
  :context ((? (status $v defective))
            (? (valve-position $v $old-p))
            (! (has-toolbox))
            (! (ope-loc $v)))
  :effects ((~> (valve-position $v $old-p))))
```

```
(defop |Go To|
  :invocation (! (ope-loc $dst))
  :call (<> (Go-To $dst:location $l:location))
  :context ((? (connex $l $dst))
            (! (ope-loc $l)))
  :effects ((~> (ope-loc $l))
            (+> (has-toolbox) (toolbox-loc $dst))
            (-> (has-toolbox) (toolbox-loc $l))))
```

A heuristic function is defined to minimise potentially dangerous human interventions (prefer |Automatic Valve Control| to |Manual Valve Control|).
Let us consider a simple domain with two sections (S1, S2) and three valves (G0, G1, G2); the topology is described in the figure (so the related facts are not enumerated). To simplify, a valve refers both to an object and the corresponding location. The initial state $(\mathcal{I})$ is defined as follows: (valve-position G$i$ cl)$_{i\in\{0,1\}}$ (valve-position G2 op)

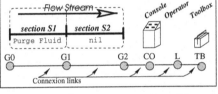

(toolbox-loc TB) (status G$i$ OK)$_{i\in\{0,2\}}$
(status G1 defective)
(in S1 purge-fluid) (in S2 nil) (ope-loc L).

▶ **OP Analysis**
When loading OPs in Propice-Plan, off-line processing makes it possible to determine OP hierarchies and cluster them according to their abstraction level. This is then used to guide plan adaptation as illustrated in this example. The following table presents the corresponding two-level OP hierarchy, with related facts.

| Higher level | Purge-Fluid-Transfer | *in, status:section* |
|---|---|---|
| **Lower level** | Manual-Valve-Ctl, Get-Toolbox, Automatic-Valve-Ctl, Go-To | *valve-position, ope-loc, status:valve, has-toolbox* |

▶ **Projections**
Suppose we simulate the execution of an OP such that the next goals are
(! (ope-loc CO)) (to control valves positions) and then (! (in S2 purge-fluid)) (with the |Minimise Human Risks| heuristic), and the initial state is $(\mathcal{I})$ described above. The projection tree is represented in Fig. 5.

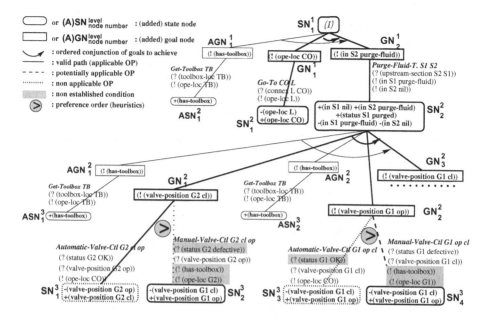

**Fig. 5.** Example of a Projection Tree.

Projection trees are layered, with two kinds of nodes appearing alternatively: *state nodes* represent the successive process states and *goal nodes*. Initially, the tree root $SN_1^1$ is a state node representing the database contents, and its children correspond to the sequence of `invocations` to be performed (here, there are therefore two goal nodes $GN_1^1$ and $GN_2^1$ to simulate (! (ope-loc CO)) and (! (in S2 purge-fluid))).

Projections are developed as follows: various OPs may be relevant for a non established goal $GN_i^l$, leading to new state nodes $SN_j^{l+1}$; the corresponding edges $\overline{GN_i^l SN_j^{l+1}}$ are labeled with the instantiated `call` and `context` fields, and ordered according to heuristics. A new state node $SN_j^{l+1}$ contents is only *the difference with its parent state node* at level $l$; '+' (resp. '-') signs indicate added (resp. deleted) effects, according to the *declarative informations* of the OP from edge $\overline{GN_i^l SN_j^{l+1}}$. Then, goal nodes at level $l + 1$ ($GN_k^{l+1}$ nodes) are obtained by developing the `body` field of the OP from edge $\overline{GN_i^l SN_j^{l+1}}$. This process repeats until no state node could be further developed (terminal node); this occurs when the `body` field of the OP leading to this node is empty or the OP is only simulated through its `effects` field (OP marked as not relevant for anticipation). Finally, a state node is *completed* if it is terminal, or all its children are *completed* (AND-node semantic); a goal node is *completed* if at least one child is *completed* (OR-node related to choice points). The simulation is considered successful if the root node is completed; then, the $\mathcal{E}m$ can be advised for all these choice points.

The result of the simulation process corresponds to **thick edges and nodes** in the projection tree. Go-To and Purge-Fluid-Transfer are the only relevant OPs for (! (ope-loc CO)) and (! (in S2 purge-fluid)), and they are applicable. For the second one, the body indicates that the goals (! (valve-position G2 cl)), (! (valve position G1 op)), and (! (valve-position G1 cl)) must be achieved. Two OPs are relevant for each of the first two ones, and the preferred one is Automatic-Valve-Ctl, which is applicable for the goal (! (valve-position G2 cl)) (goal node $GN_1^2$). Concerning the goal (! (valve-position G1 op)) ($GN_2^2$), the first relevant OP is non applicable, and the second (Manual-Valve-Ctl) is only potentially applicable. Thus, adaptation is necessary to establish the required conditions: (has-toolbox) and (ope-loc G1).

▶ **Adaptation**
The basic idea is to select a flaw (unsupported feasible condition), look for all possible fixing OPs, and add the most adequate one where it suits the best (fix selection and positioning are performed at the same time, using a cost criterion). The main step of the adaptation algorithm are:

Step 1. Let $\overline{GN_i^l SN_j^{l+1}}$ be the first potentially applicable OP appearing while developing the projection tree, and such that it is not marked as non recoverable, and the tree can still be repaired (initially, all flaws are considered recoverable).

If no such OP exists, either there is a valid path in the tree (all inconsistencies are removed: exit successfully), or it is impossible to fix the flaw (report failure).

Step 2. Let $S$ be the set of goal nodes $AGN_m^n$ that may fix the flaw $\overline{GN_i^l SN_j^{l+1}}$, and added with respect to hierarchy considerations. Set a causal link between every element $AGN_m^n$ in $S$ and $\overline{GN_i^l SN_j^{l+1}}$. For every couple of elements in $S$, set a mutual exclusion link.

Step 3. For each element $AGN_m^n$ in $S$, develop all possible edges $\overline{AGN_m^n ASN_p^{n+1}}$ corresponding to OPs relevant for $AGN_m^n$. Associate to those edges their insertion cost. Let $\overline{AGN_m^n ASN_p^{n+1}}$ be the minimal relevant OP with regards to the associated costs.

Step 4. If the insertion cost for $\overline{AGN_m^n ASN_p^{n+1}}$ is infinite, then determine the flaw $\overline{GN_i^l SN_j^{l+1}}$ that led to $AGN_m^n$ addition (use causal links), mark $\overline{GN_i^l SN_j^{l+1}}$ as non recoverable, remove all goal nodes added because of $\overline{GN_i^l SN_j^{l+1}}$ (use causal links), and go back to step 1. Otherwise, remove all goal nodes linked by a mutual exclusion relation to $AGN_m^n$, and exit to restart the anticipation process from node $AGN_m^n$.

The first selected flaw is $\overline{GN_2^2 SN_4^3}$ (Step 1), and the condition to establish is (! (has-toolbox)). Then, the set $S$ of added goal nodes (Step 2) is $\{AGN_1^1, AGN_1^2, AGN_2^2\}$ (thin edges and nodes on the Fig. 5); no such goal node can be immediately before $GN_2^1$, because they do not belong to the same hierarchic level (in is higher than has-toolbox). Causal links between a flaw and its fixes are used to suppress all added nodes if it appears that the flaw cannot be recovered; a mutual exclusion link indicates that as soon as a correct fix is found, the others are useless. Edges developed in step 3 are $\overline{AGN_1^1 ASN_1^2}$, $\overline{AGN_1^2 ASN_1^3}$ and $\overline{AGN_2^2 ASN_2^3}$. A cost is associated, estimating the OP applicability, and the interactions it may have with the rest of the plan:

- **Applicability factor.** If the added OP is applicable, this factor is zero; if it is non-applicable, it is set to $\infty$; otherwise, it is an estimate of the number of OPs needed to establish invalid conditions. Instead of solving this planning problem, we focus on a relaxed one where delete-lists are not taken into account, and conditional effects are turned into systematic ones; the resulting plan may not be correct, but it provides a rough idea of an added OP adequacy to the current situation.
- **Interaction factor.** If the effects of the inserted OP invalidate a condition required to execute another one (this is determined by using condition labels associated to edges), this factor is set to the estimate number of OP needed to re-establish it (same technique as above); otherwise, it is zero.

In this example, the global cost for the fixes is:

| | Applicability | Interaction |
|---|---|---|
| $AGN_1^1 ASN_1^2$ | 2 OPs: (Go-To TB L) + (Get-Toolbox TB) | no OP |
| $AGN_1^2 ASN_1^3$ | 3 OPs: (Go-To L CO) + (Go-To TB L) + (Get-Toolbox TB) | no OP |
| $AGN_1^2 ASN_1^3$ | 3 OPs: (Go-To L CO) + (Go-To TB L) + (Get-Toolbox TB) | no OP |

$AGN_1^1 ASN_1^2$ is therefore selected, and the two others fixes removed (step 4). The simulation process is started again to update the projection tree (indeed, new conditional effects may have been triggered); if simulation is unsuccessful, the adaptation process is run again. For example, the next flaw will thus be $AGN_1^1 ASN_1^2$ (to establish (! (ope-loc TB))). The final plan is therefore to have the operator go and get the toolbox in TB, go to CO via L to control the valve G2, and finally go to G1 (via G2) to manœuvre the defective valve.

Without anticipation planning, the $\mathcal{E}m$ would have closed G2 and failed to open G1; it would have then requested help from the $\mathcal{P}m$. However, the additional actions from the returned plan would have been more numerous than the ones added by the $\mathcal{A}m$, since the planning phase is launched *after an actual execution failure* (the operator being at CO, he would have had more moves to do to get the toolbox and go back to G1). On the opposite, anticipation provides rather light adaptations, which are integrated seamlessly in the original procedure.

## 7   Conclusion

The proposed approach addresses general issues about planning and execution in real-world domains. Our work is meant to be generic and uses two different case studies (the blast furnace and the autonomous robots domains) which showed that one single planning technique is not sufficient to cope with the wide variety of problems occurring in these domains. Therefore, Propice-Plan combines existing approaches (IPP for planning and Propice — an extended version of C-PRS — for execution) with new ones dealing with anticipation planning.

This paper presents several improvements for existing techniques to meet real-world requirements. In particular: an efficient information filtering mechanism to select only relevant facts and operators for a given planning problem; a cooperative approach between the execution and planning modules to handle possible changes concerning the initial planning state; an anticipation of OPs execution to foresee problems to come and make choices in advance; an advisable execution module that takes external recommendations into account when facing choices.

The OP formalism we adopted encompasses the notions of operational plans, operators and even constraints. This language revealed suitable to encode a large body of knowledge in a rather natural way, using classical preconditions/effects planning information but also additional operational knowledge in a formalism similar to HTN, but including programming language constructs (loops, conditional branching) and variables. This common representation ensures seamless transitions between modules. However, this raises the problem of the logical consistency between the various OP fields used by these modules. Part of this checking may be done using purely syntactic approaches, but other will require more formal logical techniques. Nevertheless, Propice-Plan provides means to avoid dangerous states with constraint OPs, which ensure that any conflict state is detected at execution time, and is not planned explicitly by the $Am$ and the $Pm$.

In the future, we intend to study how to take some additional reasoning capabilities into account. At this point, Propice-Plan handles explicit operations on dates and durations, and wait / rendezvous synchronisations, but it is unable to manage time constraints propagation as described in [Ghallab and Mounir-Aloui, 1989]. Similarly, better hypothetical reasoning techniques are needed to handle conditional branching for anticipation, and dynamic CSP techniques to cope with more general constraints. In addition, we intend to improve the current implementation, which already runs on board our mobiles robots, to use it on site for the blast furnace application (Propice-Plan runs under Solaris, VxWorks and Linux). In both application, hundreds of OPs have been encoded and are used by the $Em$. The plans produced by the $Am$ shows a real gain over the generic plans (eg. by preventing a furnace shutdown delay). Moreover, the $Pm$ is used in the robot application to synthesise multi robots plans [Botelho, 1998] and in the furnace application to handle the pipes network configuration.

# References

[Alami et al., 1998] R. Alami, R. Chatila, S. Fleury, M. Ghallab, and F. Ingrand. An Architecture for Autonomy. *International Journal of Robotics Research*, 17(4):315–337, April 1998.

[Beetz and McDermott, 1994] M. Beetz and D. McDermott. Improving Robot Plans During Their Execution. In *Proceedings of AIPS94*, June 1994.

[Blum and Furst, 1997] A. Blum and M. Furst. Fast Planning Through Planning Graph Analysis. *Artificial Intelligence*, 90:281–300, 1997. See also: http://www.cs.cmu.edu/~avrim/graphplan.html.

[Bonasso et al., 1997] R.P. Bonasso, R.J. Firby, E. Gat, D. Kortenkamp, D.P. Miller, and M.G. Slack. Experiences with an Architecture for Intelligent, Reactive Agents. *Journal of Experimental and Theoretical AI*, January 1997.

[Botelho, 1998] S. Botelho. A distributed scheme for task planning and negotiation in multi-robot systems. In *Proceedings of ECAI*, 1998.

[Currie and Tate, 1991] K. Currie and A. Tate. O-Plan: the Open Planning Architecture. *Artificial Intelligence*, 52:49–86, 1991.

[Firby, 1994] R. J. Firby. Task Networks for Controlling Continuous Processes. In *Proceedings of the Second International Conference on AI Planning Systems, Chicago IL*, June 1994.

[Geffner and Bonet, 1998] H. Geffner and B. Bonet. High-Level Planning and Control with Incomplete Information Using POMDPs. In *Proceedings AIPS-98 Workshop on Integrating Planning, Scheduling and Execution in Dynamic and Uncertain Environments*, 1998. See also: http://www.ldc.usb.ve/∼hector.

[Ghallab and Mounir-Alaoui, 1989] M. Ghallab and A. Mounir-Alaoui. The Indexed Time Table Approach for Planning and Acting. In Rodriguez and Seraji, editors, *Procs. NASA Conference on Space Telerobotics*, volume 5, pages 321–332. JPL Publications 89-7, Feb. 1989.

[Guéré and Alami, 1999] E. Guéré and R. Alami. A Possibilistic Planner That Deals With Non-Determinism and Contingency. In *Procs. IJCAI*, 1999.

[Ingrand *et al.,* 1992] F. F. Ingrand, M. P. Georgeff, and A. S. Rao. An Architecture for Real-Time Reasoning and System Control. *IEEE Expert, Knowledge-Based Diagnosis in Process Engineering*, 7(6):34–44, December 1992.

[Ingrand *et al.,* 1996] F. F. Ingrand, R. Chatila, and R. Alami F. Robert. PRS: A High Level Supervision and Control Language for Autonomous Mobile Robots. In *IEEE International Conference on Robotics and Automation, St Paul, (USA)*, 1996.

[Kambhampati, 1999] Subbarao Kambhampati. Improving Graphplan's Search with EBL & DDB Techniques. In *Procs.IJCAI*, 1999.

[Koehler *et al.,* 1997] J. Koehler, B. Nebel, J. Hoffmann, and Y. Dimopoulos. Extending Planning Graphs to an ADL Subset. In *Proceedings of European Conference on Planning*, 1997. See also: http://www.informatik.uni-freiburg.de/∼koehler/ipp.html.

[Kushmeric *et al.,* 1995] N. Kushmeric, S. Hanks., and D. Weld. An Algorithm for Probabilistic Planning. *Artificial Intelligence*, 2:239–286, 1995.

[Levinson, 1994] R. Levinson. A General Programming Langage for Unified Planning and Control. *Artificial Intelligence*, 76:281–300, 1994.

[McDermott *et al.,* 1998] D. McDermott, M. Ghallab, A. Howe, C. Knoblock, A. Ram, M. Veloso, D. Weld, and D. Wilkins. PDDL – The Planning Domain Definition Langage. Technical Report CVC TR-98-003/DCS TR-1165, Yale Center for Computational Vision and Control, October 1998. Available on request to drew.mcdermott@yale.edu.

[Myers, 1998] K. L. Myers. Towards a Framework for Continuous Planning and Execution. In *Proceedings of the AAAI Fall Symposium on Distributed Continual Planning*, 1998. See also: http://www.ai.sri.com/∼cpef.

[Nau *et al.,* 1999] D. Nau, Y. Cao, A. Lotem, and H. Muñoz-Avila. SHOP: Simple Hierarchical Ordered Planner. In *Procs. IJCAI*, 1999.

[Nebel *et al.,* 1997] B. Nebel, Y. Dimopoulos, and J. Koehler. Ignoring Irrelevant Facts and Operators in Plan Generation. In *Proceedings of European Conference on Planning*, 1997. See also: http://www.informatik.uni-freiburg.de/∼nebel.

[Pryor and Collins, 1996] L. Pryor and G. Collins. Planning for contingencies: A decision-based approach. *Artificial Intelligence Research*, 1996.

[Wilkins *et al.,* 1994] D. E. Wilkins, K. L. Myers, J. D. Lowrance, and L. P. Wesley. Planning and Reacting in Uncertain and Dynamic Environments. *Journal of Experimental and Theoretical AI*, 6:197–227, 1994. See also: http://www.ai.sri.com/people/wilkins/papers.html.

[Yang, 1997] Q. Yang. *Intelligent Planning: A Decomposition and Abstraction Based Approach*, pages 163–171. Springer-Verlag Publisher, 1997.

# What Is the Expressive Power of Disjunctive Preconditions?

Bernhard Nebel

Institut für Informatik,
Albert-Ludwigs-Universität,
D-79100 Freiburg, Germany
E-mail: nebel@informatik.uni-freiburg.de

**Abstract.** While there seems to be a general consensus about the expressive power of a number of language features in planning formalisms, one can find many different statements about the expressive power of disjunctive preconditions. Using the "compilability framework," we show that preconditions in conjunctive normal form add to the expressive power of propositional STRIPS, which confirms a conjecture by Bäckström. Further, we show that preconditions in conjunctive normal form do not add any expressive power once we have conditional effects.

## 1 Introduction

There seems to be a general consensus about the *expressive power* of a number of language features in planning formalisms. For example, everybody seems to agree that adding negative preconditions does not add very much to the expressive power of basic STRIPS, while conditional effects are considered as a significant increase in expressive power [1,5,6,10].

For *disjunctive preconditions*, however, the picture is much less clear. One can find many different statements about the expressive power of disjunctive preconditions. For instance, Anderson *et al.* [1] write that "disjunctive preconditions ... are ... essential prerequisites for handling conditional effects." This statement could be understood as stating that implementing disjunctive preconditions would make it easier to handle conditional effects. However, it also could be read as stating that once we are able to handle conditional effects, disjunctive preconditions come for free. Bäckström [2] conjectures that "disjunctive preconditions most likely increase the expressive power [of STRIPS]." Kambhampati *et al.* [6] and Gazen and Knoblock [5], finally, believe that disjunctive preconditions are easy to add to basic STRIPS.

On top of these different conjectures there is no consensus on what the term *disjunctive precondition* means. Bäckström [2] interprets it as preconditions in conjunctive normal form (CNF), Gazen and Knoblock [5] interpret it as preconditions in disjunctive normal form (DNF), and Anderson *et al.* [1] interpret it as general Boolean preconditions. This means that instead of asking for the expressive power of disjunctive preconditions (relative to basic STRIPS or STRIPS

S. Biundo and M. Fox (Eds.): ECP-99, LNAI 1809, pp. 294–307, 2000.

with conditional effects), we should ask for the expressive power of CNF-, DNF-, and general Boolean preconditions.

In order to address this problem we use the *compilability framework* [11,12], which is inspired by Bäckström's [2] approach to assess the expressive power of planning formalisms, by the work of Gazen and Knoblock [5], and by the *knowledge compilation* framework [4]. The main idea behind this approach is that a language feature does not increase the expressiveness if planning operators containing this language feature can be translated into operators that do not contain this feature without blowing up the operator description too much and without enlarging the resulting plans too much. It differs from Bäckström's [2] *ESP-reductions* in that we only consider *structured* transformations that translate the initial state, goal specification, and operators in isolation, in that we allow for arbitrary computational power in the transformation, and in that we do not require that translated plans have *exactly* the same length.[1]

Using this approach, we show that Bäckström's conjecture that CNF preconditions are more expressive than basic STRIPS is indeed correct. We also prove that CNF preconditions do not add to the expressive power if we have already conditional effects – proving a weak version of Anderson's *et al.* [1] conjecture. However, general boolean formulae as preconditions are incomparable to conditional effects as follows from earlier results [11].

Although these results are purely theoretical, they also have some significance for designing planning algorithms. Provided we can prove that a language feature can be "compiled away" easily, i.e., that it can be regarded as "syntactic sugar," a planning algorithm for the original language can be easily extended to deal with the new feature. Conversely, if we can prove that a language feature cannot be compiled away, we most probably will have significant problems when we want to extend the planning algorithm for the original language to deal with the new language feature.

The rest of the paper is structured as follows. The next section introduces general terminology and definitions. In Section 3, we introduce the notion of compilability between planning formalisms. Using this framework, we analyze the relationship between STRIPS with disjunctive preconditions and basic STRIPS in Section 4. We show that DNF preconditions can be compiled away and that Bäckström's [2] conjecture holds in the compilability framework. In Section 5 we show that CNF preconditions can be compiled away if we have conditional effects. Finally, in Section 6 we summarize and discuss the results.

## 2    Propositional Planning Formalisms with Disjunctive Preconditions and Conditional Effects

Let $\Sigma$ be a finite set of **propositional atoms** and $\widehat{\Sigma}$ be the set consisting of the constants $\top$ (denoting truth) and $\bot$ (denoting falsity) as well as atoms and negated atoms, i.e., the **literals**, over $\Sigma$. The set of all **Boolean formulae**

---

[1] For the rationales behind these requirements see [11].

over $\Sigma$ is denoted by $\mathbf{B}_\Sigma$. The set of **formulae in conjunctive normal form (CNF)** over $\Sigma$ is denoted by $\mathbf{C}_\Sigma$ and the set of **formulae in disjunctive normal form (DNF)** over $\Sigma$ by $\mathbf{D}_\Sigma$. Finally, by $\mathbf{L}_\Sigma$ we refer to the set of **formulae that are conjunctions of literals** over $\Sigma$, and the set of **formulae that are conjunctions of atoms** is denoted by $\mathbf{A}_\Sigma$. In general, we will use the symbol $\mathcal{L}_\Sigma$ to refer to a possibly restricted language over $\Sigma$.

Given a set of literals $L \subseteq \widehat{\Sigma}$, by $pos(L)$ we refer to the **positive literals** in $L$, by $neg(L)$ we refer to the **negative literals** in $L$, and $\neg L$ denotes the **element-wise negation** of the literals in $L$. **Operators** are pairs $o = \langle pre, post \rangle$. We use the notation $pre(o)$ and $post(o)$ to refer to the first and second part of an operator $o$, respectively. The **precondition** $pre$ is an element of $\mathcal{L}_\Sigma$. The set $post$, the set of **postconditions**, consists of **conditional effects** each having the form $\varphi \Rightarrow L$, where $\varphi \in \mathcal{L}_\Sigma$ is called **effect condition** and the elements of $L \subseteq \widehat{\Sigma}$ are called **effects**. If all postconditions of an operator have the form $\top \Rightarrow L$, then we say that the operator is **unconditional** and we write the postconditions as a set of literals containing all *effects*.

A **state** $S$ is a *truth-assignment* for the atoms in $\Sigma$, which is represented by the set of atoms that are true in this state. By $\bot$ we represent the **illegal state**. Given a state $S$ and an operator $o$, we define the **active effects** $A(S, o)$ as follows:

$$A(S, o) = \bigcup \{L \mid (\varphi \Rightarrow L) \in post(o), S \models \varphi\}.$$

Using this function, the result of executing operator $o$ in state $S$ can be specified as:

$$R(S, o) = \begin{cases} S - neg(A(S, o)) \cup pos(A(S, o)) & \text{if } S \neq \bot \text{ and} \\ & \qquad S \models pre(o) \text{ and} \\ & \qquad A(S, o) \not\models \bot, \\ \bot & \text{otherwise} \end{cases}$$

A **planning instance** is now a tuple $\Pi = \langle \Xi, \mathbf{I}, \mathbf{G} \rangle$, where

- $\Xi = \langle \Sigma, \mathbf{O} \rangle$ is the **domain structure** consisting of a finite set of propositional atoms $\Sigma$ and a finite set of operators $\mathbf{O}$,
- $\mathbf{I} \subseteq \Sigma$ is the **initial state**, and
- $\mathbf{G} \subseteq \widehat{\Sigma}$ is the **goal specification**.

When we talk about the **size of an instance**, symbolically $||\Pi||$, in the following, we mean the size of a (reasonable) encoding of the instance.

Let $\Delta$ be a finite sequence of operators from $\mathbf{O}$, which is called **plan**. Then $||\Delta||$ denotes the size of the plan, i.e., the number of operators in $\Delta$. We say that $\Delta$ is a $c$-**step plan** if $||\Delta|| \leq c$. The result of applying $\Delta$ to a state $S$ is recursively defined as follows:

$$Res(S, \langle \rangle) = S,$$
$$Res(S, \langle o_1, o_2, \ldots, o_n \rangle) = Res(R(S, o_1), \langle o_2, \ldots, o_n \rangle).$$

A sequence of operators $\Delta$ is said to be a **plan for** $\Pi$ or a **solution of** $\Pi$ iff

1. $Res(\mathbf{I}, \Delta) \neq \perp$ and
2. $Res(\mathbf{I}, \Delta) \models \mathbf{G}$.

The most general planning language we consider in this paper is $\text{STRIPS}_{\mathcal{C},\mathbf{B}}$, which permits conditional effects and general Boolean formulae in preconditions and effect conditions. Without the index $\mathcal{C}$, we refer to planning languages without conditional effects, i.e., all conditional effects have the form $\top \Rightarrow L$. If we have $\mathbf{C}$, $\mathbf{D}$ or $\mathbf{L}$ instead of $\mathbf{B}$, we refer to languages that permit only for CNF or DNF formulae or conjunctions of literals in preconditions and effect conditions, respectively. The language $\text{STRIPS}$ (without any index), finally, is identical to basic $\text{STRIPS}$, i.e., it requires that all preconditions are conjunctions of atoms and all effects are unconditional. In this paper, however, we assume that all formulae may contain literals, i.e., the least expressive language we consider is $\text{STRIPS}_{\mathbf{L}}$.[2] In Figure 1, the partial order induced by the syntactic restrictions is shown. In the sequel we say that $\mathcal{X}$ **is a specialization** of $\mathcal{Y}$, written $\mathcal{X} \sqsubseteq \mathcal{Y}$, iff $\mathcal{X}$ is identical to $\mathcal{Y}$ or below $\mathcal{Y}$ in the Hasse diagram depicting the partial order.

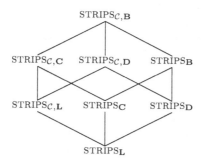

**Fig. 1.** Partial order induced by syntactic restrictions

While one would expect that planning in $\text{STRIPS}_{\mathbf{L}}$ is much easier than planning in $\text{STRIPS}_{\mathcal{C},\mathbf{B}}$, it turns out that this is not the case, provided one takes a computational complexity perspective. The plan existence problem in all the formalisms we consider is **PSPACE**-complete [11, Theorem 3]. **PSPACE**-hardness of plan existence in $\text{STRIPS}_{\mathbf{L}}$ follows from Bylander's [3] results. Membership is evident from the fact that exponentially many non-deterministic operator choices are enough for checking plan existence, where the applicability of one $\text{STRIPS}_{\mathcal{C},\mathbf{B}}$ operator can be verified using an NP-oracle.

---

[2] The reason for leaving out basic $\text{STRIPS}$ is that it has already been shown that $\text{STRIPS}$ and $\text{STRIPS}_{\mathbf{L}}$ as well as $\text{STRIPS}_{\mathcal{C}}$ and $\text{STRIPS}_{\mathcal{C},\mathbf{L}}$ are equivalent with respect to expressiveness [11, Theorem 6 and 9]. Furthermore, ignoring the case $\mathcal{L}_{\Sigma} = \mathbf{A}_{\Sigma}$ simplifies some of the proofs.

## 3   Compilation Schemes

We will consider a planning formalism $\mathcal{X}$ *as expressive as* another formalism $\mathcal{Y}$ if planning domains and plans formulated in formalism $\mathcal{Y}$ are *concisely expressible* in $\mathcal{X}$. We formalize this intuition by making use of what we call *compilation schemes*, which are *solution preserving mappings* with *polynomially sized results* from $\mathcal{Y}$ domain structures to $\mathcal{X}$ domain structures. While we restrict the size of the result of a compilation scheme, we do not require any bounds on the computational resources for the compilation. In fact, for measuring the *expressibility*, it is irrelevant whether the mapping is polynomial-time computable, exponential-time computable, or even non-recursive. If we want to use such compilation schemes in practice, they should be reasonably efficient, of course. However, if we want to prove that one formalism is *not as least as expressive as* another one, we better prove that there is no compilation scheme regardless of how many computational resources it might use. For this reason, Bäckström's [2] ESP-reductions could not be classified as compilation schemes because they are constrained to be computable in polynomial time.

So far, compilation schemes restrict only the size of domain structures. However, when measuring expressive power, the size of the generated plans should also play a role. In Bäckström's ESP-reductions [2], the plan size must be identical. Similarly, the translation from STRIPS$_{C,L}$ to STRIPS proposed by Gazen and Knoblock [5] guarantees that the plan length does not change. When comparing the expressiveness of different planning formalisms, we might, however, be prepared to accept some growth of the plans in the target formalism. For instance, we may accept an additional constant number of operators, or we may even be satisfied if the plan in the target formalism is linearly or polynomially larger. This leads to the schematic picture of compilation schemes as displayed in Figure 2.

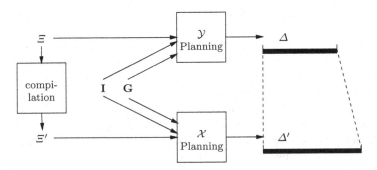

**Fig. 2.** The compilation framework

Although Figure 2 gives a good picture of the *compilation framework*, it is not completely accurate. A compilation scheme may introduce some auxiliary propositional atoms that are used to control the execution of newly introduced operators. These atoms should most likely have an initial value and may appear in the goal specification of planning instances in the target formalism. We will assume that the compilation scheme takes care of this and adds some literals to the initial state and goal specifications.

Additionally, some translations of the initial state and goal specifications may be necessary. If we want to compile a formalism that permits for literals in preconditions and goals to one that requires atoms, some trivial translations are necessary. Similarly, if we want to compile a formalism that permits us to use partial states to a formalism that requires complete state, a translation of the initial state specification is necessary. However, since we do not deal with incomplete state specifications or planning formalisms that allow only atoms in the precondition, we can ignore this issue here.

A **compilation scheme from** $\mathcal{X}$ **to** $\mathcal{Y}$ is a tuple of functions $\mathbf{f} = \langle f_\xi, f_i, f_g \rangle$ that induces a function $F$ from $\mathcal{X}$-instances $\Pi = \langle \Xi, \mathbf{I}, \mathbf{G} \rangle$ to $\mathcal{Y}$-instances $F(\Pi)$ as follows:[3]

$$F(\Pi) = \langle f_\xi(\Xi), \mathbf{I} \cup f_i(\Xi), \mathbf{G} \cup f_g(\Xi) \rangle$$

and satisfies the following conditions:

1. there exists a plan for $\Pi$ iff there exists a plan for $F(\Pi)$,
2. and the size of the results of $f_\xi, f_i$, and $f_g$ is polynomial in the size of the arguments.

Although there are no resource bounds on $f_\xi, f_i$, and $f_g$ in the general case, we are also interested in *efficient compilation schemes*. We say that $\mathbf{f}$ is a **polynomial-time compilation scheme** if $f_\xi, f_i$, and $f_g$ are polynomial-time computable functions.

In addition to that we measure the size of the corresponding plans in the target formalism. If a compilation scheme $\mathbf{f}$ has the property that for every plan $\Delta$ solving an instance $\Pi$, there exists a plan $\Delta'$ solving $F(\Pi)$ such that $||\Delta'|| \leq ||\Delta|| + k$ for some positive integer constant $k$, $\mathbf{f}$ is a **compilation scheme preserving plan size exactly** (modulo an additive constant). If $||\Delta'|| \leq c \times ||\Delta|| + k$ for positive integer constants $c$ and $k$, then $\mathbf{f}$ is a **compilation scheme preserving plan size linearly**, and if $||\Delta'|| \leq p(||\Delta||, ||\Pi||)$ for some polynomial $p$, then $\mathbf{f}$ is a **compilation scheme preserving plan size polynomially**. More generally, we say that a planning formalism $\mathcal{X}$ is **compilable** to formalism $\mathcal{Y}$ (in poly. time, preserving plan size exactly, linearly, or polynomially), if there exists a compilation scheme with the appropriate properties. We write $\mathcal{X} \preceq^x \mathcal{Y}$ in case $\mathcal{X}$ is compilable to $\mathcal{Y}$ or $\mathcal{X} \preceq_p^x \mathcal{Y}$ if the compilation can be done in polynomial time. The super-script $x$ can be 1, $c$, or $p$ depending on whether the scheme preserves plan size exactly, linearly, or polynomially, respectively. From a practical point of view, one could regard compilability preserving plan size exactly or linearly as an indication that the planning formalism we use

---

[3] We ignore the issue of *state-translation functions* as introduced in the definition of compilation schemes in an earlier paper [11,12] because they are not needed in this paper. It should be noted, however, that all results in this paper hold also for those more general compilation schemes.

as the target formalism is *at least as expressive* as the source formalism. Conversely, if a super-linear blowup of the plans in the target formalism is required by any compilation scheme, this is an indication that the source formalism *is more expressive* than the target formalism.

As is easy to see, all the notions of compilability introduced above are reflexive and transitive, i.e., compilability induces a pre-order on planning formalisms.

**Proposition 1.** [11] *The relations $\preceq^x$ and $\preceq^x_p$ are transitive and reflexive.*

Furthermore, it is obvious that when moving upwards in the diagram displayed in Figure 1, there is always a polynomial-time compilation scheme preserving plan size exactly. If $\pi_i$ denotes the projection to the $i$-th argument and $\emptyset$ the function that returns always the empty set, the generic compilation scheme for moving upwards in the partial order is $\mathbf{f} = \langle \pi_1, \emptyset, \emptyset \rangle$.

**Proposition 2.** [11] *If $\mathcal{X} \sqsubseteq \mathcal{Y}$, then $\mathcal{X} \preceq^1_p \mathcal{Y}$.*

Using (a slight generalization of) the *compilation framework* described above, it has been shown that all formalisms that we consider in this paper can be compiled to each other in polynomial time preserving plan size polynomially [11, Corollary 24]. However, if only linear growth of the plan in the target formalism is permitted, then for most of the formalisms there does not exist a compilation scheme. In particular, we have the following two results.

**Theorem 3.** [11] STRIPS$_{C,L}$ $\not\preceq^c$ STRIPS$_B$.

**Theorem 4.** [11] STRIPS$_B$ $\not\preceq^c$ STRIPS$_{C,L}$.

These results seem to invalidate the conjecture by Anderson *et al.* [1] that the expressiveness of conditional effects and disjunctive preconditions in the form of general boolean formulae are related. However, as we will see below, restricting the preconditions to be in CNF, a weak form of their conjecture is provable in the compilation framework.

## 4  Disjunctive Preconditions

Although the relative expressiveness of different planning formalisms has been extensively studied [11], syntactically restricted formulae such as CNF and DNF formulae have not been considered yet.

It is folklore in the planning community that DNF preconditions can be regarded as syntactic sugar. For each operator $o = \langle (c_1 \vee c_2 \vee \ldots \vee c_k), L \rangle$, where $c_i \in \mathbf{L}_\Sigma$ and $L \subseteq \widehat{\Sigma}$, one simply generates $k$ new operators $o_i = \langle c_i, L \rangle$. This translation is obviously a *polynomial-time compilation scheme preserving plan size exactly.*

**Proposition 5.** STRIPS$_D$ $\preceq^1_p$ STRIPS$_L$.

For CNF preconditions the situation is much less clear. As mentioned above, Bäckström [2] conjectured that CNF preconditions add to the expressiveness of STRIPS$_L$, but he was not able to prove this conjecture using his framework of ESP-reductions. Using our compilability framework for measuring expressiveness, we can, however, prove his conjecture. In order to do so, we first introduce a variation of the planning problem.

The **fixed plan-size initial-state existence problem** ($c$-FISEX) is defined as follows. Given a domain structure $\Xi = \langle \Sigma, \mathbf{O} \rangle$, a goal $\mathbf{G} \subseteq \widehat{\Sigma}$, a state $\mathbf{I} \subseteq \Sigma$, and a subset of the atoms called *choice set* $\mathbf{C} \subseteq \Sigma$, the question is whether there exists a set $C \subseteq \mathbf{C}$ such that $\langle \Xi, \mathbf{I} \cup C, \mathbf{G} \rangle$ can be solved by a plan with size $c$, where $c$ is a positive constant.

Although this problem appears to be slightly harder than the ordinary plan existence problem, fixing the plan length to the positive constant $c$ makes the problem easy, at least for the planning formalism STRIPS$_L$.

**Theorem 6.** STRIPS$_L$-$c$-FISEX *can be decided in polynomial time.*

**Proof.** Given an STRIPS$_L$-$c$-FISEX instance $(\Xi, \mathbf{I}, \mathbf{G}, \mathbf{C})$, the number of possible operator sequences is $O(|\mathbf{O}|^c)$, which is polynomial in the instance size. For each such sequence, we can do a regression analysis starting with the goal $\mathbf{G}$ computing the weakest precondition, which is a set of literals. This can be done in polynomial time. Finally, one can easily check in polynomial time, whether there is a subset $C \subseteq \mathbf{C}$ that leads to an initial state $\mathbf{I} \cup C$ that entails the weakest precondition. This means, the problem can be solved in polynomial time for any fixed $c$. ∎

If we allow for CNF preconditions, the problem becomes harder. Even if the plan length is restricted to one, 3SAT can be obviously reduced to the 1-FISEX problem in STRIPS$_C$.

**Proposition 7.** STRIPS$_C$-1-FISEX *is* NP-*complete.*

From that it follows immediately that there cannot exist any *polynomial-time* compilation scheme from STRIPS$_C$ to STRIPS$_L$ preserving plan size linearly— provided P $\neq$ NP.[4] We will show a stronger result, namely that there cannot exist *any* compilation scheme preserving plan size linearly by employing a proof technique first used by Kautz and Selman [8].

In order to prove this result, we need the notion of *advice-taking Turing machines*. These are machines with an **advice oracle**, which is a (not necessarily recursive) function $a$ from positive integers to bit strings. On input $I$, the machine loads the bit string $a(||I||)$ and then continues as usual. Note that the oracle derives its bit string only from the length of the input and not from the contents of the input. An advice is said to be a **polynomial advice** if the oracle

---

[4] Here the difference between Bäckström's [2] ESP-reductions and compilation schemes should become obvious because the former do not allow us to derive such a conclusion.

string is polynomially bounded by the instance size. Further, if X is a complexity class defined in terms of resource-bounded machines, e.g., P or NP, then X/poly (also called **non-uniform** X) is the class of problems that can be decided on machines with the same resource bounds and polynomial advice.

Because of the advice oracle, the class P/poly appears to be much more powerful than P. However, it seems unlikely that P/poly contains all of NP. In fact, one can prove that NP $\subseteq$ P/poly implies certain relationships between uniform complexity classes that are believed to be very unlikely. In particular, Karp and Lipton [7] have shown that NP $\subseteq$ P/poly implies that the polynomial hierarchy collapses on the second level,[5] which is considered to be very unlikely.

**Theorem 8.** STRIPS$_C$ $\not\preceq^c$ STRIPS$_L$, *unless the polynomial hierarchy collapses.*

**Proof.** As a first step, we construct for each $n$ a STRIPS$_C$ domain structure $\Xi_n$ and goal specification $\mathbf{G}_n$ with size polynomial in $n$ and the following properties. Satisfiability of an arbitrary 3CNF formula $\varphi_n$ of size $n$ is equivalent to 1-step plan existence for the STRIPS$_C$-1-FISEX instance $(\Xi_n, \mathbf{G}_n, \mathbf{I}_{\varphi_n}, \mathbf{C}_n)$, where $\mathbf{I}_{\varphi_n}$ can be computed in polynomial time from $\varphi_n$.

Given a set of $n$ atoms, denoted by $\mathbf{P}_n$, we define the set of clauses $\mathbf{A}_n$ to be the set containing all clauses with three literals that can be built using these atoms. The size of $\mathbf{A}_n$ is $O(n^3)$, i.e., polynomial in $n$. Let $\mathbf{D}_n$ be new atoms $p_\gamma$ corresponding one-to-one to the clauses $\gamma$ in $\mathbf{A}_n$. Finally, let s be a new atom which is not in $\mathbf{P}_n \cup \mathbf{D}_n$.

Now we construct a STRIPS$_C$ domain structure $\Xi_n = \langle \Sigma_n, \mathbf{O}_n \rangle$ goal $\mathbf{G}_n$, and the choice set $\mathbf{C}_n$ as follows:

$$\Sigma_n = \mathbf{P}_n \cup \mathbf{D}_n \cup \{\mathsf{s}\},$$
$$\mathbf{O}_n = \{o_n\},$$
$$o_n = \langle (\bigwedge_{\gamma \in \mathbf{A}_n} (p_\gamma \vee \gamma)), \{\mathsf{s}\} \rangle,$$
$$\mathbf{G}_n = \{\mathsf{s}\},$$
$$\mathbf{C}_n = \mathbf{P}_n.$$

Let $cl(\varphi)$ be the set of clauses appearing in $\varphi$. Based on that we define $\mathbf{I}_{\varphi_n}$ as follows:

$$\mathbf{I}_{\varphi_n} = \{p_\gamma \in \mathbf{D}_n \mid \gamma \notin cl(\varphi_n)\}.$$

Now it is easy to see that $\varphi_n$ is satisfiable iff for the STRIPS$_C$-1-FISEX instance $(\Xi_n, \mathbf{G}_n, \mathbf{I}_{\varphi_n}, \mathbf{C}_n)$ there exists a set of choices $C \subseteq \mathbf{C}_n$ such that the resulting planning instance $\langle \Xi_n, \mathbf{I}_{\varphi_n} \cup C, \mathbf{G}_n \rangle$ is solved by a one-step plan.

Let us now assume that there exists a compilation scheme from STRIPS$_C$ to STRIPS$_L$ preserving plan size linearly. Using this compilation scheme, we compile

---

[5] In fact, there exist even stronger collapse results [9].

the STRIPS$_C$ domain structure $\Xi_n$ into the STRIPS$_L$ domain structure $\Xi_n' = \langle \Sigma_n', \mathbf{O}_n' \rangle$. Further, $\mathbf{G}_n'$, $\mathbf{I}_{\varphi_n}'$, and $\mathbf{C}_n'$ are defined as follows:

$$\mathbf{G}_n' = \{s\} \cup f_g(\Sigma_n, \mathbf{O}_n),$$
$$\mathbf{I}_{\varphi_n}' = \mathbf{I}_{\varphi_n} \cup f_i(\Sigma_n, \mathbf{O}_n),$$
$$\mathbf{C}_n' = \mathbf{C}_n.$$

Note that all these sets can be computed in time polynomial in $n$, once we know the values of $f_g(\Sigma_n, \mathbf{O}_n)$ and $f_i(\Sigma_n, \mathbf{O}_n)$.

From the construction, it follows that the following statements are equivalent:

1. $\varphi_n$ is satisfiable,
2. for the STRIPS$_C$-1-FISEX instance $(\Xi_n, \mathbf{G}_n, \mathbf{I}_{\varphi_n}, \mathbf{C}_n)$, there exists a set of choices $C \subseteq \mathbf{C}_n$ such that $\Pi = \langle \Sigma_n, \mathbf{O}_n, \mathbf{I}_{\varphi_n} \cup C, \mathbf{G}_n \rangle$ has a one-step plan,
3. for the STRIPS$_L$-$c$-FISEX instance $(\Xi_n', \mathbf{G}_n', \mathbf{I}_{\varphi_n}', \mathbf{C}_n')$, there exists a set of choices $C' \subseteq \mathbf{C}_n'$ such that $\Pi' = \langle \Sigma_n', \mathbf{O}_n', \mathbf{I}_{\varphi_n}' \cup C', \mathbf{G}_n' \rangle$ has a $c$-step plan,

One can now construct an advice-taking Turing machine that on input of a formula $\varphi_n$ of size $n$ loads the polynomial advice $\langle \Xi_n', f_g(\Sigma_n, \mathbf{O}_n), f_i(\Sigma_n, \mathbf{O}_n) \rangle$ and then decides STRIPS$_L$-$c$-FISEX for the instance $(\Xi_n', \mathbf{G}_n', \mathbf{I}_{\varphi_n}', \mathbf{C}_n')$, which by Theorem 6 can be done in polynomial time. Since the problem 3SAT, which is solved by this deterministic, advice-taking machine in polynomial time is NP-complete, we conclude that NP $\subseteq$ P/poly. This implies by Karp and Lipton's [7] result that the polynomial hierarchy collapses on the second level, which proves the claim. ∎

The result above implies that adding CNF preconditions to STRIPS$_L$ adds to the expressiveness. However, it is not immediately obvious whether a further generalization from CNF formulae to arbitrary boolean formulae would add another level of expressiveness. We will defer this question to the next section.

## 5 Compiling Disjunctive Preconditions into Conditional Effects

As mentioned in the Introduction, sometimes the expressive power of conditional effects and of disjunctive preconditions are claimed to be related. In this section, we will analyze this claim using the compilability framework.

As in the case of unconditional actions, it is commonly agreed that DNF formulae can be regarded as "syntactic sugar." Any operator containing a DNF precondition with $k$ disjuncts can be split into $k$ new operators containing only conjunctions of literals in the precondition. Similarly, any conditional effect with a DNF effect condition $c_1 \vee \ldots \vee c_n \Rightarrow L$ can be equivalently expressed by a set of $n$ conditional effects $c_i \Rightarrow L$. Obviously, this transformation can be viewed as a polynomial-time compilation scheme preserving plan size exactly.

**Proposition 9.** STRIPS$_{C,D}$ $\preceq_p^1$ STRIPS$_{C,L}$.

Interestingly, CNF preconditions and effect conditions do not appear to add to the expressive power once we have conditional effects—provided we accept that two formalisms have the same expressive power, if they are compilable to each other preserving plan size *linearly*. The main idea behind proving this is that operators with conditional effects can be used to evaluate the truth of clauses.

**Theorem 10.** STRIPS$_{C,\mathbf{C}}$ $\preceq_p^c$ STRIPS$_{C,\mathbf{L}}$.

**Proof.** Assume that the operators of the STRIPS$_{C,\mathbf{C}}$ domain structure $\Xi = \langle \Sigma, \mathbf{O} \rangle$ contain $n$ clauses $\gamma_1, \gamma_2, \ldots, \gamma_n$ with $\gamma_i = l_{i1} \vee \ldots \vee l_{ik_i}$ in preconditions and effect conditions. For each clause $\gamma_i$, a new atom $p_{\gamma_i}$ is introduced, and the set of these new atoms is denoted by $\Gamma$. Now, the operator *eval*, which will evaluate the truth values of all the clauses in a given state, can be defined as follows:

$$eval = \langle \top, \{l_{ij} \Rightarrow p_{\gamma_i}\} \rangle.$$

If all clauses $\gamma_i$ in $\mathbf{O}$ are replaced by the new atoms $p_{\gamma_i}$—leading to the new set $\widehat{\mathbf{O}}$—the only remaining changes that are necessary are that we enforce that the *eval* operator is always executed before an operator from $\widehat{\mathbf{O}}$ is executed and that all operators from $\widehat{\mathbf{O}}$ set all the atoms from $\Gamma$ to false.

In order to enforce sequences of operators alternating between operators from $\widehat{\mathbf{O}}$ and the *eval*-operator, one can introduce a new atom $\mathsf{e}$ that is added to the initial state. In addition, we modify the *eval* operator and all operators $\widehat{o} \in \widehat{\mathbf{O}}$ as follows:

$$eval' = \langle \mathsf{e}, \, post(eval) \cup \{\top \Rightarrow \{\neg\mathsf{e}\}\} \rangle$$
$$o' = \langle \neg\mathsf{e} \wedge pre(\widehat{o}), \, post(\widehat{o}) \cup \{\top \Rightarrow \{\mathsf{e}\}\} \cup \{\top \Rightarrow \neg\Gamma\} \rangle.$$

We can now specify a compilation scheme from STRIPS$_{C,\mathbf{C}}$ to STRIPS$_{C,\mathbf{L}}$ as follows:

$$f_\xi \colon \langle \Sigma, \mathbf{O} \rangle \mapsto \langle \Sigma \cup \Gamma \cup \{\mathsf{e}\}, \{o' | \widehat{o} \in \widehat{\mathbf{O}}\} \cup \{eval'\} \rangle,$$
$$f_i \colon \langle \Sigma, \mathbf{O} \rangle \mapsto \{\mathsf{e}\},$$
$$f_g \colon \langle \Sigma, \mathbf{O} \rangle \mapsto \emptyset.$$

This is obviously a polynomial-time compilation scheme that leads to STRIPS$_{C,\mathbf{L}}$ plans that are twice as long as the original STRIPS$_{C,\mathbf{C}}$ plans. ■

This result appears to be relevant for practical planning algorithms because it suggests how to extend planning algorithms for conditional operators to algorithms for dealing with CNF preconditions and effect conditions. However, one may wonder whether we can improve on this result, coming up with a compilation scheme preserving plan size exactly. Interestingly, there does not appear to be an obvious way to do so. Further, it is completely unclear how to prove that such a compilation scheme is impossible.

Using the above result and Theorem 4, we can now easily give an answer to the question posed in the end of the previous section, namely, whether general boolean preconditions are more expressive than CNF preconditions.

**Theorem 11.** $\text{STRIPS}_\mathbf{B} \not\preceq^c \text{STRIPS}_\mathbf{C}$.

**Proof.** Assume for contradiction that there is a compilation scheme from $\text{STRIPS}_\mathbf{B}$ to $\text{STRIPS}_\mathbf{C}$ preserving plan size linearly. Since by Proposition 2 we have $\text{STRIPS}_\mathbf{C} \preceq^c \text{STRIPS}_{\mathcal{C},\mathbf{C}}$ and by Theorem 10 we have $\text{STRIPS}_{\mathcal{C},\mathbf{C}} \preceq^c \text{STRIPS}_{\mathcal{C},\mathbf{L}}$, we can conclude $\text{STRIPS}_\mathbf{B} \preceq^c \text{STRIPS}_{\mathcal{C},\mathbf{L}}$ using Proposition 1 twice. This, however, contradicts Theorem 4. ∎

This leaves us with the question whether general boolean preconditions and effect conditions are more expressive than CNF preconditions and effect conditions. However, assuming that $\text{STRIPS}_{\mathcal{C},\mathbf{B}} \preceq^c \text{STRIPS}_{\mathcal{C},\mathbf{C}}$ leads immediately to the conclusion that $\text{STRIPS}_{\mathcal{C},\mathbf{B}} \preceq^c \text{STRIPS}_{\mathcal{C},\mathbf{L}}$ (using Theorem 10 and Proposition 1), which is impossible because of Theorem 4.

**Proposition 12.** $\text{STRIPS}_{\mathcal{C},\mathbf{B}} \not\preceq^c \text{STRIPS}_{\mathcal{C},\mathbf{C}}$.

# 6  Summary and Discussion

Using the *compilability framework* [11], we analyzed the expressive power of disjunctive preconditions and conditional effects. In general, our results provide a complete classification of the relative expressiveness of STRIPS-like languages with restricted formulae and conditional effects – provided that literals are always allowed and states are always complete. Table 1 gives an overview of the results.[6] The "⊑" entries mark syntactic specialization relationships (see Figure 1). For

**Table 1.** Results

| $\preceq^x$ | $\text{STRIPS}_\mathbf{L}$ | $\text{STRIPS}_\mathbf{D}$ | $\text{STRIPS}_\mathbf{C}$ | $\text{STRIPS}_\mathbf{B}$ | $\text{STRIPS}_{\mathcal{C},\mathbf{L}}$ | $\text{STRIPS}_{\mathcal{C},\mathbf{D}}$ | $\text{STRIPS}_{\mathcal{C},\mathbf{C}}$ |
|---|---|---|---|---|---|---|---|
| $\text{STRIPS}_\mathbf{D}$ | $\preceq_p^1$ **(5)** | = | $\preceq_p^1$ (5) | ⊑ | $\preceq_p^1$ (5) | ⊑ | $\preceq_p^1$ (5) |
| $\text{STRIPS}_\mathbf{C}$ | $\not\preceq^c$ **(8)** | $\not\preceq^c$ (8,5) | = | ⊑ | $\preceq_p^c$ (10) | $\preceq_p^c$ (10) | ⊑ |
| $\text{STRIPS}_\mathbf{B}$ | $\not\preceq^c$ (11) | $\not\preceq^c$ (11,5) | $\not\preceq^c$ **(11)** | = | $\not\preceq^c$ **(4)** | $\not\preceq^c$ (4,9) | $\not\preceq^c$ (4,10) |
| $\text{STRIPS}_{\mathcal{C},\mathbf{L}}$ | $\not\preceq^c$ (3) | $\not\preceq^c$ (3) | $\not\preceq^c$ (3) | $\not\preceq^c$ **(3)** | = | ⊑ | ⊑ |
| $\text{STRIPS}_{\mathcal{C},\mathbf{D}}$ | $\not\preceq^c$ (3) | $\not\preceq^c$ (3) | $\not\preceq^c$ (3) | $\not\preceq^c$ (3) | $\preceq_p^1$ **(9)** | = | $\preceq_p^1$ (9) |
| $\text{STRIPS}_{\mathcal{C},\mathbf{C}}$ | $\not\preceq^c$ (3) | $\not\preceq^c$ (3) | $\not\preceq^c$ (3) | $\not\preceq^c$ (3) | $\preceq_p^c$ **(10)** | $\preceq_p^c$ (10) | = |
| $\text{STRIPS}_{\mathcal{C},\mathbf{B}}$ | $\not\preceq^c$ (3) | $\not\preceq^c$ (3) | $\not\preceq^c$ (3) | $\not\preceq^c$ (3) | $\not\preceq^c$ (4) | $\not\preceq^c$ (4,9) | $\not\preceq^c$ **(12)** |

all other entries we give the strongest compilation result or impossibility result. The number indicates the theorem from which the result has been derived. If the number is in bold face, it is just the statement of the theorem. Otherwise, the result can be derived from the theorem and the application of Propositions 1 and 2.

---

[6] Note that we have left out the $\text{STRIPS}_\mathbf{L}$ row and the $\text{STRIPS}_{\mathcal{C},\mathbf{B}}$ column because $\text{STRIPS}_\mathbf{L}$ is a specialization of all formalisms and all formalisms are specializations of $\text{STRIPS}_{\mathcal{C},\mathbf{B}}$.

Two particular interesting results are

1. CNF preconditions add to the power of basic STRIPS, confirming an earlier conjecture by Bäckström [2];
2. CNF preconditions and CNF effect conditions do not add anything to the expressive power if we already have conditional effects, confirming a weak version of a conjecture by Anderson *et al.* [1].

In particular the latter result may have practical value for the design of planning algorithms. It suggests that when normalizing preconditions and effect conditions it is not necessary to convert them to disjunctive normal form, but conjunctive normal form is another option that can be easily dealt with. This option may sometimes help to avoid excessive space consumption, provided the formulae are already almost CNF.

**Acknowledgments.** The research reported in this paper was started and partly carried out while the author enjoyed being a visitor at the AI department of the University of New South Wales. Many thanks go to Norman Foo, Maurice Pagnucco, and Abhaya Nayak and the rest of the AI department for the discussions and cappuccinos.

# References

1. C. R. Anderson, D. E. Smith, and D. S. Weld. Conditional effects in Graphplan. In R. Simmons, M. Veloso, and S. Smith, editors, *Proceedings of the 4th International Conference on Artificial Intelligence Planning Systems (AIPS-98)*, pages 44–53. AAAI Press, Menlo Park, 1998.
2. C. Bäckström. Expressive equivalence of planning formalisms. *Artificial Intelligence*, 76(1–2):17–34, 1995.
3. T. Bylander. The computational complexity of propositional STRIPS planning. *Artificial Intelligence*, 69(1–2):165–204, 1994.
4. M. Cadoli and F. M. Donini. A survey on knowledge compilation. *AI Communications*, 10(3,4):137–150, 1997.
5. B. C. Gazen and C. Knoblock. Combining the expressiveness of UCPOP with the efficiency of Graphplan. In Steel and Alami [13], pages 221–233.
6. S. Kambhampati, E. Parker, and E. Lambrecht. Understanding and extending Graphplan. In Steel and Alami [13], pages 260–272.
7. R. M. Karp and R. J. Lipton. Turing machines than take advice. *L' Ensignement Mathématique*, 28:191–210, 1982.
8. H. A. Kautz and B. Selman. Forming concepts for fast inference. In *Proceedings of the 10th National Conference of the American Association for Artificial Intelligence (AAAI-92)*, pages 786–793, San Jose, CA, July 1992. MIT Press.
9. J. Köbler and O. Watanabe. New collapse consequences of np having small circuits. *SIAM Journal on Computing*, 28(1):311–324, 1998.
10. J. Koehler, B. Nebel, J. Hoffmann, and Y. Dimopoulos. Extending planning graphs to an ADL subset. In Steel and Alami [13], pages 273–285.
11. B. Nebel. On the compilability and expressive power of propositional planning formalisms. Technical Report 101, Albert-Ludwigs-Universität, Institut für Informatik, Freiburg, Germany, June 1998.

12. B. Nebel. Compilation schemes: A theoretical tool for assessing the expressive power of planning formalisms. In W. Burgard, A. B. Cremers, and T. Christaller, editors, *KI-99: Advances in Artificial Intelligence*, Bonn, Germany, 1999. Springer-Verlag. To appear.
13. S. Steel and R. Alami, editors. *Recent Advances in AI Planning. 4th European Conference on Planning (ECP'97)*, volume 1348 of *Lecture Notes in Artificial Intelligence*, Toulouse, France, Sept. 1997. Springer-Verlag.

# Some Results on the Complexity of Planning with Incomplete Information

Patrik Haslum and Peter Jonsson

Department of Computer and Information Science
Linköping University, S-581 83 Linköping, Sweden
{pahas,petej}@ida.liu.se

**Abstract** Planning with incomplete information may mean a number of different things; that certain facts of the initial state are not known, that operators can have random or nondeterministic effects, or that the plans created contain sensing operations and are branching. Study of the complexity of incomplete information planning has so far been concentrated on probabilistic domains, where a number of results have been found. We examine the complexity of planning in nondeterministic propositional domains. This differs from domains involving randomness, which has been well studied, in that for a nondeterministic choice, not even a probability distribution over the possible outcomes is known. The main result of this paper is that the non-branching plan existence problem in unobservable domains with an expressive operator formalism is EXPSPACE-complete. We also discuss several restrictions, which bring the complexity of the problem down to PSPACE-complete, and extensions to the fully and partially observable cases.

## 1 Introduction

Though the planning problem, including the computational complexity of planning, has been long studied in AI, the problem of planning with incomplete information has received relatively little attention, and the computational complexity of this problem even less. Most work has been in the area of Markov decision processes, which model any uncertainty as a probability distribution over the set of alternatives. In contrast, we consider the problem of planning with *nondeterministic* operators; we assume no information about the relative probabilities of different operator outcomes.

### 1.1 Approaches to Planning with Incomplete Information

There are basically two ways in which incomplete information has been dealt with in planning research.

A *conformant* planner [15] tries to construct plans that will work in every foreseeable case, by choosing operators such that the goal is achieved no matter what the value of an initially unknown state component or the outcome of a nondeterministic operator. When the domain contains sensing operators that can

S. Biundo and M. Fox (Eds.): ECP-99, LNAI 1809, pp. 308–318, 2000.

provide information on the state of the world during plan execution, *contingent* planners [13,12,3,14] can create branching plans, plans that execute differently depending on the information sensed.

From the theory of Markov decision processes (MDPs), we borrow the terminology on observability and call a domain *fully observable* if there are sensing operators without preconditions for every proposition in the domain, *unobservable* if there are no sensing operators and *partially observable* when none of the other terms apply. Obviously, in an unobservable domain there is no point in creating a branching plan as there is no information to branch on, and thus the contingent plan existence problem coincides with the conformant version of the problem.

In *probabilistic* planning [9], all uncertainties are assigned a probability distribution. The planner seeks to create a plan that will maximize the probability of achieving the goal. Probabilistic planners may also take advantage of sensing operators to create branching plans [4].

### 1.2  Nondeterministic Operators in Planning

The problem we consider in this paper is *conformant plan existence* in unobservable domains where we lack even a probability distribution over the possible outcomes of a nondeterministic operator.

Such incompleteness naturally arises in domains where the exact process involved in an action is too complex to model, and where statistical data to create a probabilistic model is lacking. An example of this kind is domains involving agents that act independently of the planning agent, such as a game situation. Another example is synthesis of "fault-tolerant" plans.

### 1.3  Previous Results on the Complexity of Planning

There is a growing body of work on the complexity of the planning problem. We summarize the results relevant for comparison with the main result of this paper.

Deciding plan existence in a propositional domain with deterministic operators and complete information about the initial state was shown to be PSPACE-complete by Bylander [2]. Erol *et al.* [5] showed that in a general first order domain the same problem is undecidable. Restricted to a first order domain with a finite set of ground terms, the problem is EXPSPACE-complete. This representation has essentially the same expressive power as the propositional, but allows for exponentially more compact encoding. Since they consider the complete information case, all these results naturally concern only non-branching plans.

For planning with operators having context dependent and probabilistic effects, Littman [10] showed that deciding the existence of a branching plan with a probability of success greater or equal to some threshold $\epsilon$ in a fully observable propositional domain is EXPTIME-complete.

Many other results on the complexity of probabilistic planning come from the theory of Markov decision processes. The problem of finding an optimal history dependent $m$-horizon policy for a partially observable, succinctly represented MDP corresponds to the problem of finding an optimal (*i.e.* maximizing the probability of success) branching plan of depth at most $m$ in a partially observable propositional domain with context dependent operators and probabilistic effects. This problem is known to be EXPSPACE-complete [6]. Interestingly, both the unobservable and the fully observable domain versions of the problem are easier, being EXPTIME- and NEXPTIME-complete, respectively.

### 1.4   Contributions

In this paper, we consider the complexity of the problem of deciding the existence of non-branching plans in an unobservable propositional domain, and show this problem to be EXPSPACE-complete. We also discuss a number of restrictions on problem instances or solutions that bring the complexity down to PSPACE. Finally, we show briefly how a limited form of branching in plans can be encoded into a non-branching plan, taking advantage of the expressive operator formalism.

## 2   Preliminaries

The following basic definitions are used throughout the paper.

**Definition 1.** *By a* planning scenario *we mean a tuple*

$$\mathcal{P} = (\mathcal{D}, \mathcal{O}, I, G)$$

*where the domain $\mathcal{D}$ is a set of propositional symbols, $\mathcal{O}$ is a set of operators and $I$ and $G$ are the initial state and goal state descriptions, respectively. The definition of operators varies, depending on the case under consideration.*

*A state set description, or* state formula*, is any boolean combination of literals over $\mathcal{D}$. A state is represented by the subset of $\mathcal{D}$ consisting of all propositions true in the state. A state formula $\sigma$ denotes a set of possible states, in the obvious way;*

$$mod(\sigma) = \{\xi \subseteq \mathcal{D} | \xi \models \sigma\}$$

*where $\xi \models \sigma$ means that $\sigma$ holds in state $\xi$, according to the standard semantics of propositional logic.*

*For a set of states $\Xi$ and an operator $O$ we define operator semantics in terms of the* successor state set function, $S(O, \Xi)$, *denoting the set of states that may result from applying $O$ in any state $\xi \in \Xi$. A sequence of operators $O_1; \ldots; O_n$ constitutes a* plan *for scenario $\mathcal{P}$ iff $S(O_n, S(O_{n-1}, \ldots S(O_1, mod(I)) \ldots)) \models G$.*

The definition of the successor state set function depends on what operator formalism we consider. In this paper, we introduce a kind of operators, called *actions*, that may have nondeterministic and context dependent effects.

**Definition 2.** *A* reassignment *is an expression* $x{:=}v$, *where* $x{\in}\mathcal{D}$ *and* $v{\in}\{\mathcal{T},$ $\mathcal{F}\}$. *A* condition *is a boolean combination of literals over* $\mathcal{D}$. *An* action *is*

*(i) A reassignment,* $x{:=}v$.
*(ii) A sequential composition of actions* $\alpha_1;\alpha_2$, *where* $\alpha_1$ *and* $\alpha_2$ *are actions.*
*(iii) A conditional composition of actions* $\bigwedge\gamma_i \rightarrow \alpha_i$, *where each* $\gamma_i$ *is a condition and each* $\alpha_i$ *an action, and the conditions are mutually exclusive.*
*(iv) A nondeterministic composition of actions* $\alpha_1 \vee \alpha_2$, *where* $\alpha_1$ *and* $\alpha_2$ *are actions.*

*Given a set of possible states* $\Xi$ *and an action* $A$, *the successor state function is defined inductively as*

$$
\begin{aligned}
S(x{:=}\mathcal{T}, \Xi) &= \{\xi'|\xi \in \Xi \wedge \xi' = \xi \cup \{x\}\} \\
S(x{:=}\mathcal{F}, \Xi) &= \{\xi'|\xi \in \Xi \wedge \xi' = \xi - \{x\}\} \\
S(\alpha_1;\alpha_2, \Xi) &= S(\alpha_2, S(\alpha_1, \Xi)) \\
S(\bigwedge\gamma_i \rightarrow \alpha_i, \Xi) &= \bigcup S(\alpha_i, \{\xi \in \Xi|\xi \models \gamma_i\}) \cup \{\xi \in \Xi|\xi \not\models \bigvee\gamma_i\} \\
S(\alpha_1 \vee \alpha_2, \Xi) &= S(\alpha_1, \Xi) \cup S(\alpha_2, \Xi)
\end{aligned}
$$

We use sequential composition in place of conjunction to avoid the ambiguity that may otherwise arise when two subactions affect the same propositions. We may encode the unordered conjunction of two actions $\alpha_1$ and $\alpha_2$ as $(\alpha_1;\alpha_2) \vee (\alpha_2;\alpha_1)$[1], or we may include unordered conjunction into the definition of actions; it does not affect the results.

# 3 Nondeterminism, Context Dependency and Unobservable Domain

In this section, we present the main result of the paper, that the non-branching plan existence problem for a planning scenario $\mathcal{P} = (\mathcal{D}, \mathcal{O}, I, G)$ with unobservable propositional domain and containing actions is EXPSPACE-complete. In the process, we also show a bound on the length of plans (Lemma 3).

We show hardness through a reduction from the EXPSPACE-complete problem of deciding universality for regular expressions with exponentiation. To clarify the reduction, we first convert the regular expression into a nondeterministic finite automaton augmented with counters of bounded capacity. For definitions and results on regular expressions and finite automata not presented here, see for instance [7].

**Definition 3.** *A* regular expression with exponentiation *is formed as a normal regular expression from symbols of the alphabet, but using in addition to the operations of concatenation, union and closure also* exponentiation, *which is written* $r^n$, *where* $r$ *is a regular expression with exponentiation and* $n$ *a natural number written in binary (the symbols 1 and 0 are assumed not to belong to the alphabet of the expression). The expression* $r^n$ *denotes all strings consisting of* $n$ *consecutive substrings denoted by* $r$, *i.e.* $L(r^n) = \{w_1w_2...w_n|w_i \in L(r)\}$.

---

[1] Note, though, that this expression grows exponentially with the number of actions composed.

For example, the expression $(a^{10} + b^{10})^{10}$ over the alphabet $\{a, b\}$ denotes the set $\{aaaa, aabb, bbaa, bbbb\}$.

When counting the number of operators in a regular expression with exponentiation, $r$, we include alphabet symbols but count each exponentiaton as one operator, regardless of the number it is raised to. By the length of $r$, $|r|$, we mean the number of operators in $r$ plus the number of digits in the exponents in $r$ [2]. For example, the expression above consists of 6 operators and its length is 12.

**Definition 4.** *A* nondeterministic finite automaton with counters, *or NFAC, is a nondeterministic finite automaton augmented with a set of* counters $C$. *Each counter $c$ is associated with a set of states* $states_c$, *a natural number* $limit_c$, *and the two states* $loop_c$ *and* $continue_c$. *Initially, all counters are set to 0. When the automaton enters any state* $q \in states_c$, *it may increment $c$ by one and immediately change state, to* $loop_c$ *if* $c < limit_c$ *and to* $continue_c$ *if* $c \geqslant limit_c$. *The automaton terminates as usual when there is no more input, and accepts if there is some sequence of choices that makes it reach a final state.*

**Lemma 1.** *The problem of determining if a regular expression with exponentiation is universal, i.e. that it denotes all strings over its alphabet, is EXPSPACE-complete.*

*Proof.* See [7].

**Lemma 2.** *For any regular expression with exponentiation $r$, there exists a NFAC accepting the language $L(r)$ that has a number of states that is linear in the number of operators in $r$, and such that $\Sigma_{c \in C} limit_c \leqslant 2^{|r|}$.*

*Proof.* By structural induction on $r$. A similar proposition for regular expressions without exponentiation and normal NFA can be found in e.g. [7] and we refer to that for the cases concerning the normal operators.

For $r = r_0^n$, there exists a NFAC $M_0 = \langle Q, \Sigma, \delta, q_0, F, C \rangle$ accepting $L(r_0)$, by the induction assumption. Construct

$$M = \langle Q \cup \{f\}, \Sigma, \delta, q_0, \{f\}, C \cup \{c\} \rangle$$

by setting $states_c = F$, $limit_c = n$, $loop_c = q_0$ and $continue_c = f$.

Any string in $L(r)$ will consist of $n$ substrings such that each is in $L(r_0)$. $M$ starts in state $q_0$, and since $M_0$ accepts $L(r_0)$ ends in some state $f_0 \in F$ only after having read one copy of $r_0$. When, and only when, $M$ enters some $f_0 \in F$ can $c$ be incremented, and therefore $M$ will reach its final state $f$ only after having read $n$ strings in $L(r_0)$. Any string $s$ accepted by $M$ will consist of $n$ consecutive substrings accepted by $M_0$ and therefore in $L(r_0)$, so $s \in L(r)$.

---

[2] This does not exactly correspond the number of symbols in $r$ because concatenation and exponentiation are written without actual symbols for the operators.

Since the construction adds a constant number of states for each operator in $r$, the total number of states is bounded by some linear function in the size of $r$. Since the total length of all exponents in $r$ written in binary can not be greater than the length of $r$, they can not combined represent a greater number than $2^{|r|}$.

**Theorem 1.** *Deciding existence of a non-branching plan for a scenario $\mathcal{P}$ with propositional domain and containing actions, is EXPSPACE-hard.*

*Proof.* By reduction from the universality problem for regular expressions with exponentiation, via the same problem for NFACs. For a regular expression with exponentiation, $r$, we know there exists an NFAC, $M$, accepting the same language and bounded in $|r|$. We construct a scenario $\mathcal{P}$, polynomially bounded in $|r|$ and such that there exists a plan for $\mathcal{P}$ iff there exists a string in $\Sigma^*$ not accepted by $M$. This shows plan existence to be co-EXPSPACE-hard, which since EXPSPACE = co-EXPSPACE, proves the theorem.

We model the state of $M$ and its counters as the state of the scenario, and the transition function $\delta$ as the behavior of actions, with one action for each symbol in $\Sigma$. Assuming states in $Q$ are numbered $0, \ldots, |Q|$, propositions $s_0, \ldots, s_{\log |Q|}$ denote the current state in binary. Each counter $c$ is likewise represented by propositions $c_0, \ldots, c_{\log limit_c}$. We write $s = m$ for $\bigwedge_{i=0,\ldots,\log |Q|} b_i$, where $b_i$ is $s_i$ if the $i$th bit of $m$ is 1 and $\neg s_i$ if the $i$th bit of $m$ is 0, and $s:=m$ analogously. For counters $c$, we write $c = m$ and $c < m$ in the same way, and $c:=c+1$ for the conditional reassignment

$$(\neg c_0 \to c_0:=\mathcal{T}) \wedge (\neg c_1 \wedge c_0 \to c_1:=\mathcal{T} \wedge c_0:=\mathcal{F}) \wedge \ldots \wedge$$
$$(\neg c_{\log n} \wedge c_{(\log n)-1} \wedge \ldots \wedge c_0 \to c_{\log n}:=\mathcal{T}; c_{(\log n)-1}:=\mathcal{F}; \ldots ; c_0:=\mathcal{F})$$

which increments the counter by one, and which is linear in the length of the binary representation of $limit_c$. For each character $\tau \in \Sigma$, define the action

$$A_\tau = \bigwedge (s = i) \to \left( \bigvee_{q_j \in \delta(\tau, q_i)} \begin{cases} (s:=j) \vee INC(c) & \text{if } q_j \in state_c \\ s:=j & \text{if } q_j \notin state_c \end{cases} \right)$$

where $INC(c)$ is the action of incrementing $c$ and chaning state accordingly, encoded as

$$(c < limit_c - 1 \to c:=c + 1; s:=loop_c) \wedge$$
$$(c = limit_c - 1 \to c:=c + 1; s:=continue_c)$$

The initial state $I$ is $s = 0 \wedge (\bigwedge_{c \in C} c = 0)$, which is the initial state of $M$, and the goal $G$ is $\bigvee_{q_i \notin F} s = i$, the disjunction of all non-accepting states of $M$.

The construction yields a one-to-one correspondence between plans for $\mathcal{P}$ and sequences of characters that are guaranteed to take $M$ from its initial state to halt in a non-accepting state. Since $M$ accepts exactly the strings that are in $L(r)$, $r$ is universal iff no such plan exists.

The initial state and goal descriptions are bounded by $\log(|Q|) + \Sigma_{c \in C} \log (limit_c)$ and $\log(|Q|) \cdot |F|$, respectively. The number of actions is bounded by $|\Sigma|$, which is at most linear in the size of $r$, and each action will have no more than $|Q|$ branches. The effect formulas of each branch are bounded by

$$\log(limit_{c_1}) + \ldots + \log(limit_{c_k})$$

which is also linear in the size of $r$. Therefore, the size of the entire planning scenario, is bounded by $k \cdot r^3$, for some constant $k$.

To show that plan existence for a propositional scenario with actions is in EX-PSPACE, we need a bound on plan length.

**Lemma 3.** *Let $\mathcal{P} = (\mathcal{D}, \mathcal{O}, I, G)$ be a planning scenario, with $\mathcal{D}$ propositional and $|\mathcal{D}| = n$. If there exists a plan for $\mathcal{P}$, there exists a plan consisting of no more than $2^{2^n}$ steps.*

*Proof.* Consider the space of *state sets*, of which there are $2^{2^n}$. Since the successor state set function maps a set of states to another, any action is a deterministic transition in this space. Therefore, if there exists a plan for $\mathcal{P}$ there exists also a plan without loops in this space.

Plan existence for a scenario $\mathcal{P}$ with propositional domain and actions can therefore be decided by nondeterministically selecting operators and computing the resulting state set, until the goal is achieved or the length of the sequence exceeds $2^{2^n}$. The state set can be represented in space $n \cdot 2^n$ and the counter requires $2^n$. Combining this with Theorem 1, we have the following.

**Theorem 2.** *The non-branching plan existence problem for a scenario $\mathcal{P}$ with propositional domain and actions is EXPSPACE-complete.*

## 4   Bounded-Length Plans

In response to the hardness results on the complete information planning problem, some researchers have investigated the problem of finding bounded-length plans, mainly plans of polynomial length. It turns out that in a propositional domain, the problem of deciding if there for a (complete information) planning scenario $\mathcal{P}$ exists a plan no longer than $p(|\mathcal{P}|)$, where $p(x)$ is a polynomial, is NP-complete. Thus, some of the hardness of the general planning problem derives from the fact that shortest plans may be very long.

**Proposition 1.** *Let $\mathcal{P}$ be a planning scenario with propositional domain and actions, and let $p(x)$ be a polynomial. The problem of deciding if there exists a non-branching plan for $\mathcal{P}$ of length at most $p(|\mathcal{P}|)$ is solvable in PSPACE.*

The proof relies on the fact that we can write down a (nondeterministically chosen) candidate plan, then explore every possible execution of this plan depth-first. Since the depth is bounded by the length of the plan, this exploration requires no more than polynomial space (though in the worst case on the order of $p(|\mathcal{P}|)^2$).

## 5   Limited Information Incompleteness and Dependency

Another way to reduce the complexity of problem is to limit the information incompleteness and/or the context dependency of operators. In this way, we find two severely restricted classes for which the non-branching plan existence problem is PSPACE-complete[3].

**Proposition 2.** *Let $\mathcal{P}$ be a planning scenario with propositional domain, only deterministic and context independent operators[4] and an arbitrarily incomplete initial state. Then, the non-branching plan existence problem for $\mathcal{P}$ is PSPACE-complete.*

For the second class, we need some definitions.

**Definition 5.** *Let $\mathcal{D} = \{p_1, \ldots, p_n\}$ be a propositional domain and let $\Xi$ be a state set over $\mathcal{D}$. We construct a $\mathcal{T}$-$\mathcal{F}$-$\mathcal{U}$ description of $\Xi$, $\delta(\Xi)$, by mapping each $p_i$ to $\mathcal{T}$ iff $\xi(p_i) = \mathcal{T}$ for all $\xi \in \Xi$, to $\mathcal{F}$ iff $\xi(p_i) = \mathcal{F}$ for all $\xi \in \Xi$ and to $\mathcal{U}$ ("unknown") otherwise. For a given $\mathcal{T}$-$\mathcal{F}$-$\mathcal{U}$ description, $\delta$, we define the realisation of $\delta$, $\rho(\delta)$, as the state set*

$$\{\xi \mid \begin{cases} \xi(p_i) = \mathcal{T} & \text{if } \delta(p_i) = \mathcal{T} \\ \xi(p_i) = \mathcal{F} & \text{if } \delta(p_i) = \mathcal{F} \end{cases} \text{ for each } p_i \in \mathcal{D}\}$$

*We say a state set $\Xi$ is a $\mathcal{T}$-$\mathcal{F}$-$\mathcal{U}$ state if $\rho(\delta(\Xi)) = \Xi$, and that an operator $O$ preserves $\mathcal{T}$-$\mathcal{F}$-$\mathcal{U}$ states iff $S(O, \Xi)$ is a $\mathcal{T}$-$\mathcal{F}$-$\mathcal{U}$ state whenever $\Xi$ is a $\mathcal{T}$-$\mathcal{F}$-$\mathcal{U}$ state.*

What the $\mathcal{T}$-$\mathcal{F}$-$\mathcal{U}$ property means is that each proposition in the domain is either known or independent of all others. It is similar to the notion of *0-approximation* in [1].

**Proposition 3.** *Let $\mathcal{P}$ be a planning scenario with propositional domain, an initial state that is $\mathcal{T}$-$\mathcal{F}$-$\mathcal{U}$ and operators that are all $\mathcal{T}$-$\mathcal{F}$-$\mathcal{U}$ preserving. Then, the non-branching plan existence problem for $\mathcal{P}$ is PSPACE-complete.*

The proofs of propositions 2 and 3 both depend on two facts; (*i*) that we can represent the combined effect of a sequence of operators polynomially, without needing to store the actual sequence, and (*ii*) that the incompleteness is constant or decreasing, which limits the length of the shortest plan to single exponential.

   Definition 5 is rather technical, and a more intuitive characterization of $\mathcal{T}$-$\mathcal{F}$-$\mathcal{U}$ preserving operators is difficult to find. Deterministic context independent operators preserve $\mathcal{T}$-$\mathcal{F}$-$\mathcal{U}$ states, as do certain kinds of deterministic context dependent operators[5]. Nondeterministic operators that are purely information-destructive, *i.e.* that assign to the nondeterministically affected propositions any possible combination of values, also preserve $\mathcal{T}$-$\mathcal{F}$-$\mathcal{U}$ states.

---

[3] But still more general than the class of complete information scenarios, for which the plan existence problem is also PSPACE-complete.

[4] That is, essentially STRIPS operators, though with arbitrary preconditions.

[5] Note, however, that this does not imply that proposition 3 subsumes proposition 2, since the later allows arbitrary initial state.

# 6    Partially Observable Domains and Branching Plans

In the presence of incomplete information, or even nondeterminism, it may seem as a good idea to equip the executing agent with some form of sensors and to conditionalize plan execution on the results of sensing; that is, to search for branching plans.

In the case of Markov decision processes, the policy existence problems for both fully observable and unobservable domains are known to be somewhat easier than the corresponding problem for partially observable domains. By analogy, it is not unreasonable to expect the branching plan existence problem for partially observable domains with nondeterministic context dependent operators to be at least as hard as the problem corresponding non-branching plan existence problem, if not harder.

However, we can encode a limited form of "branchingness" in a non-branching plan using operator context dependency. Specifically, for a scenario $\mathcal{P}$ with propositional domain of size $n$ and actions, by adding 2 domain propositions and $2n + 3$ actions (depending only on the domain)[6], allows the construction of plans of the form

$$\textbf{IF } \gamma_1 \textbf{ THEN } A_1; \textbf{ IF } \gamma_2 \textbf{ THEN } A_2; \dots ; \textbf{ IF } \gamma_m \textbf{ THEN } A_m$$

where each $\gamma_i$ is an arbitrary formula over the domain propositions, The idea is to write each $\gamma_i$ on disjunctive normal form and evaluate it "linearly", storing the intermediate results in two control propositions.

A partially observable domain can be encoded in an unobservable domain, by duplicating propositions, letting one set represent the actual world state and the other the agents knowledge of the state. Normal operators would then depend on and effect only the "world" propositions, while sensing operators change the "knowledge" propositions depending on the corresponding world propositions.

Because both transformations are polynomial, this means that at least a limited form of the contingent plan existence problem in partially observable domains is no harder to decide than conformant plan existence in an unobservable domain.

# 7    Conclusions and Future Work

The high complexity of the planning problem in the presence of nondeterminism and context dependency should come as no surprise. This result is however interesting to compare with the corresponding problem for probabilistic domains, known to be NEXPTIME-complete and thus presumably easier[7]. That nondeterministic choice contains less information than probabilistic choice is a known

---

[6] A constant change to each action originally in $\mathcal{P}$ is also necessary.

[7] There is a class called $\bigcup_{k>0} TA[2^{n^k}, n]$ that lies inbetween NEXPTIME and EX-PSPACE, and is believed to be strictly inbetween. This fascinating class captures the decision problem for the theory of the reals with addition. For details, see [8].

fact, but these results together show that for the planning problem, the difference is significant.

This paper also leaves a large number of open questions. Is the polynomially bounded plan existence problem discussed in section 4 hard for PSPACE? To show this by reduction meets with some difficulties, since known PSPACE-complete problems involve either exponentially long sequences (for instance, the reachable deadlock problem [11]) or choice (such as the satisfiability problem for quantified Boolean formulas and various two-player games). If this problem is not hard for PSPACE, is there a more efficient algorithm that solves it?

What is the complexity of the (fully) branching plan existence problem on fully or partially observable domains? The encoding shown in section 6 depends only on the problem domain, and is linear in size. If it is also allowed to depend on operators, and increase in size proportional to a higher degree polynomial, are there then other classes of branching plans that can be encoded?

**Acknowledgments.** This research has been sponsored by the *Swedish Research Council for the Engineering Sciences* (TFR) under grant Dnr. 97-301, the *Wallenberg Foundation* and the *ECSEL/ENSYM* graduate studies program.

# References

1. C. Baral, V. Kreinovich, and R. Trejo. Computational complexity of planning and approximate planning in presence of incompleteness. In *Proc. 16th International Joint Conference on Artificial Intelligence (IJCAI'99)*, 1999.
2. T. Bylander. Complexity results for planning. In *Proc. 12th International Joint Conference on Artificial Intelligence*, 1991.
3. G. Collins and L. Pryor. Planning under uncertainty: Some key issues. In *Proc. 14th International Joint Conference on Artificial Intelligence*, 1995.
4. D. Draper, S. Hanks, and D. Weld. Probabilistic planning with information gathering and contingent execution. In *Artificial Intelligence Planning Systems: Proc. 2nd International Conference*, 1994.
5. K. Erol, D.S. Nau, and V.S. Subrahmanian. Complexity, decidability and undecidability results for domain-independent planning: A detailed analysis. Technical Report CS-TR-2797, Computer Science Department, University of Maryland, 1991.
6. J. Goldsmith, C. Lusena, and M. Mundhenk. The complexity of deterministically observable finite-horizon markov decision processes. Technical Report 268-96, Computer Science Department, University of Kentucky, 1996.
7. J.E. Hopcroft and J.D. Ullman. *Introduction to Automata theory, Languages and Computation.* Addison-Wesley, Reading, MA, 1979.
8. D.S. Johnson. A catalog of complexity classes. In J. van Leeuwen, editor, *Handbook of Theoretical Computer Science*, volume A. Elsevier, Amsterdam, 1990.
9. N. Kushmerick, S. Hanks, and D. Weld. An algorithm for probabilistic least-commitment planning. In *Proc. 12th National Conference on Artificial Intelligence*, 1994.
10. M.L. Littman. Probabilistic propositional planning: Representations and complexity. In *Proc. 14th National Conference on Artificial Intelligence*, 1997.
11. C.H. Papadimitrou. *Computational Complexity.* Addison-Wesley, Reading, MA, 1994.

12. M. Peot and D. Smith. Conditional nonlinear planning. In *Artificial Intelligence Planning Systems: Proc. International Confrence*, 1992.
13. D.H.D. Warren. Generating conditional plans and programs. In *Proceedings of the Summer Conference on AI and Simulation of Behaviour*, 1976.
14. D. Weld, C. Anderson, and D. Smith. Extending Graphplan to handle uncertainty & sensing actions. In *Proc. 15th National Conference on Artifical Intelligence (AAAI'98)*, 1998.
15. D. Weld and D. Smith. Conformant Graphplan. In *Proc. 15th National Conference on Artifical Intelligence (AAAI'98)*, 1998.

# Probabilistic Planning in the Graphplan Framework

Avrim L. Blum and John C. Langford

School of Computer Science, Carnegie Mellon University, Pittsburgh PA 15213-3891
{avrim,jcl}@cs.cmu.edu

**Abstract.** The Graphplan planner has enjoyed considerable success as a planning algorithm for classical STRIPS domains. In this paper we explore the extent to which its representation can be used for *probabilistic* planning. In particular, we consider an MDP-style framework in which the state of the world is known but actions are probabilistic, and the objective is to produce a finite horizon contingent plan with highest probability of success within the horizon.

We describe two extensions of Graphplan in this direction. The first, PGraphplan, produces an optimal contingent plan. It typically suffers a performance hit compared to Graphplan but still appears to be fast compared with other approaches to probabilistic planning problems. The second, TGraphplan, runs at essentially the same speed as Graphplan, but produces potentially sub-optimal policies: TGraphplan's policy selects the first action on the highest probability *trajectory* from its current state to the goal. Ideally, we would like an optimal planner for probabilistic domains with the same speed that Graphplan would have if the domain were made deterministic. By comparing the speed and quality of these two planners to each other and to other existing planners, we are able to estimate how far off we are from our ideal.

PGraphplan is based on a forward-chaining search, unlike the backward-chaining search of the standard Graphplan algorithm. Thus, one focus of this paper is exploring the extent to which Graphplan's representation can be used to speed up forward search in addition to the backward search for which it was originally intended.

## 1 Introduction

The Graphplan planner is based on compiling a STRIPS-style planning problem into a compact (polynomial size) graph structure in which information can be quickly propagated to aid in the search for a plan [BF97]. Empirical results indicate that this approach is often quite fast compared to other traditional methods [BF97,Byl97,KNHD97]. Since its initial formulation, this basic algorithm has been extended and improved in a number of ways, such as allowing operators with contingent effects [GK97,KNHD97,ASW97], handling certain kinds of uncertainty [SW98,WAS98], and further speed improvements [KNHD97,KLP97].

In this paper, we explore the question: to what extent can the speed of Graphplan be extended to probabilistic domains? We consider the setting in

S. Biundo and M. Fox (Eds.): ECP-99, LNAI 1809, pp. 319–332, 2000.

which initial conditions and current state are known, but actions can be probabilistic, having several possible outcomes. This falls into the framework of
Markov Decision Processes (MDPs). In particular, instead of looking for a plan
that consists of an action sequence, we will be looking for a contingent plan that
tells which action to take based on the history of outcomes so far. The specific
kind of MDP we focus on is one in which, as with STRIPS planning, our objective is to satisfy all of a given set of conjunctive goals. Our aim will be to
produce an optimal finite-horizon contingent plan, where by "optimal" we mean
maximizing the probability of success within the time window (though much of
the discussion applies to objectives such as minimizing the expected completion
time as well).

Ideally, we would like an optimal planner for probabilistic domains with the
same speed that Graphplan would have, had the domain been deterministic. Embedding probabilistic outcomes into Graphplan's graph structure is straightforward. However, how to search for a plan is less obvious. Planning in probabilistic
domains is inherently more complex than in deterministic domains, in part because a complete policy can have size exponential in the horizon time whereas
in deterministic domains, plan sizes are linear in the horizon time.

We instead present two planners: PGraphplan and TGraphplan. TGraphplan
runs at essentially the same speed as Graphplan, but produces potentially suboptimal policies. Specifically, TGraphplan finds the highest probability *trajectory*
from the start state to the goal, which can then be turned into a kind of greedy contingent plan in a natural way. In contrast, PGraphplan produces *optimal*
contingent plans but typically suffers a performance hit. Unlike the backward-
chaining search of Graphplan, PGraphplan is based on a *forward-chaining* search
through the planning graph, which is much more natural for producing optimal plans in these probabilistic settings (we discuss this further in Section 3.3).
PGraphplan is like a standard top-down Dynamic Programming algorithm, but
uses information stored in the graph to prune its search. The difficulty here is
that for forward-chaining search, Graphplan's mutual exclusion relations are not
especially helpful, and instead PGraphplan uses other kinds of information propagated *backwards* from the goals. In this sense, our objectives are similar to
those of Kambhampati and Parker [KP99].

We compare PGraphplan and TGraphplan to each other and to several other
probabilistic planners such as Buridan [KHW95], SPI [BDG95], and vanilla
dynamic programming, on a variety of domains. Comparing PGraphplan and
TGraphplan to each other allows us to estimate how far off we are from our ideal
objective. In particular, from the perspective of Graphplan, two key questions
are: "to what extent can the planning graph representation be used to speed
up *forward search?*" and "Can Graphplan's backward-chaining search produce a
near-enough-optimal solution?"

## 1.1   Other Related Work

There is a long history of work in probabilistic planning. One of the first planners
designed specifically for probabilistic STRIPS-style domains is Buridan

[KHW95], extended to the contingent planner C-Buridan by [DHW94]. Unlike our work, these planners also considered partially observability, but in general were quite slow. Closer to our motivations is Structured Policy Iteration (SPI) of [BDG95]. SPI is designed for the fully observable MDP setting, and attempts to use the propositional representation of states and actions to find an optimal policy without expanding out the entire state space. Zander [ML99] and Maxplan [ML98] by Majercik and Littman compile a planning problem onto a stochastic satisfiability problem, solving the problem in that representation. Maxplan is a blind planner, while Zander produces contingent plans.

There has also been work on probabilistic planning explicitly using representations motivated by Graphplan. In particular, Boutilier et al. [BBG98] generalize Graphplan's pairwise mutual-exclusion constraints to $k$-wise constraints, and examine the reduction in the size of the MDP that is implicitly represented by the layer at which the planning graph levels off.

The work of Dean et al. [DKKN95] has a close connection to the goals of TGraphplan. That work finds an initial trajectory using a forward search, and then attempts to interpolate (in an anytime fashion) towards an optimal solution.

## 2    Graphplan and Its Representation

Graphplan [BF97] is a planner for STRIPS domains. These domains consist of *initial conditions* that describe the starting state of the world, *operators* which describe the legal actions that may be performed, and *goals* representing those facts that we wish to be true at the end of a plan. Operators have conjunctive preconditions, add-lists, and delete-lists. For instance, in a blocks-world, the initial conditions specify the starting configuration of the blocks, the goals specify what we wish to be true about the blocks at the end of the plan, and the operators specify our legal moves.

Graphplan is based on compiling such a planning problem into a polynomial-size structure called a *planning graph*. A planning graph is a directed, leveled graph. The first level has one node for each proposition in the initial conditions. The next level has a node for each action (fully-instantiated operator) that might possibly be performed at time 1: that is, one node for each action whose preconditions all exist in the initial conditions. The next level of the graph lists all propositions that might be true at time 2: namely, the union of the add-effects of all actions in the previous action level, including the no-ops (for consistency, Graphplan also includes a *no-op* action for each proposition that simply propagates the proposition forward in time). The next level consists of actions that might possibly be performed at time 2, and so forth. Edges in a planning graph connect actions to their preconditions and their add and delete effects.

Graphplan begins by creating a planning graph forward from the initial conditions until all of the goals appear in the graph. It then searches in a backward-chaining fashion. If the recursive search does not find a plan, then the graph is extended one more time step and the process is repeated. The planner also has a mechanism for eventually halting if, in fact, the problem has no solution.

The planning graph allows for several important optimizations including:

1. Propagating pairwise *mutual exclusion* relations between propositions and between actions forward through the graph while it is being created. These relations tell the planner that, for instance, propositions $P$ and $Q$ cannot be both made true by time step 4, and are used to prune the backward search.
2. Allowing the plan to perform several operators in parallel so long as they do not conflict with each other.

Memoizing the results of unsuccessful searches is also used, and improvements to Graphplan use a number of other optimizations as well. For more details see [BF97,KNHD97,KLP97].

**Representing Probabilistic Actions**

In this paper, we consider probabilistic actions. Specifically, we consider operators with conjunctive preconditions and with several sets of add and delete effects, each set having an associated probability. For example, an "open door" action might require that the door be unlocked, and with 88% probability actually open the door, with a 10% probability do nothing, and with 2% probability pull off the handle. We assume that the outcome that actually occurred will always be known when executed (the MDP not the POMDP setting).

To represent these kinds of operators, the graph is constructed in the normal manner except that each action contains a list of possible "outcomes", each with its associated probability, and then each outgoing edge of the action indicates which outcome produced that edge. For instance, suppose that we have an operator that with probability 0.7 deletes $G0$ and adds $G1$, and with probability 0.3 just adds $G1$. Then, there would be two outcomes, one with probability 0.7 and one with probability 0.3. There would also be three outgoing edges. One edge would be a delete edge leading to $G0$ and associated with the first outcome, one would be an add edge leading to $G1$ and associated with the first outcome, and one would be an add edge leading to $G1$ and associated with the second outcome. Since each outcome contains its probability, this representation is sufficient to reconstruct the definition of the operator.

## 3   PGraphplan

A standard algorithm for solving a finite-horizon MDP is dynamic programming. One begins by computing the value of each state for a time horizon of 0, then uses that to compute values for a time horizon of 1, and so on up to the given horizon $t_{max}$. In propositional planning, because we have an initial state and the state space is not explicitly enumerated, it is more natural to do this in a recursive top-down fashion, as described in Figure 1. Notice that top-down DP, because it stores the result of each computation, explores each state at most once per time step just like bottom-up DP.

DPsolve(state $s$, time $t$): Compute $value(s, t)$ = best possible probability of success within the time window for a contingent plan starting from state $s$ at time-step $t$.

1. If $t = t_{max}$ then return 0 if goals not satisfied, else return 1.
2. If already-visited($s, t$), then return the previously-computed value.
3. For each possible action $a$,
    a) For each possible state $s'$ that could result from taking action $a$ in state $s$, recursively call DPsolve($s', t + 1$).
    b) Let $value(s, a, t)$ be the probability-weighted average of the results.
4. Let $value(s, t) = \max_a value(s, a, t)$. Return this quantity, after first storing it in case we ever visit $s$ at time $t$ in a later recursive call. Also, return $\text{argmax}_a value(s, a, t)$ as the optimal action.

**Fig. 1.** Standard top-down Dynamic Programming

PGraphplan begins with this vanilla top-down Dynamic Programming as its starting point,[1] but uses the planning graph to prune its search. In particular, PGraphplan propagates two distinct kinds of information through the graph. The first kind tells the planner how various nodes in the graph contribute to solving the problem goals, and the second focuses more on how near or remote that contribution is. Both kinds of information are used in the same way: to tell the planner when the path it is currently exploring provably cannot reach the goals within the given time horizon and therefore it may safely return failure in its recursive call. The two types of information are given below.

### 3.1   Unary and Pairwise "Needed" Nodes

One very simple optimization is to delete nodes from the planning graph that do not have any paths to the goal literals, by performing a backward sweep through the graph. This simple form of relevance analysis effectively collapses multiple states together, and is roughly equivalent to the notion of relevance used by Knoblock [Kno94]. For an even greater savings, we assign to each node not removed a vector indicating which goal literals *are* reachable from this node. The planner uses this information when at some state $S$ by looking at the vectors assigned to each literal in $S$, and checking to see if any goal is missing from all the vectors; if so, then it knows it may backtrack immediately. In fact, if there is only one non-noop action that creates some goal, then the preconditions to this action can be used instead of that goal for the purposes of defining these vectors. This can be viewed as a crude but computationally cheap version of the fact-generation trees of Nebel et al. [NDK97].

---

[1] In particular, this is a forward-chaining search, and allows only one non-noop action per time step. It would be interesting to consider parallel actions as in Graphplan, but the computation becomes quite a bit hairier in the probabilistic setting.

A more interesting version of this information is a pairwise notion of "nee-dedness". Suppose node $P$ *does* have one or more paths to goal literals, but all of these paths use node $Q$ as well (e.g., all potentially useful actions that have $P$ as a precondition also have $Q$ as a precondition); in that case, we can add a "$P$ needs $Q$" edge to the graph, allowing the planner to drop literal $P$ from its state if literal $Q$ is not also present. Specifically, we say that proposition $P$ *needs* proposition $Q$ if each action with $P$ as a precondition either (a) has $Q$ as a precondition too, or (b) needs some action that has $Q$ as a precondition. Similarly, a non-noop action $A$ needs noop $B$ if all add-effects of $A$ need the result of $B$ (and none are equal to it); a noop action $A$ needs action $B$ if the result of $A$ needs a proposition that can be created only through $B$.[2]

For example, consider a domain in which a robot, initially with key in hand, must move to the end of the hallway, unlock a door at the end, and then drop the key in order to empty its hand for some subsequent task. In this case, the fact that the move-forward action needs the noop-have-key action gets propagated back through the entire graph, ensuring that the robot does not drop the key prematurely.

It is interesting to compare the information propagated here to the "backward mutex" constraints used by Kambhampati and Parker [KP99]. They propagate the notion that "$P$ and $Q$ are redundant" in the sense that one can confidently remove one of them from any state that contains both. That information seems more difficult to propagate appropriately in a probabilistic setting and we have not attempted to do so.

### 3.2   Value Propagation

The idea for this second kind of information is to store values on nodes of the graph that allow one to compute a "permissible heuristic" for an $A^*$-style search. This method empirically produces an even greater savings than the one above.

Imagine that reaching a goal state (one in which all the problem goals are satisfied) is worth $t_{max}$ dollars, but performing a non-noop action costs \$1. In this case, the *true value* of a state that can reach the goals in $i$ steps is \$$(t_{max} - i)$. Notice that if a state $S$ at time step $t$ is worth less than \$$t$, then this means it cannot possibly reach the goals by time $t_{max}$. Thus, a magic oracle that returned the true value of any given state would be quite useful. The idea of *value propagation* is to propagate heuristic values (hvalues) through the nodes of the graph such that the heuristic value of any state $S$, defined to be the *sum* of the hvalues of the nodes of the state, is guaranteed to be greater than or equal to the true value of $S$. If the planner finds that the heuristic value of its current state at time $t$ is less than \$$t$, then it can confidently backtrack immediately.

Specifically, we begin by dividing the final value $t_{max}$ evenly among the problem goals at time $t_{max}$, breaking ties arbitrarily (all hvalues will be integral). We propagate hvalues on propositions at time $t+1$ backwards to actions at time

---

[2] This asymmetry is the result of the planner only allowing one non-noop action per time step.

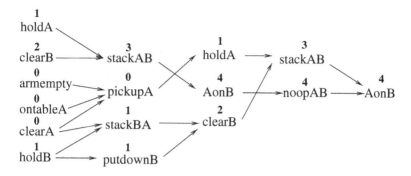

**Fig. 2.** RHS of graph for simple blocksworld domain, ending at $t = 4$. Some noops are not shown. Numbers indicate hvalues. Notice that the planner will return from any state at time $t = 2$ that doesn't have `clearB` (it will never reach the impossible state in which it is holding A and B).

$t$ as follows: the hvalue of a noop is the hvalue of its effect; the hvalue of a non-noop action $a$ is $-1 + \sum_{e \in \text{addeffect}(a)} hvalue(e)$, where preconditions that are not deleted are viewed as add-effects for this purpose. We view probabilistic actions (which may have several possible outcomes) as if they were user-controllable for this computation: that is, a probabilistic action with $k$ possible outcomes is treated as $k$ separate deterministic actions, one for each possibility, each with its own value.

Values on actions at time $t$ are then propagated to propositions at time $t$ as follows. We begin with the noops: each gives its value to its precondition. We then consider each non-noop action in turn.[3] For each action, we compare its value to the sum of the hvalues of its preconditions. If its value is larger, we give the difference to the preconditions; the semantics is that the total value of the preconditions is the maximum of performing or not performing this action. As a heuristic, we distribute the value evenly among just the preconditions deleted by the action (breaking ties arbitrarily) if any exist; if none exist, we distribute the value evenly among all preconditions. (Actions with no preconditions are given a fake "always-true" precondition that is true in the initial conditions and never deleted.) An example is given in Figure 2.

This method is *legal* in the sense that if a state at time $t$ has a sum of hvalues less than $t$, then it cannot possibly reach the goals. Unfortunately, it can at times be an over-estimate, because of a double-counting that may occur as values are propagated.

The two kinds of information described above each have a different purpose. The first kind asks *is* a node useful, and *for what*? The second focuses more on *how long* the path is from some node to its eventual use.

---

[3] The order in which the actions are examined can, in principle, affect the values, but we have not attempted to optimize the ordering in the planner.

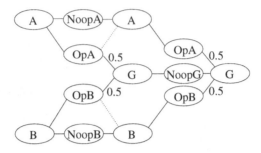

**Fig. 3.** Planning graph for the simple domain of section 3.3. Dotted lines represent deletes, and irrelevant nodes have been removed.

### 3.3   Why Forward-Chaining?

Finding an optimal policy with a planning graph appears to be considerably more difficult for a backwards search than a forwards one. Consider, for example, a domain in which the initial conditions contain the literals $A$ and $B$, and the goal is to achieve $G$. There are 2 operators: $OpA$ requires $A$, deletes $A$ and with probability 0.5 adds the goal $G$; $OpB$ requires $B$, deletes $B$ and with probability 0.5 adds $G$. See Figure 3.

For this domain, the plan $OpA, OpB$ has a 75% chance of success. However, to find this plan by reasoning backwards in the planning graph appears to require combining seemingly unrelated goal sets. In particular, to produce the optimal plan, we need at time 2 to consider the goal set $\{G\}$ (which achieves our goal via a noop) and $\{B\}$ (which has probability 0.5 of achieving the goal via $OpB$) together. We then need to realize that if $OpA$ is performed from $\{A, B\}$ at time 1, then each of the two outcomes of the action will lead to one of these goal sets at time 2. (Note: these are two distinct *goal sets*, not just two goals in the same set.) This kind of reasoning seems possible, but it also seems that it would require time quadratic in the number of goal sets at any given time step (or cubic if some action has three possible outcomes). The problem stems from the fact that we are dealing with *goal sets* (which are only subsets of states) rather than the states themselves in our backward-chaining search, unlike in bottom-up DP. Perhaps some way can be found around these difficulties, or some way to make this not too expensive in "typical" domains. In any case, this is the reason we choose to use forward chaining.

Smith and Weld [SW98] and Weld et al. [WAS98] use a different approach that allows for backward-chaining search on goal-sets in the presence of uncertainty. They handle uncertainty in the framework of Graphplan by essentially creating one graph for each "possible world". For instance, if one views the outcome of a probabilistic action as being determined by the flip of an associated coin, then there would be one graph for each possible sequence of coin flips. Ideally, one would like to use the same kind of search but at the same time handling the (possibly exponentially many) possible worlds within the context of a single planning graph.

# 4   TGraphplan

TGraphplan uses a backward chaining search (essentially the same as the original Graphplan) and finds an optimal trajectory from the initial state to the goals. A *trajectory* is a sequence of actions and outcomes leading from one state to another, such as "I turn the key in the ignition, the car starts, I drive to the airport, get there in time for my plane, I catch my flight, and arrive at my destination." An optimal trajectory is the highest probability sequence of states and actions leading to the goal. When two trajectories have the same utility, trajectories which do noops later are preferred as in Graphplan. This bias is especially important in a probabilistic setting because it maximizes the amount of time available for recovery in case the trajectory experiences a failure.

TGraphplan can be used as a subroutine to manufacture a complete policy by simulating execution of the trajectory to detect the state and time of unexpected outcomes not on the trajectory. Optimal trajectories for these unexpected outcomes can then be found and simulated forward to find new unexpected outcomes. The recursion terminates when there are no more unexpected outcomes. More naturally, TGraphplan can be used in an online fashion after the optimal trajectory is discovered. While the first action is executing (in the real world), TGraphplan can be planning for subsidiary trajectories. Thus, the important time quantity to measure for TGraphplan is the time to find the optimal trajectory rather than the time to find a complete policy.

TGraphplan starts by building a planning graph with probabilistic outcomes included. The backward-chaining search is done exactly as in Graphplan, except instead of the recursion returning a binary value (success/fail), it returns a real-valued success probability for the best sub-trajectory. In the TGraphplan search algorithm, the probability of the trajectory is determined by recursively multiplying the probability of a step succeeding by the probability of the partial trajectory already explored. The same set of optimizations that Graphplan uses are used by TGraphplan. Mutual exclusions are more complicated because it is possible for an operator to interfere with one outcome of another probabilistic operator but not with a second outcome. In TGraphplan, a pair of operators is made exclusive when any pair of outcomes interfere. Instead, we could lose the notion of "exclusive operators" and replace it with a notion of "exclusive outcomes" but we do not do that. TGraphplan can also be run in an iterative deepening mode as for Graphplan. In order to do this, the desired trajectory probability must be given in advance. The algorithm terminates as soon as a trajectory with at least the desired probability is found.

The TGraphplan algorithm is fast in comparison to PGraphplan because it only outputs *partial policies*, solves an inherently less complex problem and uses a backward chaining search which can take advantage of the mutual exclusions propagated forward in the building the graph. It is interesting to consider when TGraphplan will produce a (near) optimal policy and when the choices it makes will be substandard. This will be discussed in the following examples.

## 5  Examples and Empirical Results

We now describe several example domains and give results of running PGraphplan, TGraphplan, and other comparison planners on them. The purpose of these experiments is twofold: to examine the speed of the proposed planners, and to explore the extent to which the plans produced by TGraphplan are optimal or close to it. PGraphplan and TGraphplan are written in C. The planners we compare to are

- Top-down Dynamic Programming (Figure 1).
- Buridan [KHW95]: in compiled Lisp.
- Blackbox [KS99] (on deterministic domains): written in C.
- SPI [BDG95]: an infinite-horizon discounted MDP solver, written in C.

**Moats and Castles:** This simple domain is an adaptation of one by Majercik and Littman [ML98], in which the goal is to build a sand castle on the beach. In our version, there are two operators. Dig-moat is a deterministic action that increases the depth of a protective moat (there are 5 discrete depths), and build-castle is a probabilistic action for creating the castle, whose success probability increases with the depth of the moat. The optimal finite-horizon plan for this problem will consist of some number of dig actions, followed by a remainder of builds, where the number of digs depends on the time horizon and the specific success probabilities. The problem as stated has a very small state space; to make things more interesting, we consider having multiple castles, each with its own moat.

We consider this domain in part because it gives a simple illustration of when TGraphplan does or does not produce optimal plans. In particular, for the case of one castle, the optimal *trajectory* is to dig as many times as possible, followed by one final build operation (and then noops up to the time horizon). This may or may not be a trajectory of the optimal policy; in particular, the optimal policy may have fewer dig operations, if, for instance, deepening the moat has only a small effect on the success probability of the build operation.

We describe performance results in Table 1. For this domain, when the time horizon is large, the information propagated by PGraphplan does not provide much of a gain. That is because in this case, almost all states *can*, in principle, lead to solving the goals, and the information propagated is only intended to prune states with no chance of success. TGraphplan scales better with time horizon but worse with number of castles, compared to PGraphplan; the latter effect appears to occur because of interaction between parallel and probabilistic actions.

**Probabilistic Blocks:** In the standard blocks-world domain, we can `pickup` a block from the table, `putdown` a block onto the table, `unstack` a block from another block, and `stack` a block on another block. So, for instance, to pick up a block from the table and place it on another block takes two time steps. Imagine we augment this domain with a probabilistic operator `faststack` that can move a block from the table onto the top of another block in one time step, but it only

**Table 1.** PGraphplan (PGP), TGraphplan (TGP), top-down DP (DP), SPI, and Buridan(Bur) on the moat and castles problem with varying numbers of castles. SPI is run with discount factor 0.9. The other planners are given time horizon of 5 or 10 (for Buridan, this is done by providing a desired success probability); values for horizon of 10 are given in square brackets. Running times are given on a PII-450 xeon with 512 MBytes of memory.

|  | 1 castle | 2 castles | 3 castles | 4 castles |
|---|---|---|---|---|
| DP | .01 [.01] sec | .01 [.01] sec | .01 [.09] sec | .03 [0.54] sec |
|  | 44 [114] states | 448 [2300] states | 1977 [22113] states | 5996 [138308] states |
| PGP | .01 [.01] sec | .01 [.01] sec | .01 [.07] sec | .01 [0.36] sec |
|  | 40 [110] states | 256 [1900] states | 909 [16389] states | 1776 [89276] states |
| TGP | .01 [.01] sec | .01 [.01] sec | .01 [.05] sec | .07 [0.45] sec |
|  | 37 [97] sets | 259 [954] steps | 1959 [9974] sets | 13855 [102569] steps |
| SPI | .05 sec | 0.21 sec | 2.73 sec | 36.3 sec |
| Bur | 1 sec [> 5 min] | 6 sec [> 5 min] | > 5 min | > 5 min |
|  | 427 [>60k] plans | 4823 [>60k] plans | >60k plans | >60k plans |

**Table 2.** Results for the probabilistic blocks problem described in the text. The initial state is a tower (ABCD...) and the goal is the same tower except that the block that used to be on top is now on the bottom (BCD...A). SPI finds a policy that applies from all possible starting states, so in a sense it is unfairly penalized in this experiment (to partially alleviate this, for SPI we discarded actions not used in any reasonable trajectory from our initial state from the domain description).

|  | 2 blocks ($t=4$) | 3 blocks ($t=8$) | 7 blocks ($t=18$) | 7 blocks ($t=20$) | 7 blocks ($t=22$) | 7 blocks ($t=24$) |
|---|---|---|---|---|---|---|
| DP | .01 sec | .01 sec | 8.27 sec | 14.53 sec | 20.88 sec | 25.43 sec |
|  | 18 states | 194 states | 974k states | 1607k states | 2239k states | 2646k states |
| PGP | .01 sec | .01 sec | 1.34 sec | 4.97 sec | 10.83 sec | 14.49 sec |
|  | 6 states | 111 states | 163k states | 599k states | 1211k states | 1549k states |
| TGP | .01 sec | .01 sec | .25 sec | .42 sec | .53 sec | .61 sec |
|  | 11 sets | 29 sets | 126 sets | 718 sets | 1280 sets | 1323 sets |
| SPI | 1.17 sec | 28.54 sec | – | – | – | – |
| Bur | 1 sec | – | – | – | – | – |
|  | 783 plans | >100k plans | >100k plans | >100k plans | >100k plans | >100k plans |

succeeds with a 70% probability; with a 30% probability it has no effect. This setting is interesting because the optimal action depends on the time-to-go.

Note that the optimal trajectory will always choose the deterministic operators if there is sufficient time, otherwise it will choose faststack. Thus, in this domain, TGraphplan does, in fact, lead to an optimal policy. Results for this domain are given in Table 2.

**8-puzzle and Flat-tire:** In order to compare to deterministic planners, we also considered several deterministic domains, in particular the flat-tire problem of Russell, and the 8-puzzle problem. The 8-puzzle problem is interesting because

**Table 3.** Results on the flat-tire domain, and the easy and hard 8-puzzle problems. Blackbox was run in its default mode, with `-solver graphplan` (BlackboxGP), and with `-solver walksat` (BlackboxWS).

|  | flat-tire (19 step) | 8-puzzle (18 step) | 8-puzzle (30 step) |
|---|---|---|---|
| DP | 0.3 sec 48851 states | 0.63 sec 41242 states | 27.61 sec 1777759 states |
| PGP | .05 sec 3661 states | 0.14 sec 2069 states | 6.56 sec 437722 states |
| TGP | .04 sec 411 sets | 0.8 sec 461 sets | 43.56 sec 98946 sets |
| Blackbox | .08 sec | 0.93 sec | timeout |
| BlackboxGP | .08 sec | 0.93 sec | 45.56 sec |
| BlackboxWS | 0.18 sec | timeout | timeout |

there is no special advantage to backward-chaining on this problem. We consider the goal of achieving board state ABCDEFGH_ (reading left to right, top to bottom) from two different initial states: one in which a solution requires 18 steps and one in which a solution requires 30 steps (this is the case of initial board HGFEDCBA_). Results are given in Table 3. Note that PGraphplan is the fastest of all planners tested (even the deterministic ones) on this problem.

## 6    Discussion and Conclusions

PGraphplan performs a forward search to find an optimal contingent plan, using information stored in the graph to collapse and prune away unnecessary states. On problems such as blocks-world, flat-tire, and 8-puzzle, this pruning provides a substantial speedup. The value information seems to provide the greatest savings in general, with the information on the eventual purpose, if any, of each node in the graph providing a smaller but still significant help. PGraphplan is still in general slower than Graphplan, in part we believe because we have not yet found the best way to propagate information backwards through the planning graph. One possible avenue would be to better integrate the two kinds of information discussed above, for instance by separating the hvalues into different "flavors" depending on which goal they are intended for. It would also be helpful to propagate more probabilistic information, such as an upper-bound on the probability of reaching the goals, which could then be used to prune search. Finally, it would be interesting to explore whether knowledge of the TGraphplan's more quickly-constructed policy could guide PGraphplan's search.

The planners implemented apply to problems for which the objective is to maximize the probability of satsifying the problem goals within the given time window, or the related goal of minimizing expected completion time. More general MDP problems are often given by specifying rewards for performing certain (noop or non-noop) actions. It appears that some of the kinds of information propagated here should be useful in those more general settings, and it would be interesting to see if this generalization could be made without a sacrifice in performance.

**Acknowledgements.** This research is sponsored in part by NSF National Young Investigator grant CCR-9357793, NSF grant CCR-9732705 and an AT&T / Lucent Special Purpose Grant in Science and Technology. We would like to thank the anonymous reviewers for their detailed, thoughtful, and helpful comments. Source code is available at http://www.cs.cmu.edu/~avrim/pgp.html.

# References

[ASW97]     C. R. Anderson, D. E. Smith, and D. S. Weld. Conditional effects in Graphplan. unpublished manuscript, 1997.

[BBG98]     C. Boutilier, R. I. Brafman, and C. Geib. Structured reachability analysis for Markov Decision Processes. In *Proceedings of the Fourteenth Annual Conference on Uncertainty in Artificial Intelligence (UAI–98)*, pages 24–32, 1998.

[BDG95]     C. Boutilier, R. Dearden, and M. Goldszmidt. Exploiting structure in policy construction. In *IJCAI95*, pages 1104–1111, Montreal, 1995.

[BF97]      A. Blum and M. Furst. Fast planning through planning graph analysis. *Artificial Intelligence*, 90:281–300, 1997.

[Byl97]     T. Bylander. Linear programming heuristic for optimal planning. In *Proceedings of the Fourteenth National Conference on Artificial Intelligence*, 1997.

[DHW94]     D. Draper, S. Hanks, and D. Weld. Probabilistic planning with information gathering and contingent execution. In *Proceedings of the 2nd International Conference on Artificial Intelligence Planning Systems*, pages 31–37, 1994.

[DKKN95]    T. Dean, L. P. Kaelbling, J. Kirman, and A. Nicholson. Planning under time constraints in stochastic domains. *Artificial Intelligence*, 76, 1995.

[GK97]      B. C. Gazen and C. A. Knoblock. Combining the expressiveness of UCPOP with the efficiency of Graphplan. In *Proceedings of the Fourth European Conference on Planning*, 1997.

[KHW95]     N. Kushmerick, S. Hanks, and D. Weld. An algorithm for probabilistic planning. *Artificial Intelligence*, 76:239–286, 1995.

[KLP97]     S. Kambhampati, E. Lambrecht, and E. Parker. Understanding and extending Graphplan. In *Proceedings of the Fourth European Conference on Planning*, September 1997.

[KNHD97]    J. Koehler, B. Nebel, J. Hoffman, and Y. Dimopoulos. Extending planning graphs to an ADL subset. In *Proceedings of the Fourth European Conference on Planning*, September 1997.

[Kno94]     C. Knoblock. Automatically generating abstractions for planning. *Artificial Intelligence*, 68(2):243–302, 1994.

[KP99]      S. Kambhampati and E. Parker. Making graphplan goal-directed. In *ECP99*, September 1999.

[KS99]      H. Kautz and B. Selman. Unifying SAT-based and graph-based planning. In *IJCAI99*, 1999.

[ML98]      S. M. Majercik and M. L. Littman. MAXPLAN: a new approach to probabilistic planning. In *AIPS*, pages 86–93, 1998.

[ML99]      S. M. Majercik and M. L. Littman. Contingent planning under uncertainty via stochastic satisfiability. In *AAAI*, 1999.

[NDK97]   B. Nebel, Y. Dimopoulos, and J. Koehler. Ignoring irrelevant facts and operations in plan generation. In *Proceedings of the Fourth European Conference on Planning*, pages 338–350, September 1997.

[SW98]    D. E. Smith and D. S. Weld. Conformant graphplan. In *AAAI98*, 1998.

[WAS98]   D. S. Weld, C.R Anderson, and D. E. Smith. Extending graphplan to handle uncertainty and sensing actions. In *AAAI98*, 1998.

# Making Graphplan Goal-Directed

Eric Parker

The MITRE Corporation
1820 Dolley Madison Boulevard
McLean VA 22102
erik@mitre.org

**Abstract.** The Graphplan algorithm exemplifies the speed-up achieveable with disjunctive representations that leverage opposing directions of refinement. It is in this spirit that we introduce Bsr-graphplan, a work in progress intended to address issues of scalability and expressiveness which are problematic for Graphplan. Specifically, we want to endow the planner with intelligent backtracking and full quantification of action schemata. Since Graphplan employs a backward chaining search, it lacks the necessary state information to efficiently support these mechanisms. We hypothesize that alternatively pointing the search in the direction of the goals provides an sufficient amelioration. Further, we demonstrate that a forward chaining search strategy can be competitive by enforcing ordering constraints on the developing plan prefix. This is accomplished by using operators of a plangraph constructed in a top-down fashion to extend the plan prefix, and by introducing an additional data structure - a constraint tree - which is constructed by regressing subgoal information through the graph operators.

## 1    Introduction

Graphplan [Blum & Furst, 1995] is one of the most efficient algorithms known for automated plan synthesis in classical domains. The mental leap of Graphplan is to integrate the bottom-up and top-down refinement paradigms. The graph growing phase of the algorithm corresponds to the process of refining disjunctively represented partial plans without pushing the complexity of the refinements into the search space [Kambhampati et al, 1997]. During graph construction, data structures – i.e. sets of mutually exclusive (mutex) propositions – representing constraints are maintained and used to focus the ensuing search of the graph. Additionally, because the graph is constructed bottom-up while the search proceeds top-down, information about dead-end paths can be hashed (memoized) so that future searches can avoid redundant mistakes. Despite the algorithm's elegance, there are some problematic issues which still need to be resolved.

The problem with bottom-up graph construction is that the subsequent search must sort through all literals that might conceivably be made true starting from the current situation, rather than only those literals whose truth might be relevant to achieving the goal [McDermott, 1996]. The solution to this problem requires more than simply reversing the direction of graph construction. This is because, in general, top-down

S. Biundo and M. Fox (Eds.): ECP-99, LNAI 1809, pp. 333–346, 2000.
© Springer-Verlag Berlin Heidelberg 2000

```
(define (domain rocket)
 (:action load-pkg
    :precondition (and (at ?craft ?loc) (at ?pkg ?loc))
    :effect (and (in ?pkg ?craft) (not (at ?pkg ?loc))))
 (:action unload-pkg
    :precondition (and (at ?craft ?loc) (in ?pkg ?craft))
    :effect (and (at ?pkg ?loc) (not (in ?pkg ?craft))))
 (:action fly-craft
    :precondition (and (at ?craft ?loc1))
    :effect (and (at ?craft ?loc2) (not (at ?craft ?loc1)))))

(define (problem example-problem1)
 (:domain rocket)
 (:init (and (at package earth) (at rocket earth)))
 (:goal (at package moon)))
```

**Figure 1.** An example rocket problem.

and bottom-up approaches each have unique advantages and disadvantages [Fuchs & Fuchs, 1999]. However, in the context of planning, coupling top-down graph construction with a forward chaining search provides a mechanism for reasoning about world states. In addition to memoizing dead-end states we also get to do intelligent backtracking [Kambhampati, 1998] - a sound and complete method for anticipating and avoiding certain dead-end states in the first place.

The problem with top-down search is the intrinsic combinatorial explosion of the goal network [McDermott, 1996]. The complexity here can be handled in a manner analogous to the way that Graphplan handles the complexity of the forward state space search tree, i.e. by growing the plangraph top-down we construct a compact representation of the goal network. Additionally, since all the causal links are then explicitly documented (albeit compactly), employing a bottom-up search with state information provides support for sound, complete and full quantification of action schemata, e.g. [Burgess and Steel, 1997].

Bsr-graphplan is a planner which constructs a plangraph by backward chaining from the goal specification, and which employs a forward chaining search of the resulting graph. The current implementation uses neither intelligent backtracking nor the quantification of action schemata. Rather, the previous discussion serves to motivate our investigation of this strategy. However, analogous to Graphplan, Bsr-graphplan maintains a data structure representing constraints which are used to focus the search. This new data structure takes the form of a tree whose branches represent the possible partial orderings of the graph actions. The constraint tree is constructed by regressing subgoal information through the graph operators.

The rest of this paper is organized as follows. In Section 2 we review the Graphplan algorithm, re-stating the basic steps of the algorithm in a manner that makes the subsequent discussion more accessible. In Section 3 we describe the Bsr-graphplan algorithm, explaining in particular the way the plangraph is grown, the way the constraint tree is constructed, and the way the forward search is conducted. In Section 4 we describe the results of our experiments with Bsr-graphplan.

## 2    Review of the Graphplan Algorithm

Graphplan incrementally grows and searches a compact structure known as a planning graph. An iteration consists of extending the action-time representation by one unit of time, searching the entire representation for a solution, and finally if the search fails, memoizing information which can be used to reduce the subsequent search effort.

The rocket problem of figure 1 will serve as a running example. Figure 2 shows the planning graph structure generated by Graphplan for the rocket problem. The nodes of the planning graph are partitioned into disjoint sets (or levels) which are ordered in time, and the edges of the graph only connect nodes in adjacent levels. There are action levels and proposition levels. Each action level comes between the proposition level which contains the actions' preconditions, and the proposition level containing their effects. Levels are numbered by increasing integers beginning with zero. The level numbers for the example planning graph appear at the bottom of figure 2. The semantics of an action level 2t-1, which exists at a time t > 0, are that some subset of those actions must occur in step t of the plan. For example, the 3-step plan

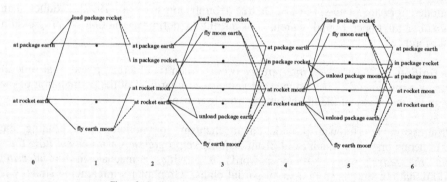

**Figure 2.** Graphplan planning graph for the example rocket problem.

<a1,{a2,a3},a4>, where {a2,a3} is a subset of actions at step 2, represents the disjunction of the 4-step plan <a1,a2,a3,a4> with the plan <a1,a3,a2,a4>.

To better understand the Graphplan algorithm, it is useful to define the notion of disjunctive progression.

**Definition 1.** *Disjunctive progression* is the progression of a set of states through a problem P. Let S be the union of the set of states, and let $operator_i()$ be a function which takes a problem and returns the $i^{th}$ operator of the problem. Let length() be a function which returns the number of operators in the problem. Further, let instantiations() be a function which takes an operator and a set of propositions, and returns all full instantiations of the operator with the propositions. Then, the result of the progression of S through P is $R = S \cup \{\cup \text{progress}(S, \text{action}) \mid \text{action} \in \{\cup \text{instantiations}(\text{operator}_i(P), S), 1 \leq i \leq \text{length}(P)\}\}$, where the function progress() takes a set of propositions and an action, and returns the classical progession of the propositions through the action. That is, $\text{progress}(S, \text{action}) = S - \text{delete\_effects}(\text{action}) \cup \text{add\_effects}(\text{action})$.

Level 2 of the graph is computed by disjunctively progressing level 0 (the set containing the initial state) through the rocket problem. In general, level 2t, $t \geq 1$, of a planning graph is computed from level 2t-2 by progressing level 2t-2 through the problem under consideration.

It is clear from the definition that, as a consequence of disjunctive progression state information is not preserved. This is due to the unioning operations performed during the computation of R. The problem is that on the next iteration, R will be progressed through the next action level, and the ensuing instantiation procedure assumes that the literals of R are conjunctively related, when the actions which participated in the computation of R are actually disjunctively related. However, because the preconditions, and the add and delete effects of the operators is known, the standard notion of interference (i.e. two actions interfere if one deletes a precondition or add effect of the other) could be used to provide a sound and complete search of the graph. Such an approach would require searching an upper bound approximation to the unioned forward state-space search tree [Kambhampati et al, 1997]. Rather than pursuing an upper bound search, it is useful to define the notion of progression interference.

**Definition 2.** *Progression interference* occurs when two literals, p and q, at time 2t cannot simultaneously be assigned the value true, regardless of the permutation of the truth values of any of the literals at any time less than 2t.

Progression interference provides a mechanism for noticing and pruning the extraneous branches which result from disjunctive progression, *without splitting them into the search space*. In other words, it provides a mechanism for *inferring* additional constraints from a set of partial plans. Graphplan's inferencing strategy is to propagate the classical interference relationships of actions to their add effects, thereby adorning the graph with progression interfering relationships. Graphplan handles this in an efficient and integrated way using persist actions. A persist action is indicated in the figure by a horizontal arc between like propositions in different levels. The dot in the middle of the arc means persist. The semantic of a persist action is that it requires and achieves the same, single proposition. Including one persist action for every proposition in level 2t means that every proposition in level 2t+2 is either the add effect of a newly instantiated action in level 2t+1, or is at least not negated by any 2t+1 action. In this way, complete state information is maintained. Since interference between persist actions and other (including other persist) actions is computed in the regular way, a procedure for finding all pairs of progression interfering propositions is well defined:

Two actions a and b in level 2t+1 are marked as progression interfering if –

1. Either action deletes a precondition or add-effect of the other.
2. There is a precondition of action a and precondition of action b that are progression interfering in level 2t.

Two propositions p and q in level 2t+2 are marked as progression interfering if –

1. Each action a in level 2t+1 which has an effect p is progression interfering with each action b in level 2t+1 which has q as an effect.

For example, in figure 2, add effects are indicated by solid arcs, and delete effects are dashed arcs.   The propositions in(package, rocket) and at(rocket, moon) are progression interfering at level 2 - since the actions <load package rocket> and <fly earth moon> interfere in level 1, and since there are no other actions which support either of these two propositions from level 1.   Therefore, <unload package moon> need not be included in the level 3 actions.   In contrast, the <unload package moon> action occurs in level 5 because its preconditions, in(package, rocket) and at(rocket, moon) are not progression interfering in level 4.   The precondition in(package, rocket) can be supported by the level 3 action <*persist* in(package, rocket)>, and the precondition at(rocket, moon) can be supported by the level 3 action <fly earth moon>, as long as these two actions are not progression interfering.   As luck would have it, the preconditions of these level 3 actions are not progression interfering in level 2, since <*persist* at(rocket, earth)> and <load package rocket> are not progression interfering in level 1, and since neither of these level 1 actions deletes a precondition or add-effect of the other.

The Graphplan algorithm is shown in figure 3.   It has three arguments:   P - the problem, n - the maximum number of levels to expand the graph, and pg - the plangraph, which initially consists of a single level containing the initial state of P, and is incrementally extended on lines 4 and 9.   After the disjunctive progression, the domain operators are fully-instantiated in all possible ways from the current situation on line 5.   Then, extraneous actions are noted in lines 6, 7 and 8, and pruned from the next action level in line 9.   Line 14 invokes a backward search of the entire plangraph.

The backward search algorithm is shown in figure 4, and when initially invoked is provided with the initial state, the problem goals, the planning graph action levels, the number of action levels, and the null plan.   A global hash_table is assumed.

The backward search is a search through the literals of the graph.   The set S is initialized to be the set of problem goals.   Lines 7 - 10 generate all the possible ways of simultaneously achieving the subgoals S, with the actions in the previous action level.   The decision at line 8 is backtrackable.   The next set of subgoals is computed by unioning the regressions of S through the A actions (line 11) and the search is recursively invoked on the next action level (line 12).   Each time backward_search() is invoked, we check to see if S has been previously encountered at level n (line 3).   If it has, the search can backtrack immediately without exploring the rest of the graph below level n.   Further, suppose the search does explore the graph below level n, and

```
GRAPHPLAN (P, n, pg)
1    if n < 1
2    then return failure
3    else S ← previous_prop_level(pg)
4         next_prop_level(pg) ← d_progress(S, P)
5         A ← {∪instantiations(operatorᵢ(P), S) | 1 ≤ i ≤ length(P)}
6         E ← {action | action ∈ A,
7                      ∃p,q ∈ preconditions(action),
8                      p,q are progression interfering}
9         next_action_level(pg) ← A – E
10        I ← initial-state(P)
11        G ← goal-formula(P)
12        A ← action-levels(pg)
13        n ← length(action-levels(pg))
14        plan ← backward_search(I,G,A,n,∅)
15        if plan
16            then return plan
17            else GRAPHPLAN(P, n-1, pg)
```

**Figure 3.** The Graphplan algorithm.

doesn't find a plan. The unproven paths are marked as a dead-end on line 15, and this information will be used by the subsequent searches.

## 3    The Bsr-Graphplan Algorithm

The construction of the Bsr-graphplan planning graph is based on the notion of regression. The regression problem is to compute the weakest precondition Q, given an action A and a *proposition* p, such that if Q is true, then A is applicable in Q, and p will be true after A is applied to Q. $Q = <A>^R(p)$ is called the regression of p through A [McDermott, 1996]. For Strips operators, this involves seeing if A has p in its list of add-effects. If so, then p regresses to the preconditions of A. If, however, p is one of A's delete-effects, then p regresses through A to False. Otherwise, p does not appear in A's effects list - in this case, p regresses to p through A. Regressing a *formula* through an action A consists of individually regressing the literals of the formula through the action, and unioning the results. The Bsr-graphplan algorithm is somewhat complicated by the fact that, in general, regressing literals through action terms containing variables means that *Q can contain variables as well*. Because of this, the following definitions will be useful:

```
BACKWARD_SEARCH (I, S, action_levels, n, plan)
1   if n < 1
2       then return failure
3   if S ∈ hash_table[n]
4       then backtrack
5   if S ⊂ I
6       then return plan
7   while unexplored combinations of actions still remain
8           A ← ∀p ∈ S,
9                   nondeterministically choose a ∈ action_levels[n]
10                  such that a has p as an add-effect.
11          next_S ← {∪regress(S,a) | a ∈ A}
12          solution ← backward_search(I, next_S, action_levels, n-1, A+plan)
13          if solution
14              then return solution
15  hash_table[n] ← S
16  return failure
```

**Figure 4.** The backward search algorithm.

**Definition 3.** Two literals $p = (p_1, p_2, \ldots, p_m)$ and $q = (q_1, q_2, \ldots, q_n)$ are *unifiable* if $m = n$, and for all pairs $(p_i, q_i)$, $1 \leq i \leq m$, either $p_i$ and $q_i$ are equal, or one of them is a variable.

**Definition 4.** Two literals $p = (p_1, p_2, \ldots, p_m)$ and $q = (q_1, q_2, \ldots, q_n)$ are *unify_equal* if $m = n$, and for all pairs $(p_i, q_i)$, $1 \leq i \leq m$, either $p_i$ and $q_i$ are equal, or both of them are variables.

To better understand the Bsr-graphplan algorithm, it is useful to define the notion of disjunctive regression.

**Definition 5.** *Disjunctive regression* is the regression of a set of goal formulas through a problem P. Let G be the union of the propositions (literals) of the goal formulas, and let operator$_i$() and length() be as in Definition 1. Let partial_instantiations() be a function which takes an operator and a set of propositions, and returns all partial instantiations of the operator with the propositions (i.e. all *unifiable* subsets of the action's add-effects with the propositions). Then, the result of the regression of G through P is $Q = G \cup \{\cup regress(G, action) \mid action \in \{\cup partial\_instantiations(operator_i(P), G), 1 \leq i \leq length(P)\}\}$, where the function regress() takes a set of propositions and an action, and returns the classical regression of the propositions through the action. That is, regress(Q, action) = Q $-$ unifiable(add_effects(action),Q) $\cup$ preconditions(action). (note: the $\cup$ operator uses *unify_equal* instead of '=' to compare set elements.)

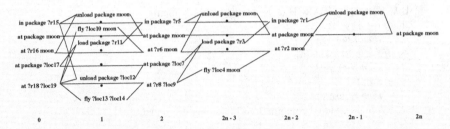

**Figure 5:** Bsr-graphplan planning graph for the example rocket problem.

Level 2n-2 of the graph is computed by disjunctively regressing level 2n (the set containing the problem goals) through the rocket problem. Figure 5 shows the planning graph structure generated by Bsr-graphplan for the rocket problem of figure 1. Notice that the Bsr-graphplan planning graph and the Graphplan planning graph, figure 2, are both numbered in ascending order from left to right. This seems awkward at first since the Bsr-graphplan planning graph is actually grown from right to left. To keep the discussion intuitive, the orientation is with respect to the intended direction of plan execution. The graph in figure 5 is both drawn and labeled in directions opposite to that of its growth. This is because it is grown in the backward direction. Further, the graph in figure 2 is both drawn and labeled in directions opposite to that in which it will be searched. This is because it is searched in the backward direction.

**Figure 6.** Constraint tree for the example rocket problem.

It is interesting to note that, unlike disjunctive progression, disjunctive regression maintains information about the goals. The unioning operations performed during the computation of Q do not cause problems for the next iteration, when Q will be regressed through the next action level. This is because the ensuing partial instantiation procedure assumes that the literals of Q are disjunctively related (it only looks for a non-empty subset of each action's add-effects). The idea here is that, whereas Graphplan needs to reason about disjunctive relationships, Bsr-graphplan needs to reason about conjunctive relationships. To put it another way, Graphplan spends its life separating literals, whereas the cross of Bsr-graphplan is to assemble literals.

The regression operator is used to aggregate literals into the nodes of a tree whose branches represent the possible partial orderings of the graph actions. Figure 6 shows the constraint tree generated by Bsr-graphplan for the rocket problem of figure 1. Each node is made up of two parts – the action which caused the generation of the node appears in the upper part, and the regression of the node's parent appears in the lower part. The root node contains the null action as its upper part and the goal formula as its lower part. Each time an action level is added to the planning graph, the partially-instantiated operators from that level are each considered in turn as possible successors of the leaf nodes of the tree. The idea is that we want as children of a node only those actions which must be ordered with respect to the other actions along that branch. In other words, to qualify as a child, an action must support one of the preconditions of one of its ancestors, and the action must interfere with all of its ancestors. Alternatively, if the action doesn't support any of its ancestors' preconditions, then the action is not required in the partial ordering under consideration. If the action doesn't interfere with all of its ancestors but is required in the final solution, then the action occurs in another branch of the constraint tree. This is not a problem because all of the branches are considered when checking the plan prefix against the constraint tree. During search, as new actions are added to the plan prefix the constraint tree is consulted to see if the newly added action is in violation of the ordering constraints. If it is, we say that some subset of the literals of the world state which exists after the application of the action are *regression interfering* with the problem goals.

**Definition 6.** *Regression interference* occurs when the truth of a literal p, at time 2t, removes the possibility of assigning true to a literal q, at some time greater than 2t, whose truth is required for the problem solution.

Formally, a literal p in level 2t of the graph is marked as regression interfering with a literal q in the graph if –

1 The literal q occurs in a level $m \geq 2t+2$ of the graph.
2 Either no action in a level $n \geq 2t+1$ of the graph, containing q as an add-effect, is unifiable with any action in the constraint tree, or for every node in the tree containing a literal r for which q is unifiable, one of the following holds:
   a. Every action in level 2t-1 of the graph containing p as an add-effect, also contains a literal s as a delete effect, s is unify_equal to a literal along the branch containing r.
   b. Some action in level 2t-1 of the graph containing p as an add-effect can be executed in parallel with one of the actions along the branch containing r.

BSR-GRAPHPLAN (P, n, pg, tree)
1   *if* n < 1
2       *then* return failure
3       *else* G ← previous_prop_level(pg)
4           next_prop_level(pg) ← d_regress(G, P)
5           A ← {∪partial_instantiations(operator$_i$(P), G) | 1 ≤ i ≤ length(P)}
6           next_action_level(pg) ← A
7           *while* more branches in constraint tree
8               branch ← next branch
9               E ← {action | action ∈ A,
10                          ∃p ∈ effects(action),
11                          ∃q ∈ literals(branch),
12                          p,q are regression interfering}
13              l ← literals( leaf( branch))
14              children( leaf( branch)) ← {(action, Q) | action ∈ A – E,
15                                          Q = regress(l, action)}
16          I ← initial-state(P)
17          G ← goal-formula(P)
18          A ← action-levels(pg)
19          n ← 1
20          plan ← forward_search(G,I,A,n,tree,∅)
21          *if* plan
22              *then* return plan
23              *else* BSR-GRAPHPLAN(P,n-1,pg)

**Figure 7.** The Bsr-graphplan algorithm.

For example, suppose either of the actions <fly ?loc10 moon> or <fly ?loc13 ?loc14> is chosen by the search as the first step of the plan for achieving the problem goal of figure 1. Then the subsequent state will contain the literal at(package, earth) - since at(package, earth) is part of the initial state and flying an empty rocket around has no affect on packages. However by condition 2a above, at(package, earth) is regression interfering with in(package, ?r1) at level 2n-2 of the plangraph of figure 5 (which matches the literal in(package, ?r1) of the action <unload package moon> of the constraint tree in figure 6). This is because the only (non-no-op) action of level 1 of the plangraph which gives at(package, earth) is <unload package ?loc12>, and this action has a delete effect which unifies with in(package, ?r1).

The Bsr-graphplan algorithm is shown in figure 7. It has four arguments: P – the problem, n – the maximum number of levels to expand the graph, pg – the plangraph – initially consisting of a single level containing the goal formula of P, and tree - the partial ordering constraints – which initially consists of a single node containing the null action as its upper part, and the goal formula as its lower part. The plangraph is incrementally extended on lines 4 and 6. After the disjunctive regression (line 4), the domain operators' add-effects are partially-instantiated in all possible ways with the current set of subgoals on line 5. Lines 7-15 are used to incrementally expand the next level of the constraint tree, considering the partially-instantiated operators which result from the disjunctive regression. Actions containing regression interfering

```
FORWARD_SEARCH (G, S, action_levels, n, tree, plan)
1   if n > length(action_levels)
2     then return failure
3   if G ⊂ S
4     then return plan
5   actions ← ∪ instantiations(action_level[n], S)
6   while unexplored combinations of actions still remain
7         A ← nondeterministically choose {a₁, a₂, ..., aₘ} ⊂ actions
8         if ∀a∃B s.t. a ∈ A, B ∈ branches(tree),
9                 b ∈ actions(B),
10                a & b are unifiable,
11                ∀c∃D s.t. c ∈ ancestors(b),
12                      D ∈ plan, d ∈ D,
13                      c & d are unifiable
14        then next_S ← {∪progress(S,a) | a ∈ A}
15            if next_S ∈ hash_table[n]
16               then backtrack
17               else solution ← forward_search(G, next_S, action_levels, n+1, plan+A)
18                   if solution
19                      then return solution
20                      else hash_table[n] ← next_S
21     return failure
```

**Figure 8.** The forward search algorithm.

effects are noted in lines 9-12 and pruned from the partial orderings tree in lines 13-15. Line 20 invokes a forward search of the entire plangraph.

The forward search algorithm is shown in figure 8, and when initially invoked is provided with the problem goals, the initial state, the planning graph action levels, the index of the first action level, the tree of partial ordering constraints, and the null plan. A global hash table is used to store states from which it is not possible to reach the goals (given the current plan-graph). The set S is initialized to be the problem initial state. Lines 5 – 7 generate all the possible combinations of actions from the current action-level which can be fully-instantiated from the current state S. The decision at line 7 is backtrackable. Although, as written it seems to consider all subsets of actions, subsets containing interfering actions will automatically be pruned by the constraint tree analysis. The next state is computed by unioning the progressions of S through the A actions (line 14) and the search is recursively invoked on the next action level (line 17). Upon the computation of each new state, we check to see if S has previously been encountered at level n (line 15), and has been found to be failing. If it has, the search can backtrack immediately without exploring the rest of the graph above level n. Further, suppose the search does explore the graph above level n, and doesn't find a plan. The dead-end paths are noted on line 20, and these memos will be used by the subsequent searches. Notice that the memos, i.e., the states that do not lead to goal states, are possible even in domains like blocks world where every state is reachable from every other state—because the memos are with respect to a specific bounded plan-graph, and only state that goal state cannot be reached within a specific number of steps.

## 4    Empirical Evaluation

We have implemented the Bsr-graphplan algorithm described in this paper, and have evaluated its performance in comparison to Graphplan on a set of problems from blocks world (containing, pickup, putdown, stack and unstack actions and the flexibility to use multiple robot arms). Our current experiments have concentrated on highlighting the differences between Bsr-graphplan and Graphplan. More extensive evaluation of the implementation on the standard benchmark problems is underway. Our implementation of Bsr-graphplan is based on the common lisp implementation of Graphplan due to Mark Peot. To make the comparisons meaningful, we ensured that both Graphplan and Bsr-graphplan share many of the same datastructures.

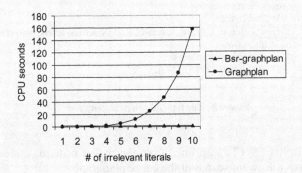

Figure 9. Effect of irrelevant initial state literals.

In the first set of experiments, we compared the performance of the two planners on a set of blocksworld problems, all of which share the same goal state - that of inverting a 2-block stack. The problems however contain varying amount of irrelevant literals (i.e., other 2-block stacks) in the initial state. As the results in figure 9 show, Bsr-graphplan, which is goal directed, is unaffected by the irrelevant literals while Graphplan worsens in performance as the amount of irrelevant literals in the initial state increase.

In the second set of experiments we compared the performance of Bsr-graphplan and Graphplan in solving a stack inversion problem (with four 2-block stacks) as the number of robot arms is varied from 1 to 4. Here, the ability of Bsr-graphplan to handle partially instantiated actions is expected to help since there are many search branches that fail regardless of the identity of the robot arm being used, and Graphplan repeatedly discovers these failures with different identities of robot arms. As the results in figure 10 show, Bsr-graphplan once again performs significantly better than Graphplan. The shape of the graph for Graphplan can be explained by the fact that as the number of robot arms increase, there are two factors that come into play: the "width" of the plan-graph increases, and the "length" of the solution bearing plan-graph reduces. The search time, which is of the order $w^l$ first increases and then reduces. The length starts at 16 (for the single robot case) and stabilizes at 4, which is the number of operations required to invert a single stack once we have four robots (since each robot can independently invert one of the stacks). Although not shown in the plot, beyond 4 robots, Graphplan again starts worsening in performance while

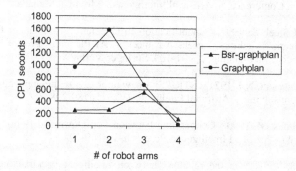

Figure 10.  Effect of the number of robot arms.

Bsr-graphplan remains immune.

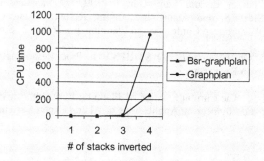

Figure 11.  Effect of the problem size.

In the final set of experiments, we compared the performance of Bsr-graphplan and Graphplan on a set of stack inversion problems. Each problem of size k contains k 2-block stacks in the initial state, all of which must be inverted in the goal state. A single robot hand is assumed to be available. Here, although there are no obvious irrelevant literals anywhere in the initial state, Graphplan considers many fruitless actions involving shifting blocks from one stack to another. Bsr-graphplan, being goal directed, is able to avoid much of this inefficiency. Figure 11 shows the results of this set of experiments. The lengths of the plans produced here go from 4 to 16. We note, once again, that Bsr-graphplan is able to dominate Graphplan in terms of performance.

# References

1. Blum, A. and Furst, M.L. 1995, "Fast Planning Through Planning Graph Analysis". In Proc. International Joint Conference on Artificial Intelligence. Montreal, Canada.

2. Burgess, A. and Steel, S. 1997, "Quantification in Generative Refinement Planning". In Steel, S. and Alami, R., editors, Recent Advances in AI Planning: Proceedings of the 4th European Conference on Planning, pp. 91-103. Springer-Verlag.

3. Fikes, R.E. and Nilsson, N.J. 1971, "STRIPS: a new approach to the application of theorem proving to problem solving". Artificial Intelligence, 2(3-4):189-208.

4. Fuchs, D. and Fuchs, M. 1999, "Cooperation between Top-Down and Bottom-Up Theorem Provers". Journal of Artificial Intelligence Research, 10:169-198.

5. Kambhampati, S. 1998, "On the relations between intelligent backtracking and failure-driven explanation based learning in constraint satisfaction and planning". Artificial Intelligence, 105(1-2):161-208.

6. Kambhampati, S. 1997, "Refinement Planning as a Unifying Framework for Plan Synthesis". AI Magazine, 18(2):67-97.

7. Kambhampati S., Parker E. and Lambrecht E. 1997, "Understanding and Extending Graphplan". In Steel, S. and Alami, R., editors, Recent Advances in AI Planning: Proceedings of the 4th European Conference on Planning, pp. 260-272. Springer-Verlag.

8. Lifschitz, V. 1986, "On the semantics of STRIPS". In Proc. Reasoning about Actions and Plans, pp. 1-9. Morgan Kaufmann.

9. McDermott, D. 1996, "The Current State of AI Planning Research". Invited paper, 9th Intl. Conf. on Industrial and Engineering Applications of AI and Expert Systems.

# GRT: A Domain Independent Heuristic for STRIPS Worlds Based on Greedy Regression Tables

Ioannis Refanidis  and  Ioannis Vlahavas

Department of Informatics, Aristotle University of Thessaloniki
54006, Thessaloniki, Greece
yrefanid, vlahavas@csd.auth.gr

**Abstract.** This paper presents Greedy Regression Tables (GRT), a new domain independent heuristic for STRIPS worlds. The heuristic can be used to guide the search process of any state-space planner, estimating the distance between each intermediate state and the goals. At the beginning of the problem solving process a table is created, the records of which contain the ground facts of the domain, among with estimates for their distances from the goals. Additionally, the records contain information about interactions that occur while trying to achieve different ground facts simultaneously. During the search process, the heuristic, using this table, extracts quite accurate estimates for the distances between intermediate states and the goals. A simple best-first search planner that uses this heuristic has been implemented in C++ and has been tested on several "classical" problem instances taken from the bibliography and on some new taken from the AIPS-98 planning competition. Our planner has proved to be faster in all of the cases, finding also in most (but not all) of the cases shorter solutions.

## 1 Introduction

It is broadly accepted that plan-space planning is more efficient than state-space planning [1, 10]. Recent results have shown that the performance of planners like GRAPHPLAN and SATPLAN seems to be even better [2, 8]. However, experience suggests that state-space planners with appropriate heuristic functions outperform all the others. For example, by using domain-dependent heuristics, very large blocks world problems can be solved in a few seconds [6, 12].

In this paper a new domain independent heuristic for state-space planners is presented. The heuristic has been inspired by the one used in the ASP planner [3] and we believe that it copes successfully with the main inefficiencies of the former. The proposed heuristic algorithm works as follows: At the beginning of the problem solving process it creates a table, the records of which contain the ground facts of the domain, together with estimates for their distances from the goals. Additionally, this table contains information on the interactions occurring while trying to achieve different ground facts simultaneously. During the planning process, the heuristic uses this table to extract quite accurate estimates for the distances between the intermediate states and the goal state. We name the proposed heuristic algorithm GRT, from the acronym of the words "Greedy Regression Tables", since its tables are obtained through greedy regression on the goals of each problem instance.

In order to have a notion of the heuristic performance, we developed, using C++, a simple best-first planner that uses the GRT heuristic to guide the search process (we

S. Biundo and M. Fox (Eds.): ECP-99, LNAI 1809, pp. 347–359, 2000.

call it 'GRT planner' henceforth). We ran this planner on several classical problem instances, taken from the bibliography, such as some blocks world instances [8], some "logistics" instances [13] and some "rocket" instances [2]. Moreover, we ran GRT planner on some new problem instances originally presented at the AIPS-98 planning competition. Our planner was faster in all of the cases, finding also in most (but not all) of the cases shorter solutions.

The rest of the paper is organized as follows: Section 2 presents briefly the ASP heuristic, while section 3 presents in detail the GRT heuristic. Section 4 presents comparative performance results of the GRT planner with other known planners and, finally, section 5 concludes the paper and poses future directions.

## 2   The ASP Heuristic

This section gives a brief description of the ASP heuristic, while the next section presents in detail the greedy regression tables approach.

In STRIPS [5], each ground action $\bullet$ is represented by three sets of facts: the precondition list $P(\bullet)$, the add list $A(\bullet)$ and the delete list $D(\bullet)$, where $D(\bullet) \subset P(\bullet)$. A state $s$ is defined as a collection of ground facts and an action $\bullet$ determines a mapping from a state $s$ to a new state $s' = res(s,\bullet)$. In the formalization being used henceforth, it is assumed a finite set of constants and no function symbols, so that the set of ground actions is finite. An action $\bullet$ can be applied to a state $s$, if $P(\bullet) \subseteq s$. Then, the new state $s'$ is defined as:

$$s' = res(s,\bullet) = s - D(\bullet) + A(\bullet) . \tag{1}$$

For each ground action $\bullet$ and for each ground fact $p \in A(\bullet)$, a rule of the form $C \rightarrow p$ can be written, where $C = P(\bullet)$. Assuming a set of rules $C \rightarrow p$ resulting from the actions to be considered, it is said that a fact $p$ is *reachable* from a state $s$ if $p \in s$ or there is a rule $C \rightarrow p$ such that each fact $q$ in $C$ is reachable from $S$.

So, a function $g(p,s)$ is defined, which inductively assigns each fact $p$ a number $i$, which is an estimate of the number of steps that are needed to achieve $p$ from $s$. That is, $g(p,s)$ is set to 0 for every fact $p$ that holds in $s$, while $g(p,s)$ is set to $i+1$, $i \geq 0$, for each fact $p$ for which a rule $C \rightarrow p$ exists, such that $\bullet_{r \in C} g(r,s) = i$. So:

$$g(p,s) \stackrel{def}{=} \begin{cases} 0 & if \ \ p \in s \\ i+1 & if \ \ for \ \ some \ \ C \rightarrow p, \ \sum_{r \in C} g(r,s) = i \\ \infty & if \ \ p \ \ is \ \ not \ \ reachable \ \ from \ \ s \end{cases} \tag{2}$$

It should be noted that $g(p,s)$ does not define a unique value for a domain fact $p$. Since it is possible to exist more than one rules of the form $C \rightarrow p$, while the same may happen for the facts included in $C$, the computed value depends strongly on the rule selection strategy. Moreover, $g(p,s)$ may assign a finite distance to a fact that can not be achieved.

For a set of facts $P$, we define $g(P,s) \stackrel{def}{=} \sum_{q \in P} g(q,s)$ and the function $H_{ASP}(s)$ as:

$$H_{ASP}(s) \stackrel{def}{=} g(G,s) . \tag{3}$$

where $G$ are the goals of a problem. The complexity of computing $H_{ASP}(s)$ is linear in both the number of ground actions and the number of ground facts. The function always overestimates the cost to the goal and hence is not admissible.

# 3   The Greedy Regression Tables Heuristic

The ASP heuristic presented above has two main drawbacks. The first is that it is recomputed from scratch for each intermediate state. The second is that it does not take into account the interactions that occur while trying to achieve several conjunctive goals simultaneously, resulting as a consequence in overestimates. In the following subsections the GRT heuristic, denoted henceforth also as $H_{GRT}$, is presented. The new heuristic, solves the first problem and copes well with the second one.

## 3.1   Estimating Distances Backwards

Instead of estimating the distance of the goal state forwards from an intermediate state, $H_{GRT}$ estimates the distance of an intermediate state backwards from the goal. At the beginning of the planning process, the heuristic estimates once the distances of all the domain's ground facts backwards from the goal state, as it is shown in the next sub-section. During the planning process, the above estimates are used to compute the distance of each intermediate state from the goal. Calculating the distance between two states either backwards or forwards often results in different values, since no heuristic is precise. Yet there is no reason why any of the two directions should be preferable.

In order to compute the distances backwards, the domain operators have to be inverted. Suppose that we have the operator • and two states $s$ and $s'$, such that • is applicable in $s$ and $s' = res(s,•)$. The *inverted operator* $a'$ of • is an operator, such that $s = res(s', a')$. The inverted operator is defined from the original operator as follows:

$$P(•')=A(•) + P(•) - D(•) .$$
$$D(•')=•(•) .$$
$$•(•)=D(•) .$$

(4)

The inverted ground operators are applied to the goal state, assigning inductively each ground fact $p$ an integer $i$ that estimates the steps needed to achieve $p$ from the goal state, as it is described in the next sub-section.

A problem when calculating $H_{GRT}$ backwards is that the goal state in most of the problems is incomplete. For example, in the "logistics" domains [13] the goal state does not determine where are the trucks and the planes. If the goal state is incomplete, it is impossible to apply the inverted operators to it and hence to estimate the distances of the domain's facts from the goal. The solution adopted is to enrich the incomplete goal state with all the missing facts that are not in contradiction with the existing ones. For example, since the goal state of the "logistics.a" problem instance does not determine at which airports are the two planes, it is considered that each one of the two planes could be at any of the tree airports. So, the ground facts:

$$at(plane1, pgh\_air), at(plane1, bos\_air), at(plane1, la\_air),$$
$$at(plane2, pgh\_air), at(plane2, bos\_air), at(plane2, la\_air) .$$

are added to the new goal state and are assigned a distance equal to 0. This new goal state will be referenced henceforth as the *enriched goal state*. It is obvious that the enriched goal state is not a valid state, since, for example, a plane cannot be simultaneously both in *pgh_air* and *bos_air* or *la_air* airports. Including these facts in the enriched goal state, it is possible to apply the inverted operators to it and hence compute the distances of the domain's facts from the goal. Currently, the enriched goal state is defined manually, together with the definition of each problem. The issue of its semi- or fully-automated definition will be addressed in the future.

## 3.2 Interactions among Conjunctive Goals

The second innovation of the proposed heuristic is that it takes into account the interactions that occur while trying to achieve several conjunctive goals simultaneously, leading to more accurate estimates of the distances between the intermediate states and the goal state. GRT assigns to each ground fact $p$ of a domain not only an estimate of its distance from the goal state, but also a list of other ground facts $\{r_1, r_2, ...\}$, which are also achieved while trying to achieve $p$. This list will be called the list of *related facts* of $p$ and will be denoted as $rel(p)$. The distance between a fact $p$ and the goal state will be denoted as $dist(p)$. Henceforth, when it is referred that a fact has been achieved, this will mean that its distance is finite. The calculation of the distances and the lists of related facts is done by algorithm 1. Algorithm 2, which is used by algorithm 1 (step 4b), is presented next.

**ALGORITHM 1:** Finding the distances from the goal state and the lists of related facts for all the ground facts of a domain

1.  Calculate the set Inv of all the inverted ground actions of the domain.
2.  Let F be the set of all the ground facts of the domain. For each f ∈ F set dist(f)= +∞ .
3.  Let G' be the enriched goal state of the problem. For each f ∈ G' set dist(f)= 0 and rel(f)= ∅ .
4.  While (Inv ≠ ∅ ∧ ∃ f ∈ F : dist(f)= +∞) do:
    a)  Select an action   from Inv, such that ∀ q_i ∈ P( ), dist(q_i)< +∞ .
    b)  Call algorithm 2, passing to it as input parameters the set of q_is, together with their distances dist(q_i) and their lists of related facts rel(q_i). Let Cost be the returned value of algorithm 2.
    c)  Let A'( ) be the subset of the ground facts of A( ) that have not already been achieved:
        A'( )= { p_i ∈ A( ): dist(p_i)= +∞ }
        For every p_i ∈ A'( ), set dist(p_i) = Cost +1.
    d)  For each fact p_i ∈ A'( ) compute the list rel(p_i) as the union of the elements of the precondition list of action   , the lists of related facts of  's precondition elements, the other elements in A'( ), excluding the elements of the delete list of   and those facts whose distance is equal to zero. More formally:
        
        rel(p_i) = { f ∈ (∪_i q_i) ∪ (∪_i rel(q_i)) ∪ A'( )- {D( ), p_i}: 
        
        $$dist(f)>0 \ \}$$
        
    e)  Remove from Inv all the actions, whose all facts in their add list have distances less than +∞.

As it follows by the presentation of algorithm 1, at each iteration the algorithm selects an inverted action, whose preconditions have all been achieved (step 4a), computes the cost of achieving all the preconditions simultaneously (step 4b), adds 1 to this cost and assigns the new cost to those facts of the action's add list, that have not already been achieved (step 4c), calculates their list of related facts (step 4d) and finally removes from the set of the inverted actions those actions, whose facts in their add lists have all been achieved. This is repeated until all the ground facts of the domain have been achieved, or until the set of the inverted actions becomes empty (which means that there are facts with infinite distances).

The number of iterations that the above algorithm performs is bounded both by the number of the domain's ground facts (since at each iteration at least one fact is achieved) and by the number of domain's ground actions (since at each iteration at least one inverted action is removed from *Inv*). The selection strategy for the inverted actions used in step 4a affects the results. A good strategy is to select the operator with the minimum cost preconditions.

It should be noted that in step 4d, the facts that have zero distances are excluded from $rel(p_i)$. This is not obligatory, but it is done for efficiency, since these facts do not play any role while calculating distances by the second algorithm that follows.

The next algorithm estimates the total cost (number of steps) needed to achieve simultaneously a set of facts, provided that all these facts have finite distances. The input to this algorithm is a set of facts $\{p_1, p_2, ....p_n\}$, together with their individual costs $dist(p_i)$ and their lists of related facts $rel(p_i)$. This algorithm is used by the 4b step of the algorithm 1, in order to compute the cost of achieving simultaneously the preconditions of an operator, but is also used during the planning process, in order to compute the distances between the intermediate states and the goal state (i.e. the cost of achieving simultaneously all the facts of each intermediate state, which is the estimate of the distance of the intermediate state from the goal).

**ALGORITHM 2:** Estimating the cost of achieving a set of ground facts simultaneously

```
1.  Let M₁ = { p₁, p₂, ..., pₙ }, such that ∀pᵢ ∈ M₁: dist(pᵢ)<+∞.
    Let Cost = 0 .
2.  While (M₁ ≠ ∅) do:
    a) Let M₂ be the set of facts p ∈ M₁ that are not included in
       any list of related facts of another fact q of M₁, without
       q being also included in their lists of related facts.
       More formally:

       M₂ = { p: p ∈ M₁, ∀ q ∈ M₁, p ∈ rel(q) ⇒ q ∈ rel(p) }

    b) Let M₃ be the set of those facts of M₁ that are not
       included in M₂, but are included in at least one of the
       lists of related facts of the elements of M₂.

       M₃ = { p: p ∈ M₁ - M₂, ∃ q ∈ M₂, p ∈ rel(q) }

    c) Sum the distances of the facts of M₂. For groups of facts
       where each one is contained in the list of related facts
       of the others, consider their common distance once. Add
       the result to the Cost.
    d) Let M₁ = M₁ - M₂ - M₃.
3.  Return Cost
```

Let us explain how algorithm 2 works. Suppose that $M_1$ initially consists of three facts, $M_1 = \{p_1, p_2, p_3\}$, such that $dist(p_1) = 1$, $rel(p_1) = \emptyset$, $dist(p_2) = 2$, $rel(p_2) = \{p_1\}$, $dist(p_3) = 3$, $rel(p_3) = \{p_2\}$. Such a situation can arise if an action $\alpha_1$ achieves first $p_1$, then another action $\alpha_2$, which has $p_1$ as precondition achieves $p_2$ without deleting $p_1$, and finally a third action $\alpha_3$, which has $p_2$ and $p_1$ as precondition but which deletes $p_1$, achieves $p_3$.

In step 2a of the first iteration, $M_2$ is set equal to $\{p_3\}$, since $p_3$ is not included in any list of related facts. The other two facts are not added to $M_2$, since for example $p_2$ is included in $rel(p_3)$, while $p_3$ is not included in $rel(p_2)$. Similarly, $p_1$ is included in $rel(p_2)$, while $p_2$ is not included in $rel(p_1)$. In step 2b, $M_3$ is set equal to $\{p_2\}$, since $p_2$ is the only fact that is included in $rel(p_3)$. In step 2c, $dist(p_3)$ is added to $Cost$, so $Cost$ becomes 3. Finally, in step 2d, $M_1$ is set to $M_1-M_2-M_3$, so $M_1$ becomes equal to $\{p_1\}$.

The second iteration begins with $M_1 = \{p_1\}$, so obviously $M_2 = \{p_1\}$ (step 2a), $M_3 = \varnothing$ (step 2b), *Cost* becomes $Cost + dist(p_1) = 4$ (step 2c) and $M_1$ becomes empty (step 2d). This is the last iteration and value 4 is returned.

It should be noted that $M_3$ is partitioned in groups, such that the facts of each group contain one another in their lists of related facts. The facts in each group have all been achieved by the same inverted action and have obviously the same distance. So, step 2c sums the costs of these groups.

The number of iterations that the algorithm 2 performs is bounded by the initial size of $M_1$, but usually the algorithm performs at most two iterations.

### 3.3 An Example

Consider the 3-blocks problem situation shown in figure 1, which has been taken from [3]. In the following paragraphs, we will show how the above two algorithms calculate the distances and the lists of related facts for all the ground facts of this domain, creating the greedy regression table for this problem. Next, we will use this table to select one of the three applicable actions to the initial state. Finally, we will compare these results with that produced by the ASP heuristic.

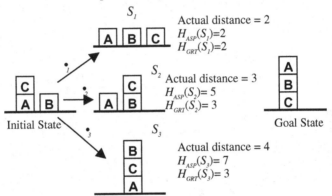

**Fig. 1.** Comparison between $H_{ASP}$ and $H_{GRT}$

The goal state for the above problem consists of the following ground facts:

$$on(C, table), on(B, C), on(A, B), clear(A)$$

where *on(X,Y)* means that object *X* is on top of *Y* and *clear(X)* means that object *X* has no other object on top of it (we prefer a Prolog-style representation, although we use capitals for the names of the blocks). The goal state is complete, so it is not needed to enrich it with more ground facts. Moreover, because in the blocks world domain, it is possible to completely cancel any action by performing another valid action, the set of the inverted actions *Inv* is identical with the set of the original ground actions (step 1 of the first algorithm).

Algorithm 1 assigns distances equal to 0 and empty lists of related facts to each ground fact of the goal state (step 3), as it is shown in table 1, rows 1 through 4. The only "applicable" action to the goal state is "Move_A_from_B_onto_table" (step 4a). The precondition list of this action consists of the facts:

$$on(A, B), clear(A)$$

which have zero distances and empty lists of related facts. Using the first algorithm, a distance equal to 1 is assigned to the ground facts that are included in the add list of the selected action, i.e. to the facts *on(A, table)* and *clear(B)* (the second algorithm computes a cost equal to 0 -step 4b-, and step 4c of the first algorithm adds 1 to this cost). Since the facts in the precondition list of the action have zero distances and empty lists of related facts, the lists of related facts for the newly achieved facts contain (step 4d ) one another, as it is shown in table 1 (rows 5 and 6). Finally, some actions, like "Move_A_from_Table_onto_B" etc., are removed from *Inv*, because all the facts in their add lists have already been achieved (step 4e).

Suppose that the first algorithm selects next (step 4a) to apply the action "Move_B_from_C_to_table" (although it could select to apply the action "Move_B_from_C_to_A"). The precondition list of this action consists of the facts:

$$on(B, C), clear(B)$$

The cost of achieving these facts simultaneously, as it is computed by the second algorithm, is 1. So the distances for the facts in the add list of the selected action, i.e. for the facts *on(B, table)* and *clear(C)*, are set equal to 2 (table 1, rows 7 and 8). Let us calculate the list of related facts for the fact *on(B, table)*. The only fact from the action's precondition list that has non-zero distance and non-empty list of related facts is *clear(B)*. The union of this fact and its list of related facts is:

$$\{clear(B)\} \cup rel(clear(B)) = \{clear(B)\} \cup \{on(A, table)\} = \{ clear(B), on(A, table)\}$$

From this set the facts of the delete list of the selected action are removed:

$$\{ clear(B), on(A, table) \} - \{on(B, C)\} = \{ clear(B), on(A, table) \}$$

Finally, the other facts that the selected action achieves are added to the above set, resulting in (table 1, row 7):

$$rel(on(B,table)) = \{ clear(B), on(A, table), clear(C) \}$$

Similarly, the distances and the lists of related facts for the other facts of the domain are computed.

Having created the Greedy Regression Table for the problem of figure 1, we can now compute, using algorithm 2, the distances from the goal state of the three states that result by applying all the applicable actions to the initial state of this problem instance. Let us compute for example the distance of state $S_2$ from the goal state. This state consists of the following ground facts:

$$on(A, table), clear(A), on(B, table), on(C, B), clear(C)$$

According to the second algorithm, $M_1$ contains initially all the above facts. $M_2$ contains only the fact *on(C,B)*, since all the other facts of $M_1$ are contained in *rel(on(C,B))*, while *on(C,B)* is not contained in any list of related facts of the other facts (step 2a). $M_3$ contains all the other facts (step 2b). So, the algorithm adds $dist(on(C,B)) = 3$ to *Cost* (step 2c). Algorithm 2 does not perform a second iteration, since $M_1$ becomes empty (step 2d), so the algorithm terminates and returns the value 3, as the distance of $S_2$ from the goal state. Similarly, the distances for $S_1$ and $S_3$ are computed.

For each of the three resulting states $S_1$, $S_2$ and $S_3$, figure 1 shows their actual distance from the goal state together with the estimates of these distances using $H_{ASP}$ and $H_{GRT}$. For $S_1$, both the heuristics find the exact distance. For $S_2$, only $H_{GRT}$ finds the exact distance, while $H_{ASP}$ has an overestimate by 2. And for $S_3$, $H_{ASP}$ overestimates the distance by 3, while $H_{GRT}$ underestimates the distance only by 1. It is worth noting that most of the times when $H_{GRT}$ does not find the exact distance, it results in an underestimate of it.

In the example of figure 1, both heuristics select the same operator, i.e. $\bullet_T$="Move_C_from_A_to_table", but in more complicated situations, there are a lot of cases where $H_{ASP}$ does not select the best one. This happens sometimes with $H_{GRT}$ too, but more rarely.

**Table 1.** The Greedy Regression Table for the problem of figure 1

|    | Ground facts | Distance from Goal | Related facts |
|----|--------------|--------------------|---------------|
| 1  | on(C, table) | 0 | { } |
| 2  | on(B,C)      | 0 | { } |
| 3  | on(A,B)      | 0 | { } |
| 4  | clear(A)     | 0 | { } |
| 5  | clear(B)     | 1 | {on(A, table)} |
| 6  | on(A, table) | 1 | {clear(B)} |
| 7  | on(B, table) | 2 | {clear(B), on(A, table), clear(C)} |
| 8  | clear(C)     | 2 | {on(B, table), clear(B), on(A, table)} |
| 9  | on(B,A)      | 2 | {clear(C), clear(B), on(A, table)} |
| 10 | on(A,C)      | 3 | {on(B, table), clear(B), on(A, table)} |
| 11 | on(C,B)      | 3 | {clear(C), on(B, table), on(A, table), clear(A)} |
| 12 | on(C,A)      | 3 | {clear(C), on(B, table), clear(B), on(A, table)} |

## 4 Performance Results

In order to measure the performance of the proposed heuristic, we have implemented, using C++, a simple planner (we name it 'GRT planner') that uses a best-first search algorithm and the GRT heuristic to guide the search process. We ran the GRT planner on two sets of domains. The first set consists of some "classical" problem instances that can be found in the bibliography in the last decade. The second set of problem instances is taken by the AIPS-98 planning competition. In the following sub-sections we give comparative results (solution time and length) of GRT, against other known planners.

### 4.1 Classical Problem Instances

The classical problem instances selected for testing are some "blocks world" instances [8], some "logistics" instances [13] and some "rocket" instances [2]. Table 2 compares the performance, in terms of solution time and solution length, of GRT with other known planners, i.e. GRAPHPLAN, BLACKBOX [9] and HSP [4]. We thank Blum, Furst, Kautz, Selman, Bonet, Loerincs and Geffner for making the code of their planners available[1].

---

[1] GRAPHPLAN is available at
http://www.cs.cmu.edu/afs/cs.cmu.edu/usr/avrim/www/graphplan.html.
BLACKBOX is available at http://www.research.att.com/~kautz/blackbox/index.html.
HSP is available at http://www.ldc.usb.ve/~hector/.
GRT is available through mail at yrefanid@csd.auth.gr. Also, a Prolog version of the planner is available.

BLACKBOX is an evolution of SATPLAN. BLACKBOX automates the transformation of a planning problem to a satisfiability one, using some of the GRAPHPLAN features. The version of BLACKBOX we have tested is 3.4. HSP is an evolution of ASP that uses hill-climbing instead of B-LRTA. The version of HSP we have tested is 1.1.

The measurements for all planners have been taken on a SUN Enterprise 3000 Unix server, running at 167 MHz and having 64MB memory. All planners have been implemented in C/C++. The results are shown in table 2. Short dashes mean that no solution was found after 10 minutes, or the planner exhausted the available memory. The time measurements for GRT include both the time for the creation of the table and the time of the search process. As it is shown by table 2, GRT performs better than all the other planners, with respect both to the time needed to find a solution and to the solution length.

**Table 2.** Comparison with several known planners (time in msecs)

| problem | GRAPHPLAN | | BLACKBOX | | HSP | | GRT | |
|---|---|---|---|---|---|---|---|---|
| | steps | time | steps | time | steps | time | steps | time |
| bw_large_a | 10 | 1810 | 23 | 2400 | 11 | 600 | 8 | 310 |
| bw_large_b | 15 | 9570 | 18 | 206000 | 16 | 1400 | 12 | 770 |
| bw_large_c | - | - | - | - | 17 | 6800 | 16 | 2310 |
| bw_large_d | - | - | - | - | 31 | 46000 | 21 | 5900 |
| logistics_a | - | - | 69 | 67000 | 61 | 1400 | 55 | 450 |
| logistics_b | - | - | 64 | 80000 | 49 | 1200 | 47 | 470 |
| logistics_c | - | - | 77 | 101000 | 74 | 2200 | 56 | 650 |
| rocket_a | - | - | 31 | 105000 | - | - | 24 | 100 |
| rocket_b | - | - | 28 | 104000 | 35 | 580 | 28 | 220 |

## 4.2 AIPS-98 Problem Instances

During the 4th Artificial Intelligence Planning Systems conference, which held in June 1998 at the Carnegie Mellon University, Pittsburgh, Pennsylvania, it took place a planning competition. During this competition, some new problem domains and a lot of new problem instances have been presented. Four contestants participated at the STRIPS track of this competition: BLACKBOX, HSP, IPP (by Jana Koehler, Freiburg University) and STAN (by Derek Long and Maria Fox, Durham University). The platform of the competition consisted of Pentium-II machines at 300 MHz and 128 MB main memory, running Linux operating system. All planners were compiled programs written in C/ C++[2].

In order to have comparable results, we ran GRT on the same problem instances and on a similar machine. Actually, our machine had a Pentium Celeron 300 processor, with 64 MB main memory.

From the domains used in the competition, we present in this paper the "Gripper", "Logistics" and "Movie" domains. With the other domains, i.e. the "Grid", "Mystery" and "Mprime", GRT do not cope quite successfully. These domains embody the notion of the universal quantification. For example, in the "Mystery" and "Mprime" domains, each vehicle can contain more than one cargoes. The problem with the STRIPS representation of these domains is that if a cargo is contained within a vehicle, the fact that determines the location of the cargo is removed from the state

---

[2] The official competition results and other information about this competition can be found at ftp://ftp.cs.yale.edu/pub/mcdermott/aipscomp-results.html.

description. But this fact is necessary for the correct application of the GRT heuristic. In general, a pure STRIPS representation is not suitable for this type of domains. What is needed is a representation that permits operators to have universal quantification in their effects (which of course is not supported currently by GRT, but it is subject of future work). Moreover, the "Mystery" and "Mprime" domains embody the notion of the resource usage, something that is not currently faced by the GRT heuristic effectively.

The absence of universal quantification in pure STRIPS operators affects also the effectiveness of the GRT heuristic in the "Logistics" domain. But in this domain, the possibility that have the airplanes and the trucks to go from any location to any other valid location with only one action decreases the consequences of the above problem.

### 4.2.1 Gripper Domain

This domain consists of a robot that has two grippers, with the ability to carry a ball in each one. The goal is to take N balls from one room to another. This domain was created by Jana Koehler. There were 20 problem instances. Table 3 shows comparative results for this domain.

**Table 3.** Comparative results for the Gripper domain (time in msecs)

| Problems | BLACKBOX | | HSP | | IPP | | STAN | | GRT | |
|---|---|---|---|---|---|---|---|---|---|---|
| | steps | time | steps | time | steps | time | steps | time | steps | time |
| prob01 | 11 | 113 | 13 | 2007 | 15 | 50 | 11 | 46 | 11 | 10 |
| prob02 | 17 | 7820 | 21 | 2150 | 23 | 380 | 17 | 1075 | 17 | 25 |
| prob03 | - | - | 31 | 2485 | 31 | 3270 | 23 | 54693 | 23 | 50 |
| prob04 | - | - | 37 | 3060 | 39 | 26680 | 29 | 3038381 | 29 | 50 |
| prob05 | - | - | 47 | 3320 | 47 | 226460 | - | - | 35 | 110 |
| prob06 | - | - | 53 | 3779 | - | - | - | - | 41 | 160 |
| prob07 | - | - | 63 | 4797 | - | - | - | - | 47 | 220 |
| prob08 | - | - | 71 | 5565 | - | - | - | - | 53 | 280 |
| prob09 | - | - | 79 | 6675 | - | - | - | - | 59 | 440 |
| prob10 | - | - | 85 | 7583 | - | - | - | - | 65 | 550 |
| prob11 | - | - | 93 | 9060 | - | - | - | - | 71 | 770 |
| prob12 | - | - | 101 | 10617 | - | - | - | - | 77 | 1040 |
| prob13 | - | - | 109 | 12499 | - | - | - | - | 83 | 1260 |
| prob14 | - | - | 119 | 15050 | - | - | - | - | 89 | 1600 |
| prob15 | - | - | 125 | 16886 | - | - | - | - | 95 | 2030 |
| prob16 | - | - | 135 | 20084 | - | - | - | - | 101 | 2470 |
| prob17 | - | - | 143 | 23613 | - | - | - | - | 107 | 3080 |
| prob18 | - | - | 151 | 26973 | - | - | - | - | 113 | 3741 |
| prob19 | - | - | 157 | 29851 | - | - | - | - | 119 | 4560 |
| prob20 | - | - | 165 | 33210 | - | - | - | - | 125 | 5390 |

Except for the first problems of this domain (which are the easiest), the other problems were solved only by HSP. As it is shown in table 3, GRT heuristic solved all problems, it needed less time and found shorter solutions.

### 4.2.2 Logistics Domain

This is a classical domain, with several cities, each containing several locations, some of which are airports. There are also trucks, which can drive within a single city, and airplanes, which can fly between airports. The goal is to get some packages from various locations to various new locations. In the competition there have been presented 30 new problem instances, most of them were really hard and have not been solved by any contestant. As it is shown in table 4, GRT solved most of the unsolved problem instances, while in the solved instances it was faster than the contestants. Concerning solution length, BLACKBOX and IPP found shorter solutions in some instances.

**Table 4.** Comparative results for the Logistics domain (time in msecs)

| Problems | AIPS-98 CONTESTANTS | | | | | | | | GRT | |
|---|---|---|---|---|---|---|---|---|---|---|
| | BLACKBOX | | HSP | | IPP | | STAN | | | |
| | steps | time | steps | time | steps | time | steps | time | steps | time |
| prob01 | 27 | 2062 | 43 | 79682 | 26 | 900 | 27 | 767 | 30 | 280 |
| prob02 | 32 | 6436 | 44 | 97114 | - | - | 32 | 4319 | 34 | 1320 |
| prob03 | - | - | - | - | - | - | - | - | 60 | 5550 |
| prob04 | - | - | - | - | - | - | - | - | 69 | 19280 |
| prob05 | - | - | 26 | 14413 | 24 | 2400 | 29 | 364932 | 26 | 390 |
| prob06 | - | - | - | - | - | - | - | - | 80 | 14390 |
| prob07 | - | - | 112 | 788914 | - | - | - | - | 37 | 1760 |
| prob08 | - | - | - | - | - | - | - | - | 48 | 16370 |
| prob09 | - | - | - | - | - | - | - | - | 98 | 50480 |
| prob10 | - | - | - | - | - | - | - | - | 117 | 23130 |
| prob11 | 32 | 6544 | 30 | 86195 | 33 | 6940 | 34 | 12806 | 36 | 1540 |
| prob12 | - | - | - | - | - | - | - | - | 48 | 43060 |
| prob13 | - | - | - | - | - | - | - | - | 79 | 85580 |
| prob14 | - | - | - | - | - | - | - | - | 104 | 60200 |
| prob15 | - | - | - | - | - | - | - | - | 106 | 6750 |
| prob16 | - | - | - | - | - | - | - | - | 62 | 31580 |
| prob17 | - | - | - | - | - | - | - | - | 53 | 12190 |
| prob18 | - | - | - | - | - | - | - | - | 193 | 335050 |
| prob19 | - | - | - | - | - | - | - | - | 174 | 238980 |
| prob20 | - | - | - | - | - | - | - | - | 169 | 324120 |
| prob21 | - | - | - | - | - | - | - | - | 120 | 294230 |
| prob22 | - | - | - | - | - | - | - | - | - | - |
| prob23 | - | - | - | - | - | - | - | - | 118 | 16860 |
| prob24 | - | - | - | - | - | - | - | - | 49 | 98540 |
| prob25 | - | - | - | - | - | - | - | - | - | - |
| prob26 | - | - | - | - | - | - | - | - | - | - |
| prob27 | - | - | - | - | - | - | - | - | - | - |
| prob28 | - | - | - | - | - | - | - | - | - | - |
| prob29 | - | - | - | - | - | - | - | - | - | - |
| prob30 | - | - | - | - | - | - | - | - | - | - |

### 4.2.3 Movie Domain

This is a relatively simple domain. The goal is always the same: to have lots of snacks in order to watch a movie. The various instances differ in the number of problem objects. As it is shown in table 5, the optimal solution in all of the 30 problem instances consists of 7 steps, something that has been found by all planners.

**Table 5.** Comparative results for the Movie domain (time in msecs)

| Problems | AIPS-98 CONTESTANTS | | | | | | | | GRT | |
| | BLACKBOX | | HSP | | IPP | | STAN | | | |
| | steps | time | steps | time | steps | time | steps | time | steps | time |
|---|---|---|---|---|---|---|---|---|---|---|
| prob01 | 7 | 11 | 7 | 2121 | 7 | 10 | 7 | 19 | 7 | 10 |
| prob02 | 7 | 12 | 7 | 2104 | 7 | 10 | 7 | 18 | 7 | 10 |
| prob03 | 7 | 14 | 7 | 2144 | 7 | 10 | 7 | 19 | 7 | 10 |
| prob04 | 7 | 16 | 7 | 2188 | 7 | 10 | 7 | 20 | 7 | 10 |
| prob05 | 7 | 18 | 7 | 2208 | 7 | 10 | 7 | 21 | 7 | 20 |
| prob06 | 7 | 20 | 7 | 2617 | 7 | 10 | 7 | 22 | 7 | 20 |
| prob07 | 7 | 22 | 7 | 2316 | 7 | 20 | 7 | 22 | 7 | 20 |
| prob08 | 7 | 24 | 7 | 2315 | 7 | 20 | 7 | 23 | 7 | 20 |
| prob09 | 7 | 26 | 7 | 2357 | 7 | 20 | 7 | 25 | 7 | 20 |
| prob10 | 7 | 29 | 7 | 2511 | 7 | 10 | 7 | 26 | 7 | 20 |
| prob11 | 7 | 30 | 7 | 2427 | 7 | 30 | 7 | 27 | 7 | 20 |
| prob12 | 7 | 32 | 7 | 2456 | 7 | 30 | 7 | 28 | 7 | 20 |
| prob13 | 7 | 36 | 7 | 3070 | 7 | 20 | 7 | 29 | 7 | 20 |
| prob14 | 7 | 35 | 7 | 2573 | 7 | 30 | 7 | 31 | 7 | 20 |
| prob15 | 7 | 38 | 7 | 2577 | 7 | 30 | 7 | 32 | 7 | 20 |
| prob16 | 7 | 39 | 7 | 2699 | 7 | 10 | 7 | 34 | 7 | 30 |
| prob17 | 7 | 41 | 7 | 2645 | 7 | 30 | 7 | 35 | 7 | 30 |
| prob18 | 7 | 43 | 7 | 2686 | 7 | 10 | 7 | 37 | 7 | 30 |
| prob19 | 7 | 45 | 7 | 2727 | 7 | 30 | 7 | 39 | 7 | 30 |
| prob20 | 7 | 47 | 7 | 2787 | 7 | 20 | 7 | 40 | 7 | 30 |
| prob21 | 7 | 49 | 7 | 2834 | 7 | 20 | 7 | 42 | 7 | 30 |
| prob22 | 7 | 51 | 7 | 2834 | 7 | 20 | 7 | 45 | 7 | 30 |
| prob23 | 7 | 53 | 7 | 2866 | 7 | 20 | 7 | 48 | 7 | 40 |
| prob24 | 7 | 55 | 7 | 3341 | 7 | 20 | 7 | 50 | 7 | 40 |
| prob25 | 7 | 57 | 7 | 2997 | 7 | 30 | 7 | 52 | 7 | 40 |
| prob26 | 7 | 58 | 7 | 3013 | 7 | 40 | 7 | 54 | 7 | 40 |
| prob27 | 7 | 60 | 7 | 3253 | 7 | 50 | 7 | 57 | 7 | 40 |
| prob28 | 7 | 63 | 7 | 3049 | 7 | 40 | 7 | 62 | 7 | 50 |
| prob29 | 7 | 64 | 7 | 3384 | 7 | 50 | 7 | 64 | 7 | 50 |
| prob30 | 7 | 66 | 7 | 3127 | 7 | 40 | 7 | 67 | 7 | 50 |

## 5  Conclusions

In this paper, a new domain-independent heuristic for STRIPS-style problems has been presented. The heuristic is suitable for state-space planners, providing quite accurate estimates of the distance between any two states. Starting from the goal state, the heuristic creates a table, the records of which contain the ground facts of the domain, including estimates of their distances from the goal state and information on the interactions occurring while trying to achieve different ground facts simultaneously. This table is created once, at the beginning of the planning process, and is used during planning to guide the action selection. We call this heuristic GRT, from the acronym of the words "Greedy Regression Tables", since its tables are obtained through greedy regression on the goal state of each problem instance.

Using C++, a simple planner, consisting of a best-first algorithm and the proposed heuristic has been implemented and tested on several classical problem instances, taken by the bibliography, and on some new that have been originally presented at the AIPS-98 planning competition. The benchmarks show that the combination of the proposed heuristic with a simple best-first search strategy is quite competitive, with respect to the time needed to find a solution. Moreover, our planner finds the shortest plans in many cases.

Our future work is to develop an integrated planning system that will use the Greedy Regression Tables to guide search. This system will be able to cope with more sophisticated domains, that embody universal quantification, resource usage, constraint satisfaction and temporal characteristics. The goal is that the system will be able to handle as many as possible aspects of the Planning Domain Definition Language (PDDL). Strictly concerning the heuristic, we will try to convert it to an admissible one, without losing any of its efficiency. Finally, we are about to develop an algorithm that will automate the detection and completion of incomplete goal states.

# References

1. Barett A., Weld D.S.: Partial order planning: Evaluating possible efficiency gains. J. Artificial Intelligence, 67 (1994) 71-112
2. Blum A.L., Furst M.L.: Fast planning through planning graph analysis. 14th International Joint Conference on Artificial Intelligence (IJCAI-95), Montreal, Canada (1995) 636-1642
3. Bonet B., Loerincs G., Geffner H.: A robust and fast action selection mechanism for planning. 14th International Conference of the American Association of Artificial Intelligence (AAAI-97), Providence, Rhode Island (1997) 714-719
4. Bonet B., Geffner H.: HSP: Heuristic Search Planner. Entry at the Artificial Intelligence Planning Systems (AIPS-98) Planning Competition, Pittsburgh (1998).
5. Fikes R.E. , Nilsson N.J.: Strips: A new approach to the application of theorem proving to problem solving. Artificial Intelligence, 2 (1971) 189-208
6. Gupta N. , Nau D.S.: On the complexity of blocks world planning. Artificial Intelligence, 56(2-3) (1992) 223-254
7. Hart P.E., Nilsson N.J., Raphael B.: A formal basis for the heuristic determination of minimum cost paths. IEEE Trans. on Systems Science and Cybernetics, 4(2) (1968) 100-107
8. Kautz H., Selman B.: Pushing the envelope: Planning, propositional logic and stochastic search. 13th International Conference of the American Association of Artificial Intelligence (AAAI-96), Portland, Oregon (1996) 1194-1201
9. Kautz H., Selman B.: BLACKBOX: A New Approach to the Application of Theorem Proving to Problem Solving. Working notes of the Workshop on Planning as Combinatorial Search, held in conjunction with Artificial Intelligence Planning Systems (AIPS-98), Pittsburgh (1998)
10. Minton S., Bresina J., Drummond M.: Total-order and partial-order planning: A comparative analysis. J. of Artificial Intelligence Research, 2, (1994) 227—261
11. Pearl J.: Heuristics: intelligent strategies for computer problem solving. MA: Addison-Wesley (1984)
12. Slaney J., Thiebaux S.: Linear-time near-optimal planning in the blocks world. 13th International Conference of the American Association of Artificial Intelligence (AAAI-96), Portland, Oregon (1996)
13. Veloso M.: Learning by Analogical Reasoning in General Problem Solving. PhD Dissertation, Computer Science Dept., CMU Tech. Report, CMU-CS-92-174 (1992)

# Planning as Heuristic Search: New Results

Blai Bonet and Héctor Geffner

Depto. de Computación
Universidad Simón Bolívar
Aptdo. 89000, Caracas 1080-A, Venezuela
{bonet,hector}@usb.ve

**Abstract.** In the recent AIPS98 Planning Competition, the hsp planner, based on a forward state search and a domain-independent heuristic, showed that heuristic search planners can be competitive with state of the art Graphplan and Satisfiability planners. hsp solved more problems than the other planners but it often took more time or produced longer plans. The main bottleneck in HSP is the computation of the heuristic for every new state. This computation may take up to 85% of the processing time. In this paper, we present a solution to this problem that uses a simple change in the direction of the search. The new planner, that we call HSPr, is based on the same ideas and heuristic as HSP, but searches backward from the goal rather than forward from the initial state. This allows HSPr to compute the heuristic estimates only *once*. As a result, HSPr can produce better plans, often in less time. For example, HSPr solves each of the 30 logistics problems from Kautz and Selman in less than 3 seconds. This is two orders of magnitude faster than BLACKBOX. At the same time, in almost all cases, the plans are substantially smaller. HSPr is also more robust than HSP as it visits a larger number of states, makes deterministic decisions, and relies on a single adjustable parameter than can be fixed for most domains. HSPr, however, is not better than HSP across all domains and in particular, in the blocks world, HSPr fails on some large instances that HSP can solve. We discuss also the relation between HSPr and Graphplan, and argue that Graphplan can also be understood as a heuristic search planner with a precise heuristic function and search algorithm.

## 1   Introduction

The last few years have seen a number of promising new approaches in Planning. Most prominent among these are Graphplan [BF95] and Satplan [KS96]. Both work in stages by building suitable structures and then searching them. In Graphplan, the structure is a graph, while in Satplan, it is a set of clauses. Both planners have shown impressive performance and have attracted a good deal of attention. Recent implementations and significant extensions have been reported in [KNHD97, LF99, KS99, ASW98].

In the recent AIPS98 Planning Competition [McD98], three out of the four planners in the Strips track, IPP [KNHD97], STAN [LF99], and BLACKBOX [KS99],

S. Biundo and M. Fox (Eds.): ECP-99, LNAI 1809, pp. 360–372, 2000.

were based on these ideas. The fourth planner, HSP [BG98], was based on heuristic search [Nil80,Pea83]. In HSP, the search is assumed to be similar to the search in problems like the 8-puzzle, the sole difference being in the heuristic: while in problems like the 8-puzzle the heuristic is normally given (e.g., as the sum of Manhattan distances), in planning it has to be extracted automatically from the representation of the problem. HSP thus appeals to a simple scheme for computing the heuristic from Strips encodings and uses the heuristic to guide the search. Similar ideas have been used recently in [McD96] and [BLG97].

In this paper, we review the heuristic and search algorithm used in HSP, discuss the problem of having to compute the heuristic from scratch in every new state, and introduce a reformulation of HSP that avoids this problem. We also analyze the relation between the resulting planner and Graphplan, and argue that Graphplan can also be understood as an heuristic search planner with a precise heuristic and search algorithm.

# 2  HSP: Heuristic Search Planner

HSP casts planning problems as problems of heuristic search. A Strips problem $P = \langle A, O, I, G \rangle$ is a tuple where $A$ is a set of atoms, $O$ is a set of operators, and $I \subseteq A$ and $G \subseteq A$ encode the initial and goal situations. The operators $op \in O$ are all assumed grounded, and each has precondition, add, and delete lists denoted as $Prec(op)$, $Add(op)$, and $Del(op)$, given by sets of atoms from $A$. A Strips problem $P = \langle A, O, I, G \rangle$ defines a state-space $\mathcal{S} = \langle S, s_0, S_G, A(\cdot), f, c \rangle$ where

S1. the states $s \in S$ are collections of atoms from $A$
S2. the initial state $s_0$ is $I$
S3. the goal states $s \in S_G$ are such that $G \subseteq s$
S4. the actions $a \in A(s)$ are the operators $op \in O$ such that $Prec(op) \subseteq s$
S5. the transition function $f$ maps states $s$ into states $s' = f(s, a)$, such that $s' = s - Del(a) + Add(a)$ for $a \in A(s)$.

HSP search this state-space, starting from $s_0$, with the aid of an heuristic extracted from the Strips representation of the problem (see also [McD96, BLG97]).

## 2.1  Heuristic

The heuristic function $h$ for a problem $P$ is obtained by considering a 'relaxed' problem $P'$ in which all *delete lists* are ignored. From any state $s$, the optimal cost $h'(s)$ for solving $P'$ is a lower bound on the optimal cost $h^*(s)$ for solving $P$. The heuristics $h(s)$ could thus be set to $h'(s)$. Solving optimally the relaxed problem $P'$ and hence obtaining $h'(s)$, however, is still exponentially hard. We thus settle for an approximation: we set $h(s)$ to an estimate of the optimal value function $h'(s)$ of the relaxed problem. This estimate is computed as follows.

For a given state $s$ and every atom $p$ we compute a measure $g(p)$ that provides an estimate of the cost (number of steps) of achieving $p$ from $s$. These measures are initialized to 0 if $p \in s$ and to $\infty$ otherwise. Then, every time an operator $op$ with preconditions $C = Prec(op)$ is applicable in $s$, each atom $p \in Add(op)$ is added to $s$ and $g(p)$ is updated to

$$g(p) := \min[g(p), 1 + g(C)] \tag{1}$$

where $g(C)$ stands for the cost of achieving the conjunction of atoms $C$ (see below). The expansions and updates continue until the measures $g(p)$ do not change. This is basically a shortest path algorithm over the graph that connects action preconditions with their effects.

In HSP, the costs $g(C)$ for conjunctions of atoms $C$ is defined as the sum of the costs $g(r)$ of the individual atoms $r$ in $C$:

$$g(C) \stackrel{\text{def}}{=} \sum_{r \in C} g(r) \quad \text{(additive costs)} \tag{2}$$

The heuristic function $h(s)$ that provides an estimate of the number of steps required to make the goal $G$ true from $s$, is then defined as:

$$h(s) \stackrel{\text{def}}{=} g(G) \tag{3}$$

Note that the $g$ measures depend on $s$, which is the state from which they are computed. In HSP, these measures are recomputed from scratch in every new state. This is expensive and may take up to 85% of the processing time.

The definition of the cost of *conjunctions* in (2) assumes that 'subgoals' are *independent*. Namely, that plans that achieve one subgoal have no effect on the plans that achieve other subgoals. In general, this is not true, and as a result the heuristic function defined by 1–3 is not *admissible* (it may overestimate the true optimal costs).

An admissible heuristic function *could* be defined by simply changing the formula (2) to

$$g(C) = \max_{r \in C} g(r) \quad \text{(max costs)} \tag{4}$$

The heuristic function that would result from this change, that we call the 'max heuristic', is admissible but is less informative than the *additive* function defined above. This is the reason that it is not used in HSP. In fact, while the 'additive' heuristic combines the costs of *all* subgoals, the 'max' heuristic focuses only on the most difficult subgoals ignoring all others. In Sect. 5 we will see that a refined version of the 'max heuristic' is used in Graphplan.

## 2.2  Algorithm

Provided with the heuristic $h$, HSP approaches planning as a problem of heuristic search. While standard search algorithms like A\* or IDA\* [Kor93] could be used, HSP uses a variation of hill-climbing. This is due to several reasons, the most important being that the computation of $h(s)$ for *every* new state is expensive.

HSP thus tries to get to the goal with as few node evaluations as possible. Surprisingly, the hill-climbing search plus a few simple enhancements, like a limited number of 'plateau' moves, restarts, and memory of past states, performs relatively well. We have obtained similar results before using Korf's LRTA* [Kor90] in place of hill-climbing [BLG97]. This is probably due to the fact that the heuristic $h$, while not admissible, is informative and tends to drive the search in a good direction.

Table 1 from [McD98], shows the results for the Strips track of the recent AIPS Competition. As can be seen, HSP solved more problems than the other planners but it often took more time or produced longer plans (see [McD98] for details).

**Table 1.** Results of the AIPS98 Competition (Strips track). Columns show the number of problems solved by each planner, and the number of problems in which each planner was fastest or produced shortest plans (from McDermott)

| Round | Planner | Avg. Time | Solved | Fastest | Shortest |
|-------|---------|-----------|--------|---------|----------|
| Round 1 | BLACKBOX | 1.49 | 63 | 16 | 55 |
|  | HSP | 35.48 | 82 | 19 | 61 |
|  | IPP | 7.40 | 63 | 29 | 49 |
|  | STAN | 55.41 | 64 | 24 | 47 |
| Round 2 | BLACKBOX | 2.46 | 8 | 3 | 6 |
|  | HSP | 25.87 | 9 | 1 | 5 |
|  | IPP | 17.37 | 11 | 3 | 8 |
|  | STAN | 1.33 | 7 | 5 | 4 |

## 3  HSPr: Heuristic Regression Planning

In HSP, the main bottleneck is the computation of the heuristic in every new state. The main innovation in HSPr is a simple change in the direction of the search that avoids this problem. As a result, the search proceeds faster, more states can be visited, and often better plans can be found in less time.

HSPr searches backward from the goal rather than forward from the initial state. This is an old idea in planning known as *regression search* [Nil80, Wel94]. In regression search, the states can be thought as *subgoals*; i.e., the 'application' of an action in a goal yields a situation in which the execution of the action achieves the goal. The crucial point in HSPr is that the measures $g(p)$ computed from $s_0$ that estimate the cost of achieving each atom $p$ from $s_0$ (Sect. 2.1), can be used with *no recomputation* to estimate the heuristic of *any* state arising during the regression. For example, if $s = \{p, q, r\}$ is one such state, the cost of reaching $s_0$ from $s$ by regression can be estimated in terms of the costs $g(p)$, $g(q)$, and $g(r)$ of achieving $p$, $q$, $r$ from $s_0$. The same applies to any other state. In other words, in HSPr, the cost estimates $g(p)$ are computed *once* from $s_0$ and are used to define the heuristic of *any* state. This is the key change from HSP from HSPr which is formalized below.

## 3.1   State Space

HSPr and HSP search two different state spaces. HSP searches the *progression space* $S$ defined by [S1]–[S5] above, where the actions are the available Strips operators. HSPr, on the other hand, searches a *regression space* $R$ where the actions correspond to 'reverse' application of the Strips operators. The regression space $R$ associated with a Strips problem $P = \langle A, O, I, G \rangle$ can be defined in analogy to the progression space $S$ defined by [S1]–[S5] above, as the tuple $\langle S, s_0, S_G, A(\cdot), f, c \rangle$ where

R1. the states s are sets of atoms from $A$
R2. the initial state $s_0$ is the goal $G$
R3. the goal states $s \in S_G$ are such $s \subseteq I$
R4. the set of actions $A(s)$ applicable in s are the operators $op \in O$ that are
     *relevant* and *consistent*; namely, $Add(op) \cap s \neq \emptyset$, and $Del(op) \cap s = \emptyset$
R5. the state $s' = f(a, s)$ that follows the application of $a \in A(s)$ is such that
     $s' = s - Add(a) + Prec(a)$.

The solution of this state space is, like the solution of any space $\langle S, s_0, S_G, A(\cdot), f, c \rangle$, a finite sequence of actions $a_0, a_1, \ldots, a_n$ such that for a sequence of states $s_0, s_1, \ldots, s_{n+1}$, $s_{i+1} = f(a_i, s_i)$, for $i = 0, \ldots, n, a_i \in A(s_i)$, and $s_{n+1} \in S_G$. The solution of the progression and regression spaces are related in the obvious way; one is the inverse of the other.

## 3.2   Heuristic

HSPr searches the regression space [R1]–[R5] using an heuristic based on the additive cost estimates $g(p)$ described in Sect 2.1. *These estimates are computed only once from the initial state* $s_0 \in S$. The heuristic $h(s)$ associated with *any* state s is then defined as

$$h(s) = \sum_{p \in s} g(p)$$

While in HSP, the heuristic $h(s)$ combines the cost estimates $g(p)$ of a fixed set of goal atoms computed from each state $s$, in HSPr, $h(s)$ combines the estimates $g(p)$ computed from a fixed state $s_0$ over the set of subgoals $p \in s$.

## 3.3   Mutexes

The regression search often leads to states s that are not reachable from the initial state $s_0$ in $S$ and cannot lead to a solution. Graphplan [BF95] recognized this problem and showed a way to deal with it based on the notion of *mutual exclusivity relations or mutexes*. In HSPr, we use the same idea with a slightly different formulation.

Roughly, two atoms $p$ and $q$ form a mutex $\langle p, q \rangle$, when they cannot be both true in a state reachable from $s_0$. For example, in block problems, atoms like $on(a.b)$ and $on(a, c)$ will normally form a mutex. States containing mutex pairs can be safely pruned as they cannot be part of a solution. We are thus interested in recognizing as many mutexes as possible.

A tentative definition is to have a pair of atoms $\langle p, q \rangle$ as a mutex when $p$ and $q$ are not both true in $s_0$ and every action that asserts $p$ deletes $q$, and every action that asserts $q$ deletes $p$. This definition, however, while sound, is too weak. In particular, it does not recognize a pair like $\langle on(a, b), on(a, c) \rangle$ as a mutex, since actions like $move(a, d, b)$ add the first atom but do not delete the second.

We thus use a different definition in which a pair $\langle p, q \rangle$ is recognized as mutex when the actions that add $p$ and do not delete $q$ can guarantee through their preconditions that $q$ will not be true after the action (symmetrically for actions that add $q$). For that we consider *sets* of mutexes.

**Definition 1.** *A set $M$ of atom pairs is a mutex set given a set of operators $O$ and an initial state $s_0$ iff for all $\langle p, q \rangle$ in $M$*

1. *$p$ and $q$ are not both true in $s_0$,*
2. *for every $op \in O$ that adds $p$, $q \in Del(op)$, or $q \notin Add(op)$ and for some $r \in Prec(op)$, $\langle r, q \rangle \in M$, and*
3. *for every $op \in O$ that adds $q$, $p \in Del(op)$, or $p \notin Add(op)$ and for some $r \in Prec(op)$, $\langle r, p \rangle \in M$.*

It is easy to verify that if $\langle p, q \rangle$ belongs to a mutex set, then $p$ and $q$ are really mutually exclusive, i.e., no reachable state will contain them both. Also if $M_1$ and $M_2$ are two mutex sets, $M_1 \cup M_2$ will be a mutex set as well, and hence according to this definition, there is a single *largest* mutex set. Rather than computing this set, however, that is difficult, we compute an approximation as follows.

We say that $\langle p, q \rangle$ is a 'bad pair' in $M$ when it does not comply with one of the conditions 1–3 above. The procedure for approximating the largest mutex set starts with a set of pairs $M := M_0$ and iteratively removes all bad pairs from $M$ until no bad pair remains. The initial set $M_0$ can be chosen as the set of all pairs $\langle p, q \rangle$ for atoms $p$ and $q$. In practice, however, we've found that we could speed up the computation without losing mutexes by considering a smaller initial set defined as the union of two sets $M_A$ and $M_B$ where

- $M_A$ is the set of pairs $\langle p, q \rangle$ such that for some $op \in O$, $p \in Add(op)$ and $q \in Del(op)$,
- $M_B$ is the set of pairs $\langle r, q \rangle$ such that for some $\langle p, q \rangle$ in $M_A$ and some $op \in O$, $r \in Prec(op)$ and $p \in Add(op)$

The structure of this definition mirrors that of Def. 1. We treat pairs symmetrically and hence assume that $\langle q, p \rangle$ is in a set when $\langle p, q \rangle$ is, and vice versa (in the implementation, atoms are ordered so only one pair is stored).

A mutex in HSPr refers to a pair in the set $M^*$ obtained from $M_0 = M_A + M_B$ by sequentially removing all 'bad' pairs. Like the analogous definition in Graphplan, the set $*$ does not capture all actual mutexes, yet it can be computed fast, and in the domains we have considered appears to prune the obvious unreachable states. A difference with Graphplan is that this definition identifies non-temporal mutexes while Graphplan identifies time-dependent mutexes. On the other hand, because of the fixed point construction, it can identify mutexes that Graphplan cannot. For example, in the complete TSP domain [LF99], pairs like $\langle at(city_1), at(city_2) \rangle$ would be recognized as a mutex by this definition but not by Graphplan, as the actions of going to different cities are not mutually exclusive for Graphplan.[1]

## 3.4 Algorithm

The search algorithm in HSPr uses the heuristic $h$ to guide the search in the regression space and the mutex set $M^*$ to prune states. The choice of the actual search algorithm follows from several considerations:

1. We want HSPr to solve problems with large state-spaces (e.g., spaces of $10^{20}$ states are not uncommon).
2. Node generation in HSPr, while faster than in HSP, is slow in comparison with domain-specific search algorithms.[2]
3. The heuristic function, while often quite informative, is not admissible and often overestimates the true optimal costs.
4. State spaces in planning are quite redundant with many paths leading to the same states.

These constraints are common to HSP where node generation is several times slower because of the computation of the heuristic. Slow node generation with large state-spaces practically rules out optimal search algorithms such as s* and IDA* that visit too many states before finding a solution. HSP uses instead a Hill-Climbing search with a few extensions like a memory of past states, a limited number of 'plateau' moves, and multiple restarts. In HSPr, we wanted to take advantage of the faster node generation to use a more systematic algorithm. After some experimentation, we settled on a simple algorithm that we call *Greedy Best First*. GBFS is a BFS algorithm that uses a cost function $f(n) = g(n) + W \cdot h(n)$, where $g(n)$ is the accumulated cost (number of steps),[3] $h(n)$ is the estimated cost, and $W \geq 1$ is a real parameter.[4] As any BFS algorithm, GBFS maintains the nodes for expansion in an OPEN list, and the nodes already expanded in a CLOSED

---

[1] An additional distinction is that Graphplan focuses on parallel macro-actions, while we focus on sequential primitive actions. See Sect. 5.

[2] HSPr generates a few thousand nodes per second while Korf's algorithms for the Rubik's Cube and the 24-puzzle generate in the order of a million nodes per second [Kor98,KT96].

[3] Not to be confused with the cost measures $g(p)$ defined in Sect. 2.1 for *atoms*.

[4] For the role of $W$ in BFS, see [Kor93].

list. The only difference between GBFS and BFS is that before selecting a node $n$ with minimum $f$-cost from OPEN, GBFS checks first to see whether the $f$-best children of the *last node* expanded are 'good enough'. If so, GBFS selects one such child; else it selects $n$ from OPEN as BFS. In HSPR, a child $n'$ of a node $n$ is 'good enough' when it appears closer to the goal; i.e., when $h(n') < h(n)$. GBFS thus performs some sort of Hill-Climbing search but 'backtracks' in a BFS mode when the heuristic is not improved. We also explored schemes in which 'discrepancies' are counted and used to control a LDS algorithm [HG95] but didn't get as good results. The results reported below for HSPR are all based on the GBFS algorithm described above with $W = 5$. Small changes in $W$ have little effect in HSPR, but some problems are not solved with $W = 1$.

# 4   Experiments

We report results on two domains for which there is a widely used body of difficult but solvable instances. We use the 30 logistic instances and the 5 largest block-world instances included in the BLACKBOX distribution (that we refer to as log-$i$ and bw-$k$, $i = 1 \ldots, 30$, $k = 1, \ldots, 5$)[5] and 5 hard block-world instances of our own (that we refer to as bw-$i$, $i = 6, \ldots, 10$).

In logistics, we compare HSPR with HSP, BLACKBOX, and TLPLAN. TLPLAN [BK98] is not a domain-independent planner (it uses control information tailored for each domain) but provides a reference for the results achievable in the domain. The results are shown in Table 2. The 'time' column measures CPU time in seconds, while 'steps' displays the number of actions in the solutions found.[6]

As the table shows, HSPR solves each of the 30 logistics problems in less than 3 seconds. This is faster (and less erratic) than HSP, and two orders of magnitude faster than BLACKBOX. At the same time, in almost all cases, the plans produced by HSPR are substantially smaller than those produced by BLACKBOX and HSP. The speed of HSPR is comparable with TLPLAN that includes fine-tuned knowledge for the domain. The plans obtained by TLPLAN, however, are shorter (with one exception).

HSPr is not better than HSP accross all domains, and there are problems, including a number of 'grid', 'mystery' and 'mprime' instances from the AIPS

---

[5] The logistics instances are from the 'logistics-strips' directory; while the block-world instances are from the prodigy-bw directory (they contain 7, 9, 11, 15, and 19 blocks, respectively).

[6] The results for BLACKBOX and TLPLAN were taken from [BK98]. The results for BLACKBOX in [BK98] are compatible with the results in the distribution but are slightly more detailed. From [BK98], BLACKBOX was run on a SPARC Ultra 2 with a 296MHz clock and 256M of RAM, while TLPLAN was run on a Pentium Pro with a 200MHz clock and 128M of RAM. The results for HSP and HSPR were obtained on a SPARC Ultra 5 running at 333MHz with 128M of RAM. In the case of HSP and hspr, individual problems are converted into C programs that are then compiled and run. This takes on the order of a couple of seconds for each instance. This time is not included in the table.

**Table 2.** Performance over `logistics` problems. A dash indicates that the planner was not able to solve the problem in 10 mins or 100 Mb.

| Problem | Time | | | | Steps | | | |
|---|---|---|---|---|---|---|---|---|
| | HSPR | HSP | BBOX | TLPLAN | HSPR | HSP | BBOX | TLPLAN |
| log-01 | 0.09 | 0.63 | 0.57 | 0.26 | 27 | 39 | 25 | 25 |
| log-02 | 0.08 | 0.48 | 95.97 | 0.28 | 28 | 31 | 31 | 27 |
| log-03 | 0.08 | 0.47 | 98.99 | 0.24 | 29 | 27 | 28 | 27 |
| log-04 | 0.23 | 0.95 | 130.74 | 1.37 | 67 | 58 | 71 | 51 |
| log-05 | 0.19 | 1.12 | 231.93 | 1.10 | 51 | 53 | 69 | 42 |
| log-06 | 0.33 | 1.51 | 321.27 | 1.91 | 69 | 60 | 82 | 51 |
| log-07 | 1.12 | — | 264.04 | 5.54 | 81 | — | 96 | 70 |
| log-08 | 1.49 | 9.43 | 317.42 | 6.84 | 82 | 170 | 110 | 70 |
| log-09 | 0.76 | — | 1609.45 | 3.79 | 77 | — | 121 | 70 |
| log-10 | 0.99 | 4.77 | 84.04 | 2.42 | 46 | 109 | 71 | 41 |
| log-11 | 0.64 | 1.61 | 137.93 | 2.24 | 54 | 47 | 68 | 46 |
| log-12 | 0.43 | 1.30 | 136.22 | 1.93 | 41 | 36 | 49 | 38 |
| log-13 | 1.26 | 5.78 | 165.84 | 6.54 | 74 | 102 | 85 | 66 |
| log-14 | 1.46 | 8.48 | 77.74 | 9.34 | 82 | 141 | 89 | 73 |
| log-15 | 1.10 | 3.53 | 424.36 | 5.36 | 69 | 76 | 91 | 63 |
| log-16 | 0.34 | 1.09 | 926.96 | 1.14 | 44 | 43 | 85 | 39 |
| log-17 | 0.36 | — | 758.47 | 1.24 | 48 | — | 83 | 43 |
| log-18 | 2.36 | 5.88 | 152.35 | 9.27 | 56 | 57 | 105 | 46 |
| log-19 | 0.74 | 2.97 | 149.22 | 2.66 | 50 | 68 | 78 | 45 |
| log-20 | 1.77 | 7.46 | 538.22 | 10.18 | 99 | 144 | 113 | 89 |
| log-21 | 1.51 | 5.09 | 190.49 | 6.83 | 69 | 81 | 87 | 59 |
| log-22 | 1.22 | 4.12 | 846.84 | 6.40 | 87 | 99 | 111 | 75 |
| log-23 | 0.95 | — | 173.93 | 4.69 | 70 | — | 85 | 62 |
| log-24 | 1.05 | 5.00 | 74.83 | 4.71 | 73 | 113 | 87 | 64 |
| log-25 | 1.04 | — | 73.99 | 4.09 | 67 | — | 84 | 57 |
| log-26 | 0.80 | 3.66 | 233.40 | 3.64 | 52 | 77 | 80 | 55 |
| log-27 | 1.08 | 4.45 | 145.16 | 5.52 | 76 | 115 | 97 | 70 |
| log-28 | 2.96 | — | 867.34 | 14.53 | 88 | — | 118 | 74 |
| log-29 | 2.40 | 8.40 | 89.51 | 5.99 | 50 | 105 | 84 | 45 |
| log-30 | 0.89 | — | 495.37 | 3.42 | 52 | — | 92 | 51 |

competition (see [McD98]), were neither planner does well. In the blocks-world, in particular, HSP appears to do better than HSPr over some hard instances (bottom of Table 3). This appeared surprising to us as HSP and HSPr use similar heuristics and HSPr can visit many more nodes than HSP. The problem we found is that the additive heuristic may not work as well for HSPr as for HSP due to the changes in the size of the states during the regression search. Since $h(s)$ is defined as the sum of the costs $g(p)$ for $p \in s$, this means that the heuristics favors smaller states over large states. This is not necessarily bad but often makes the definition of 'good children' in the Greedy BFS algorithm inadequate. Indeed, sometimes perfectly good 'children' do not appear closer to the goal than their 'parents' simply because they contain more atoms. We are currently

**Table 3.** Performance over `blocks-world` problems. A dash indicates that a planner was not able to solve the problem in 10 mins or 100 Mb.

| Problem | Blocks | HSPR Time | HSPR Steps | HSP Time | HSP Steps |
|---------|--------|-----------|------------|----------|-----------|
| bw-01 | 7 | 0.07 | 6 | 0.16 | 6 |
| bw-02 | 9 | 0.15 | 6 | 0.59 | 10 |
| bw-03 | 11 | 0.33 | 10 | 0.77 | 12 |
| bw-04 | 15 | 8.11 | 18 | 5.88 | 19 |
| bw-05 | 19 | 6.96 | 18 | 15.80 | 28 |
| bw-06 | 8 | 0.12 | 19 | 0.42 | 12 |
| bw-07 | 10 | 1.10 | 13 | 0.65 | 12 |
| bw-08 | 12 | 3.73 | 14 | 1.97 | 18 |
| bw-09 | 14 | – | – | 3.83 | 17 |
| bw-10 | 16 | – | – | 12.63 | 20 |

exploring possible ways around this problem. This problem does not happen in HSP because in the forward search all states $s$ have the same size. Namely, atoms that are not in $s$ can be safely assumed to be false then. The same is not true for the regression search.

## 5   Related Work

HSPr and HSP are descendants of the real-time planner ASP reported in [BLG97]. All three planners perform planning as heuristic search and use the same heuristic but search in different directions or with different algorithms. UNPOP is a planner based on similar ideas developed independently by Drew McDermott [McD96].

The search in HSPr can be understood as consisting of two phases. In the first, a forward propagation is used to compute the measures $g(p)$ that estimate the cost of achieving each of the atoms $p$ from the initial state $s_0$; in the second, a regression search is performed using those measures. These two phases are in correspondence with the two phases in Graphplan [BF95], where a plan graph is built forward in the first phase, and is searched backward in the second. The two planners are also related in the use of mutexes that HSPr borrows from Graphplan. For the rest, HSPr and Graphplan look quite different. However, we argue below that Graphplan, like HSPr, can also be understood as a heuristic search planner with a precise heuristic function and search algorithm. From this point of view, the main innovation in Graphplan is the implementation of the search algorithm, that is quite efficient, and the derivation of the heuristic function making use of the mutex information. Basically, we argue that the main features of Graphplan can be understood as follows:

1. **Graph:** The graph encodes an admissible heuristic $h_G$ that is a refined version of the 'max' heuristic discussed in Sect. 2.1, where $h_G(\mathsf{s}) = j$ iff $j$

is the index of the first level in the graph that contains (the set of atoms) s without a mutex, and in which s is not memoized (note that memoizations are updates on the heuristic function; see 4).

2. **Mutex:** Mutexes are used to prune states in the regression search (as in HSPr) and to refine the 'max' heuristic into $h_G$

3. **Algorithm:** The search algorithm is a version of Iterative Deepening sc a* (IDA*) [Kor93], where the *sum* of the accumulated cost $g(n)$ and the estimated cost $h_G(n)$ is used to prune nodes $n$ whose cost exceed the current threshold. Actually, Graphplan never generates such nodes.

4. **Memoization:** Memoizations are updates on the heuristic function $h_G$ (see 1). The resulting algorithm is a memory-extended version of IDA* that closely corresponds to the MREC algorithm [SB89]. In MREC, the heuristic of a node $n$ is updated and stored in a hash-table after the search below the children of $n$ completes without a solution (given the current threshold).

5. **Parallelism:** Graphplan searches a regression space $\mathcal{R}_p$ that is slightly different than $\mathcal{R}$ above, in which the actions are macro-actions (parallel-actions) composed of compatible primitive actions. Solutions costs in $\mathcal{R}_p$ are given by the number of macro-actions used (the number of time steps). The heuristic $h_G(\cdot)$ is admissible in this space as well, where it provides a *tighter* lower bound than in $\mathcal{R}$. While the branching factor in $\mathcal{R}_p$ can be very high (the number of macro-actions applicable in a state), Graphplan makes smart use of the information in the graph to generate only the children that are 'relevant' and whose cost does not exceed the current threshold.

This interpretation can be tested by extracting the heuristic $h_G$ from the graph and plugging it into a suitable version of IDA* (MREC) running over the parallel-regression space $\mathcal{R}_p$. The same care as Graphplan has to be taken for generating the applicable 'parallel' actions in each state, avoiding redundant or unnecessary effort. Our claim is that the performance of the resulting algorithm should be close to the basic Graphplan system [BF95] except for a small constant factor.

We haven't done this experiment ourselves but hope others may want to do it. If the account is correct, it will show that Graphplan, like HSP and HSPr, is best understood as a heuristic search planner. This would provide a different perspective to evaluate the strengths and limitations of Graphplan, and may suggest ways in which Graphplan can be improved. For another perspective on Graphplan, see [KLP97].

## 6   Discussion

We have presented a reformulation of HSP that makes 'planning as heuristic search' more competitive with state of art planners. A number of problem and challenges remain open. Among those that we think are the most important are the following:

- **better heuristics:** the 'additive' heuristic used in HSP and HSPr is too simple and often overestimates widely the true costs. The 'max' heuristic is too simple in the other direction, and focuses only on the most difficult subgoals ignoring all others. Better heuristics are needed.

- **node generation:** the node generation rates in HSP and HSPr are orders of magnitude slower than in specialized search algorithms [KT96,Kor98]. Implementations need to be improved substantially if domain-independent planning approaches are to compete with domain-specific search algorithms.

- **use of memory:** in hard problems, HSPr may run out of memory. A number of approaches for searching with limited memory are discussed in the literature but none is currently incorporated in the planners discussed.

- **optimal search:** so far we have been concerned with finding 'reasonable' plans, yet if better *admissible* heuristics can be obtained and node generation rates are improved, it'll be feasible to use standard optimal search algorithms such as IDA*.

- **modeling languages:** HSP and HSPr are tied to the Strips language, yet the ideas can be generalized to more expressive planning languages. In [Gef99], we describe Functional Strips, a language that extends Strips with first-class function symbols, and allows the codification of a number of problems (e.g., Hanoi, 8-puzzle, etc) in a way that mimics more specialized representations. In the current implementation, the heuristic function is given by the user, but we expect to be able to exploit the richer language for extracting better heuristics.

Except for the modeling point, the rest of the issues are common to the problems encountered in the application of domain-specific heuristic search methods [JS99]. Indeed, the only thing that distinguishes planning as heuristic search from classical heuristic search is the use of a general declarative language for encoding problems and getting the heuristic information. The biggest challenge is to make the declarative approach competitive with specialized methods while being more general.

**Acknowledgements:** Part of this work was done while H. Geffner was visiting IRIT, Toulouse. He thanks J. Lang, D. Dubois, H. Prade, and H. Farreny, for their hospitality and for making the visit possible. Both authors also thank R. Korf and D. Furcy for comments related to this work. Partial funding is due to Conicit. B. Bonet is currently at UCLA under a USB-Conicit fellowship. The code for HSPr is available at www.ldc.usb.ve/~hector.

# References

[ASW98]   C. Anderson, D. Smith, and D. Weld. Conditional effects in graphplan. In *Proc. AIPS-98*. AAAI Press, 1998.

[BF95]    A. Blum and M. Furst. Fast planning through planning graph analysis. In *Proceedings of IJCAI-95*, Montreal, Canada, 1995.

[BG98]    B. Bonet and H. Geffner. HSP: Planning as heuristic search. http://www.ldc.usb.ve/~hector, 1998.

372    B. Bonet and H. Geffner

[BK98]    F. Bacchus and F. Kabanza. Using temporal logics to express search control knowledge for planning, 1998. Submitted. At `http://www.lpaig.uwaterloo.ca/~fbacchus`.

[BLG97]   B. Bonet, G. Loerincs, and H. Geffner. A robust and fast action selection mechanism for planning. In *Proceedings of AAAI-97*, 1997.

[Gef99]   H. Geffner. Functional strips: a more general language for planning and problem solving. Presented at the Logic-based AI Workshop, Washington D.C., June 1999. At `http://www.ldc.usb.ve/~hector`.

[HG95]    W. Harvey and M. Ginsberg. Limited discrepancy search. In *Proc. IJCAI-95*, 1995.

[JS99]    A. Junghanns and J. Schaeffer. Domain-dependent single-agent search enhancements. In *Proc. IJCAI-99*. Morgan Kaufmann, 1999.

[KLP97]   S. Kambhampati, E. Lambrecht, and E. Parker. Understanding and extending Graphplan. In *Proc. 4th European Conf. on Planning*, LNAI 1248. Springer, 1997.

[KNHD97]  J. Koehler, B. Nebel, J. Hoffman, and Y. Dimopoulos. Extending planning graphs to an ADL subset. In *Proc. 4th European Conf. on Planning*, volume LNAI 1248. Springer, 1997.

[Kor90]   R. Korf. Real-time heuristic search. *Art. Intelligence*, 42:189–211, 1990.

[Kor93]   R. Korf. Linear-space best-first search. *Art. Intelligence*, 62:41–78, 1993.

[Kor98]   R. Korf. Finding optimal solutions to to Rubik's cube using pattern databases. In *Proceedings of AAAI-98*, pages 1202–1207, 1998.

[KS96]    H. Kautz and B. Selman. Pushing the envelope: Planning, propositional logic, and stochastic search. In *Proceedings of AAAI-96*.

[KS99]    H. Kautz and B. Selman. Unifying SAT-based and Graph-based planning. *Proceedings IJCAI-99*.

[KT96]    R. Korf and L. Taylor. Finding optimal solutions to the twenty-four puzzle. In *Proceedings of AAAI-96*, pages 1202–1207. MIT Press, 1996.

[LF99]    D. Long and M. Fox. The efficient implementation of the plan-graph. *JAIR*, 1999.

[McD96]   D. McDermott. A heuristic estimator for means-ends analysis in planning. In *Proc. Third Int. Conf. on AI Planning Systems (AIPS-96)*, 1996.

[McD98]   D. McDermott. AIPS-98 Planning Competition Results. `http://ftp.cs.yale.edu/mcdermott/aipscomp-results.html`, 1998.

[Nil80]   N. Nilsson. *Principles of Artificial Intelligence*. Tioga, 1980.

[Pea83]   J. Pearl. *Heuristics*. Morgan Kaufmann, 1983.

[SB89]    A. Sen and A. Bagchi. Fast recursive formulations for BFS that allow controlled used of memory. In *Proc. IJCAI-89*, pages 297–302, 1989.

[Wel94]   D. Weld. An introduction to least commitment planning. *AI Magazine*, 1994.

# Author Index

# Lecture Notes in Artificial Intelligence (LNAI)

# Lecture Notes in Computer Science